Graduate Texts in Mathematics 229

Graduate Texts in Mathematics

(continued after index)

David Eisenbud

The Geometry of Syzygies

A Second Course in Commutative Algebra and Algebraic Geometry

With 27 Figures

 Springer

David Eisenbud
Mathematical Sciences Research Institute
Berkeley, CA 94720
USA
de@msri.org

Mathematics Subject Classification (2000): 13Dxx 14-xx 16E05

Library of Congress Cataloging-in-Publication Data

A C.I.P. Catalogue record for this book is available from the Library of Congress.

ISBN 0-387-22215-4 (hardcover)   Printed on acid-free paper.
ISBN 0-387-22232-4 (softcover)

Printed in the United States of America.   (MVY)

9 8 7 6 5 4 3 2 1   SPIN 10938621 (hardcover)   SPIN 10946992 (softcover)

springeronline.com

Contents

Preface: Algebra and Geometry

Syzygy [from] Gr. συζυγία
yoke, pair, copulation, conjunction

— Oxford English Dictionary (etymology)

Implicit in the name "algebraic geometry" is the relation between geometry and equations. The qualitative study of systems of polynomial equations is the chief subject of commutative algebra as well. But when we actually study a ring or a variety, we often have to know a great deal about it before understanding its equations. Conversely, given a system of equations, it can be extremely difficult to analyze its qualitative properties, such as the geometry of the corresponding variety. The theory of syzygies offers a microscope for looking at systems of equations, and helps to make their subtle properties visible.

This book is concerned with the qualitative geometric theory of syzygies. It describes geometric properties of a projective variety that correspond to the numbers and degrees of its syzygies or to its having some structural property — such as being determinantal, or having a free resolution with some particularly simple structure. It is intended as a second course in algebraic geometry and commutative algebra, such as I have taught at Brandeis University, the Institut Poincaré in Paris, and the University of California at Berkeley.

What Are Syzygies?

In algebraic geometry over a field $\mathbb{K}$ we study the geometry of varieties through properties of the polynomial ring

$$S = \mathbb{K}[x_0, \ldots, x_r]$$

and its ideals. It turns out that to study ideals effectively we we also need to study more general graded modules over S. The simplest way to describe a module is by generators and relations. We may think of a set $A \subset M$ of generators for an S-module M as a map from a free S-module $F = S^A$ onto M, sending the basis element of F corresponding to a generator $m \in A$ to the element $m \in M$.

Let M_1 be the kernel of the map $F \to M$; it is called the *module of syzygies* of M corresponding to the given choice of generators, and a *syzygy* of M is an element of M_1 — a linear relation, with coefficients in S, on the chosen generators. When we give M by generators and relations, we are choosing generators for M and generators for the module of syzygies of M.

The use of "syzygy" in this context seems to go back to Sylvester [1853]. The word entered the language of modern science in the seventeenth century, with the same astronomical meaning it had in ancient Greek: the conjunction or opposition of heavenly bodies. Its literal derivation is a yoking together, just like "conjunction", with which it is cognate.

If $r = 0$, so that we are working over the polynomial ring in one variable, the module of syzygies is itself a free module, since over a principal ideal domain every submodule of a free module is free. But when $r > 0$ it may be the case that any set of generators of the module of syzygies has relations. To understand them, we proceed as before: we choose a generating set of syzygies and use them to define a map from a new free module, say F_1, onto M_1; equivalently, we give a map $\phi_1 : F_1 \to F$ whose image is M_1. Continuing in this way we get a *free resolution* of M, that is, a sequence of maps

$$\cdots \longrightarrow F_2 \xrightarrow{\phi_2} F_1 \xrightarrow{\phi_1} F \longrightarrow M \longrightarrow 0,$$

where all the modules F_i are free and each map is a surjection onto the kernel of the following map. The image M_i of ϕ_i is called the *i-th module of syzygies* of M.

In projective geometry we treat S as a graded ring by giving each variable x_i degree 1, and we will be interested in the case where M is a finitely generated graded S-module. In this case we can choose a *minimal* set of homogeneous generators for M (that is, one with as few elements as possible), and we choose the degrees of the generators of F so that the map $F \to M$ preserves degrees. The syzygy module M_1 is then a graded submodule of F, and Hilbert's Basis Theorem tells us that M_1 is again finitely generated, so we may repeat the procedure. Hilbert's Syzygy Theorem tells us that the modules M_i are free as soon as $i \geq r$.

The free resolution of M appears to depend strongly on our initial choice of generators for M, as well as the subsequent choices of generators of M_1, and so on. But if M is a finitely generated graded module and we choose a minimal set of generators for M, then M_1 is, up to isomorphism, independent of the minimal set of generators chosen. It follows that if we choose minimal sets of generators at each stage in the construction of a free resolution we get a *minimal free resolution of M* that is, up to isomorphism, independent of all the choices made. Since, by the Hilbert Syzygy Theorem, M_i is free for $i > r$, we see that in the minimal free resolution $F_i = 0$ for $i > r+1$. In this sense the minimal free resolution is finite: it has length at most $r+1$. Moreover, any free resolution of M can be derived from the minimal one in a simple way (see Section 1B).

The Geometric Content of Syzygies

The minimal free resolution of a module M is a good tool for extracting information about M. For example, Hilbert's motivation for his results just quoted was to devise a simple formula for the dimension of the d-th graded component of M as a function of d. He showed that the function $d \mapsto \dim_{\mathbb{K}} M_d$, now called the *Hilbert function* of M, agrees for large d with a polynomial function of d. The coefficients of this polynomial are among the most important invariants of the module. If $X \subset \mathbb{P}^r$ is a curve, the Hilbert polynomial of the homogeneous coordinate ring S_X of X is

$$(\deg X)\, d + (1 - \operatorname{genus} X),$$

whose coefficients $\deg X$ and $1 - \operatorname{genus} X$ give a topological classification of the embedded curve. Hilbert originally studied free resolutions because their discrete invariants, the *graded Betti numbers*, determine the Hilbert function (see Chapter 1).

But the graded Betti numbers contain more information than the Hilbert function. A typical example is the case of seven points in $\mathbb{P}^3$, described in Section 2C: every set of 7 points in $\mathbb{P}^3$ in linearly general position has the same Hilbert function, but the graded Betti numbers of the ideal of the points tell us whether the points lie on a rational normal curve.

Most of this book is concerned with examples one dimension higher: we study the graded Betti numbers of the ideals of a projective curve, and relate them to the geometric properties of the curve. To take just one example from those we will explore, Green's Conjecture (still open) says that the graded Betti numbers of the ideal of a canonically embedded curve tell us the curve's Clifford index (most of the time this index is 2 less than the minimal degree of a map from the curve to $\mathbb{P}^1$). This circle of ideas is described in Chapter 9.

Some work has been done on syzygies of higher-dimensional varieties too, though this subject is less well-developed. Syzygies are important in the study of embeddings of abelian varieties, and thus in the study of moduli of abelian varieties (for example [Gross and Popescu 2001]). They currently play a part in the study of surfaces of low codimension (for example [Decker and Schreyer 2000]), and other questions about surfaces (for example [Gallego and Purnaprajna 1999]). They have also been used in the study of Calabi–Yau varieties (for example [Gallego and Purnaprajna 1998]).

What Does Solving Linear Equations Mean?

A free resolution may be thought of as the result of fully solving a system of linear equations with polynomial coefficients. To set the stage, consider a system of linear equations $AX = 0$, where A is a $p \times q$ matrix of elements of $\mathbb{K}$, which we may think of as a linear transformation

$$F_1 = \mathbb{K}^q \xrightarrow{\ A\ } \mathbb{K}^p = F_0.$$

Suppose we find some solution vectors $X_1, \ldots, X_n$. These vectors constitute a complete solution to the equations if every solution vector can be expressed as a linear combination of them. Elementary linear algebra shows that there are complete solutions consisting of $q - \operatorname{rank} A$ independent vectors. Moreover, there is a powerful test for completeness: A given set of solutions $\{X_i\}$ is complete if and only if it contains $q - \operatorname{rank} A$ independent vectors.

A set of solutions can be interpreted as the columns of a matrix X defining a map $X : F_2 \to F_1$ such that

$$F_2 \xrightarrow{\ X\ } F_1 \xrightarrow{\ A\ } F_0$$

is a complex. The test for completeness says that this complex is exact if and only if $\operatorname{rank} A + \operatorname{rank} X = \operatorname{rank} F_1$. If the solutions are linearly independent as well as forming a complete system, we get an exact sequence

$$0 \to F_2 \xrightarrow{\ X\ } F_1 \xrightarrow{\ A\ } F_0.$$

Suppose now that the elements of A vary as polynomial functions of some parameters $x_0, \ldots, x_r$, and we need to find solution vectors whose entries also vary as polynomial functions. Given a set $X_1, \ldots, X_n$ of vectors of polynomials that are solutions to the equations $AX = 0$, we ask whether every solution can be written as a linear combination of the X_i with polynomial coefficients. If so we say that the set of solutions is complete. The solutions are once again elements of the kernel of the map $A : F_1 = S^q \to F_0 = S^p$, and a complete set of solutions is a set of generators of the kernel. Thus Hilbert's Basis Theorem implies that there do exist finite complete sets of solutions. However, it might be that every complete set of solutions is linearly dependent: the syzygy module $M_1 = \ker A$ is not free. Thus to understand the solutions we must compute the dependency relations on them, and then the dependency relations on these. This is precisely a free resolution of the cokernel of A. When we think of solving a system of linear equations, we should think of the whole free resolution.

One reward for this point of view is a criterion analogous to the rank criterion given above for the completeness of a set of solutions. We know no simple criterion for the completeness of a given set of solutions to a system of linear equations over S, that is, for the exactness of a complex of free S-modules $F_2 \to F_1 \to F_0$. However, if we consider a whole free resolution, the situation is better: a complex

$$0 \longrightarrow F_m \xrightarrow{\phi_m} \cdots \xrightarrow{\phi_2} F_1 \xrightarrow{\phi_1} F_0$$

of matrices of polynomial functions is exact if and only if the ranks r_i of the ϕ_i satisfy the conditions $r_i + r_{i-1} = \operatorname{rank} F_i$, as in the case where S is a field, and the set of points

$$\{p \in \mathbb{K}^{r+1} \mid \text{the evaluated matrix } \phi_i|_{x=p} \text{ has rank} < r_i\}$$

has codimension $\geq i$ for each i. (See Theorem 3.4.)

This criterion, from joint work with David Buchsbaum, was my first real result about free resolutions. I've been hooked ever since.

Experiment and Computation

A qualitative understanding of equations makes algebraic geometry more accessible to experiment: when it is possible to test geometric properties using their equations, it becomes possible to make constructions and decide their structure by computer. Sometimes unexpected patterns and regularities emerge and lead to surprising conjectures. The experimental method is a useful addition to the method of guessing new theorems by extrapolating from old ones. I personally owe to experiment some of the theorems of which I'm proudest. Number theory provides a good example of how this principle can operate: experiment is much easier in number theory than in algebraic geometry, and this is one of the reasons that number theory is so richly endowed with marvelous and difficult conjectures. The conjectures discovered by experiment can be trivial or very difficult; they usually come with no pedigree suggesting methods for proof. As in physics, chemistry or biology, there is art involved in inventing feasible experiments that have useful answers.

A good example where experiments with syzygies were useful in algebraic geometry is the study of surfaces of low degree in projective 4-space, as in work of Aure, Decker, Hulek, Popescu and Ranestad [Aure et al. 1997]. Another is the work on Fano manifolds such as that of of Schreyer [2001], or the applications surveyed in [Decker and Schreyer 2001, Decker and Eisenbud 2002]. The idea, roughly, is to deduce the form of the equations from the geometric properties that the varieties are supposed to possess, guess at sets of equations with this structure, and then prove that the guessed equations represent actual varieties. Syzygies were also crucial in my work with Joe Harris on algebraic curves. Many further examples of this sort could be given within algebraic geometry, and there are still more examples in commutative algebra and other related areas, such as those described in the *Macaulay 2 Book* [Decker and Eisenbud 2002].

Computation in algebraic geometry is itself an interesting field of study, not covered in this book. It has developed a great deal in recent years, and there are now at least three powerful programs devoted to computation in commutative algebra, algebraic geometry and singularities that are freely available: CoCoA, Macaulay 2, and Singular.[1] Despite these advances, it will always be easy to give sets of equations that render our best algorithms and biggest machines useless, so the qualitative theory remains essential.

A useful adjunct to this book would be a study of the construction of Gröbner bases which underlies these tools, perhaps from [Eisenbud 1995, Chapter 15], and the use of one of these computing platforms. The books [Greuel and Pfister 2002, Kreuzer and Robbiano 2000] and, for projective geometry, the forthcoming book [Decker and Schreyer ≥ 2004], will be very helpful.

What's In This Book?

The first chapter of this book is introductory: it explains the ideas of Hilbert that give the definitive link between syzygies and the *Hilbert function*. This is the origin of the modern theory of syzygies. This chapter also introduces the basic discrete invariants of resolution, the *graded Betti numbers*, and the convenient Betti diagrams for displaying them.

At this stage we still have no tools for showing that a given complex is a resolution, and in Chapter 2 we remedy this lack with a simple but very effective idea of Bayer, Peeva, and Sturmfels for describing some resolutions in terms of *labeled simplicial complexes*. With this tool we prove the Hilbert Syzygy Theorem and we also introduce Koszul homology. We then spend some time on the example of seven points in $\mathbb{P}^3$, where we see a deep connection between syzygies and an important invariant of the positions of the seven points.

In the next chapter we explore a case where we can say a great deal: sets of points in $\mathbb{P}^2$. Here we characterize all possible resolutions and derive some invariants of point sets from the structure of syzygies.

The following Chapter 4 introduces a basic invariant of the resolution, coarser than the graded Betti numbers: the *Castelnuovo–Mumford regularity*. This is a topic of central importance for the rest of the book, and a very active one for research. The goal of Chapter 4, however, is modest: we show that in the setting of sets of points in $\mathbb{P}^r$ the Castelnuovo–Mumford regularity is the degree needed to interpolate any function as a polynomial function. We also explore different characterizations of regularity, in terms of local or Zariski cohomology, and use them to prove some basic results used later.

Chapter 5 is devoted to the most important result on Castelnuovo–Mumford regularity to date: the theorem by Castelnuovo, Mattuck, Mumford, Gruson, Lazarsfeld, and Peskine bounding the regularity of projective curves. The techniques introduced here reappear many times later in the book.

---

[1]These software packages are freely available for many platforms, at cocoa.dima.unige.it, www.math.uiuc.edu/Macaulay2 and www.singular.uni-kl.de, respectively. These web sites are good sources of further information and references.

The next chapter returns to examples. We develop enough material about linear series to explain the free resolutions of all the curves of genus 0 and 1 in complete embeddings. This material can be generalized to deal with nice embeddings of any hyperelliptic curve.

Chapter 7 is again devoted to a major result: Green's Linear Syzygy theorem. The proof involves us with exterior algebra constructions that can be organized around the Bernstein–Gelfand–Gelfand correspondence, and we spend a section at the end of Chapter 7 exploring this tool.

Chapter 8 is in many ways the culmination of the book. In it we describe (and in most cases prove) the results that are the current state of knowledge of the syzygies of the ideal of a curve embedded by a complete linear series of *high degree* — that is, degree greater than twice the genus of the curve. Many new techniques are needed, and many old ones resurface from earlier in the book. The results directly generalize the picture, worked out much more explicitly, of the embeddings of curves of genus 0 and 1. We also present the conjectures of Green and Green–Lazarsfeld extending what we can prove.

No book on syzygies written at this time could omit a description of Green's conjecture, which has been a wellspring of ideas and motivation for the whole area. This is treated in Chapter 9. However, in another sense the time is the worst possible for writing about the conjecture, since major new results, recently proven, are still unpublished. These results will leave the state of the problem greatly advanced but still far from complete. It's clear that another book will have to be written some day...

Finally, I have included two appendices to help the reader: Appendix 1 explains local cohomology and its relation to sheaf cohomology, and Appendix 2 surveys, without proofs, the relevant commutative algebra. I can perhaps claim (for the moment) to have written the longest exposition of commutative algebra in [Eisenbud 1995]; with this second appendix I claim also to have written the shortest!

Prerequisites

The ideal preparation for reading this book is a first course on algebraic geometry (a little bit about curves and about the cohomology of sheaves on projective space is plenty) and a first course on commutative algebra, with an emphasis on the homological side of the field. Appendix 1 proves all that is needed about local cohomology and a little more, while Appendix 2 may help the reader cope with the commutative algebra required.

How Did This Book Come About?

This text originated in a course I gave at the Institut Poincaré in Paris, in 1994. The course was presented in my imperfect French, but this flaw was corrected by three of my auditors, Freddy Bonnin, Clément Caubel, and Hèléne Maugendre. They wrote up notes and added a lot of polish.

I have recently been working on a number of projects connected with the exterior algebra, partly motivated by the work of Green described in Chapter 7. This led me to offer a course on the subject again in the Fall of 2001, at the University of California, Berkeley. I rewrote the notes completely and added many topics and results, including material about exterior algebras and the Bernstein–Gelfand–Gelfand correspondence.

Other Books

Free resolutions appear in many places, and play an important role in books such as [Eisenbud 1995], [Bruns and Herzog 1998], and [Miller and Sturmfels 2004]. The last is also an excellent reference for the theory of monomial and toric ideals and their resolutions. There are at least two book-length treatments focusing on them specifically, [Northcott 1976] and [Evans and Griffith 1985]. The books [Cox et al. 1997] and [Schenck 2003] give gentle introductions to computational algebraic geometry, with lots of use of free resolutions, and many other topics. The notes [Eisenbud and Sidman 2004] could be used as an introduction to parts of this book.

Thanks

I've worked on the things presented here with some wonderful mathematicians, and I've had the good fortune to teach a group of PhD students and postdocs who have taught me as much as I've taught them. I'm particularly grateful to Dave Bayer, David Buchsbaum, Joe Harris, Jee Heub Koh, Mark Green, Irena Peeva, Sorin Popescu, Frank Schreyer, Mike Stillman, Bernd Sturmfels, Jerzy Weyman, and Sergey Yuzvinsky, for the fun we've shared while exploring this terrain.

I'm also grateful to Eric Babson, Baohua Fu, Leah Gold, George Kirkup, Pat Perkins, Emma Previato, Hal Schenck, Jessica Sidman, Greg Smith, Rekha Thomas, Simon Turner, and Art Weiss, who read parts of earlier versions of this text and pointed out infinitely many of the infinitely many things that needed fixing.

Notation

Throughout the text $\mathbb{K}$ denotes an arbitrary field; $S = \mathbb{K}[x_0, \ldots, x_r]$ denotes a polynomial ring; and $\mathfrak{m} = (x_0, \ldots, x_r) \subset S$ denotes its homogeneous maximal ideal. Sometimes when r is small we rename the variables and write, for example, $S = \mathbb{K}[x, y, z]$.

1

Free Resolutions and Hilbert Functions

A *minimal free resolution* is an invariant associated to a graded module over a ring graded by the natural numbers $\mathbb{N}$ or by $\mathbb{N}^n$. In this book we study minimal free resolutions of finitely generated graded modules in the case where the ring is a polynomial ring $S = \mathbb{K}[x_0, \ldots, x_r]$ over a field $\mathbb{K}$, graded by $\mathbb{N}$ with each variable in degree 1. This study is motivated primarily by questions from projective geometry. The information provided by free resolutions is a refinement of the information provided by the Hilbert polynomial and Hilbert function. In this chapter we define all these objects and explain their relationships.

The Generation of Invariants

As all roads lead to Rome, so I find in my own case at least that all algebraic inquiries, sooner or later, end at the Capitol of modern algebra, over whose shining portal is inscribed The Theory of Invariants.

— J. J. Sylvester (1864)

In the second half of the nineteenth century, invariant theory stood at the center of algebra. It originated in a desire to define properties of an equation, or of a curve defined by an equation, that were invariant under some geometrically defined set of transformations and that could be expressed in terms of a polynomial function of the coefficients of the equation. The most classical example is the discriminant of a polynomial in one variable. It is a polynomial function of the coefficients that does not change under linear changes of variable and whose vanishing is the condition for the polynomial to have multiple roots. This example had been studied since Leibniz's work: it was part of the motivation for his invention of matrix notation and determinants (first attested in a letter to l'Hôpital of April 1693; see [Leibniz 1962, p. 239]). A host of new examples had become important with the rise of complex projective plane geometry in the early nineteenth century.

The general setting is easy to describe: If a group G acts by linear transformations on a finite-dimensional vector space W over a field $\mathbb{K}$, the action extends uniquely to the ring S of polynomials whose variables are a basis for W. The fundamental problem of invariant theory was to prove in good cases — for example when $\mathbb{K}$ has characteristic zero and G is a finite group or a special linear group — that the ring of invariant functions S^G is finitely generated as a $\mathbb{K}$-algebra, that is, every invariant function can be expressed as a polynomial in a finite generating set of invariant functions. This had been proved, in a number of special cases, by explicitly finding finite sets of generators.

Enter Hilbert

The typical nineteenth-century paper on invariants was full of difficult computations, and had as its goal to compute explicitly a finite set of invariants generating all the invariants of a particular representation of a particular group. David Hilbert changed this landscape forever with his papers [Hilbert 1978] or [Hilbert 1970], the work that first brought him major recognition. He proved that the ring of invariants is finitely generated for a wide class of groups, including those his contemporaries were studying and many more. Most amazing, he did this by an existential argument that avoided hard calculation. In fact, he did not compute a single new invariant. An idea of his proof is given in [Eisenbud 1995, Chapter 1]. The really new ingredient was what is now called the *Hilbert Basis Theorem*, which says that submodules of finitely generated S-modules are finitely generated.

Hilbert studied syzygies in order to show that the generating function for the number of invariants of each degree is a rational function [Hilbert 1993]. He also showed that if I is a homogeneous ideal of the polynomial ring S, the "number of independent linear conditions for a form of degree d in S to lie in I" is a polynomial function of d [Hilbert 1970, p. 236]. (The problem of counting the number of conditions had already been considered for some time; it arose both in projective geometry and in invariant theory. A general statement of the problem, with a clear understanding of the role of syzygies (but without the word yet — see page x) is given by Cayley [1847], who also reviews some of the earlier literature and the mistakes made in it. Like Hilbert, Cayley was interested in syzygies (and higher syzygies too) because they let him count the number of forms in the ideal generated by a given set of forms. He was well aware that the syzygies form a module (in our sense). But unlike Hilbert, Cayley seems concerned with this module only one degree at a time, not in its totality; for instance, he did not raise the question of finite generation that is at the center of Hilbert's work.)

1A   The Study of Syzygies

Our primary focus is on the homogeneous coordinate rings of projective varieties and the modules over them, so we adapt our notation to this end. Recall that the *homogeneous coordinate ring* of the projective r-space $\mathbb{P}^r = \mathbb{P}^r_{\mathbb{K}}$ is the polynomial ring $S = \mathbb{K}[x_0, \ldots, x_r]$ in $r+1$ variables over a field $\mathbb{K}$, with all variables of degree 1. Let $M = \bigoplus_{d \in \mathbb{Z}} M_d$ be a finitely generated graded S-module with d-th graded component M_d. Because M is finitely generated, each M_d is a finite-dimensional vector space, and we define the *Hilbert function of M* to be

$$H_M(d) = \dim_{\mathbb{K}} M_d.$$

Hilbert had the idea of computing $H_M(d)$ by comparing M with free modules, using a *free resolution*. For any graded module M, denote by $M(a)$ the module M shifted (or "twisted") by a:

$$M(a)_d = M_{a+d}.$$

(For instance, the free S-module of rank 1 generated by an element of degree a is $S(-a)$.) Given homogeneous elements $m_i \in M$ of degree a_i that generate M as an S-module, we may define a map from the graded free module $F_0 = \bigoplus_i S(-a_i)$ onto M by sending the i-th generator to m_i. (In this text a map of graded modules means a degree-preserving map, and we need the shifts m_i to make this true.) Let $M_1 \subset F_0$ be the kernel of this map $F_0 \to M$. By the Hilbert Basis Theorem, M_1 is also a finitely generated module. The elements of M_1 are called *syzygies* on the generators m_i, or simply *syzygies of M*.

Choosing finitely many homogeneous syzygies that generate M_1, we may define a map from a graded free module F_1 to F_0 with image M_1. Continuing in this way we construct a sequence of maps of graded free modules, called a *graded free resolution of M*:

$$\cdots \longrightarrow F_i \xrightarrow{\varphi_i} F_{i-1} \longrightarrow \cdots \longrightarrow F_1 \xrightarrow{\varphi_1} F_0.$$

It is an exact sequence of degree-0 maps between graded free modules such that the cokernel of φ_1 is M. Since the φ_i preserve degrees, we get an exact sequence of finite-dimensional vector spaces by taking the degree d part of each module in this sequence, which suggests writing

$$H_M(d) = \sum_i (-1)^i H_{F_i}(d).$$

This sum might be useless, or even meaningless, if it were infinite, but Hilbert showed that it can be made finite.

Theorem 1.1 (Hilbert Syzygy Theorem). *Any finitely generated graded S-module M has a finite graded free resolution*

$$0 \longrightarrow F_m \xrightarrow{\varphi_m} F_{m-1} \longrightarrow \cdots \longrightarrow F_1 \xrightarrow{\varphi_1} F_0.$$

Moreover, we may take $m \leq r+1$, the number of variables in S.

We will prove Theorem 1.1 in Section 2B.

As first examples we take, as did Hilbert, three complexes that form the beginning of the most important, and simplest, family of free resolutions. They are now called Koszul complexes:

$$\mathbf{K}(x_0) : 0 \longrightarrow S(-1) \xrightarrow{(x_0)} S$$

$$\mathbf{K}(x_0,x_1) : 0 \longrightarrow S(-2) \xrightarrow{\binom{x_1}{-x_0}} S^2(-1) \xrightarrow{(x_0\ x_1)} S$$

$$\mathbf{K}(x_0,x_1,x_2) : 0 \longrightarrow S(-3) \xrightarrow{\begin{pmatrix} x_0 \\ x_1 \\ x_2 \end{pmatrix}} S^3(-2) \xrightarrow{\begin{pmatrix} 0 & x_2 & -x_1 \\ -x_2 & 0 & x_0 \\ x_1 & -x_0 & 0 \end{pmatrix}} S^3(-1) \xrightarrow{(x_0\ x_1\ x_2)} S$$

The first of these is obviously a resolution of $S/(x_0)$. It is quite easy to prove that the second is a resolution — see Exercise 1.1. It is also not hard to prove directly that the third is a resolution, but we will do it with a technique developed in the first half of Chapter 2.

The Hilbert Function Becomes Polynomial

From a free resolution of M we can compute the Hilbert function of M explicitly.

Corollary 1.2. *Suppose that $S = \mathbb{K}[x_0,\dots,x_r]$ is a polynomial ring. If the graded S-module M has finite free resolution*

$$0 \longrightarrow F_m \xrightarrow{\varphi_m} F_{m-1} \longrightarrow \cdots \longrightarrow F_1 \xrightarrow{\varphi_1} F_0,$$

with each F_i a finitely generated free module $F_i = \bigoplus_j S(-a_{i,j})$, then

$$H_M(d) = \sum_{i=0}^{m}(-1)^i \sum_j \binom{r+d-a_{i,j}}{r}.$$

Proof. We have $H_M(d) = \sum_{i=0}^{m}(-1)^i H_{F_i}(d)$, so it suffices to show that

$$H_{F_i}(d) = \sum_j \binom{r+d-a_{i,j}}{r}.$$

Decomposing F_i as a direct sum, it even suffices to show that $H_{S(-a)}(d) = \binom{r+d-a}{r}$. Shifting back, it suffices to show that $H_S(d) = \binom{r+d}{r}$. This basic combinatorial identity may be proved quickly as follows: a monomial of degree d is specified by the sequence of indices of its factors, which may be ordered to make a weakly increasing sequence of d integers, each between 0 and r. For example, we could specify $x_1^3 x_3^2$ by the sequence $1,1,1,3,3$. Adding i to the i-th element of the sequence, we get a d element subset of $\{1,\dots,r+d\}$, and there are $\binom{r+d}{d} = \binom{r+d}{r}$ of these. □

Corollary 1.3. *There is a polynomial $P_M(d)$ (called the* Hilbert polynomial *of M) such that, if M has free resolution as above, then $P_M(d) = H_M(d)$ for $d \geq \max_{i,j}\{a_{i,j} - r\}$.*

Proof. When $d + r - a \geq 0$ we have

$$\binom{d+r-a}{r} = \frac{(d+r-a)(d+r-1-a)\cdots(d+1-a)}{r!},$$

which is a polynomial of degree r in d. Thus in the desired range all the terms in the expression of $H_M(d)$ from Proposition 1.2 become polynomials.   □

Exercise 2.15 shows that the bound in Corollary 1.3 is not always sharp. We will investigate the matter further in Chapter 4; see, for example, Theorem 4A.

1B   Minimal Free Resolutions

Each finitely generated graded S-module has a *minimal free resolution*, which is unique up to isomorphism. The degrees of the generators of its free modules not only yield the Hilbert function, as would be true for any resolution, but form a finer invariant, which is the subject of this book. In this section we give a careful statement of the definition of minimality, and of the uniqueness theorem.

Naively, minimal free resolutions can be described as follows: Given a finitely generated graded module M, choose a minimal set of homogeneous generators m_i. Map a graded free module F_0 onto M by sending a basis for F_0 to the set of m_i. Let M' be the kernel of the map $F_0 \to M$, and repeat the procedure, starting with a minimal system of homogeneous generators of M'. . . .

Most of the applications of minimal free resolutions are based on a property that characterizes them in a different way, which we will adopt as the formal definition. To state it we will use our standard notation $\mathfrak{m}$ to denote the homogeneous maximal ideal $(x_0, \ldots, x_r) \subset S = \mathbb{K}[x_0, \ldots, x_r]$.

Definition. *A complex of graded S-modules*

$$\cdots \longrightarrow F_i \overset{\delta_i}{\longrightarrow} F_{i-1} \longrightarrow \cdots$$

is called minimal *if for each i the image of δ_i is contained in $\mathfrak{m}F_{i-1}$.*

Informally, we may say that a complex of free modules is minimal if its differential is represented by matrices with entries in the maximal ideal.

The relation between this and the naive idea of a minimal resolution is a consequence of Nakayama's Lemma. See [Eisenbud 1995, Section 4.1] for a discussion and proof in the local case. Here is the lemma in the graded case:

Lemma 1.4 (Nakayama). *Suppose M is a finitely generated graded S-module and $m_1, \ldots, m_n \in M$ generate $M/\mathfrak{m}M$. Then $m_1, \ldots, m_n$ generate M.*

Proof. Let $\bar{M} = M/(\sum Sm_i)$. If the m_i generate $M/\mathfrak{m}M$ then $\bar{M}/\mathfrak{m}\bar{M} = 0$ so $\mathfrak{m}\bar{M} = \bar{M}$. If $\bar{M} \neq 0$, since $\bar{M}$ is finitely generated, there would be a nonzero element of least degree in $\bar{M}$; this element could not be in $\mathfrak{m}\bar{M}$. Thus $\bar{M} = 0$, so M is generated by the m_i. □

Corollary 1.5. *A graded free resolution*

$$\mathbf{F}: \quad \cdots \longrightarrow F_i \xrightarrow{\ \delta_i\ } F_{i-1} \longrightarrow \cdots$$

is minimal as a complex if and only if for each i the map δ_i takes a basis of F_i to a minimal set of generators of the image of δ_i.

Proof. Consider the right exact sequence $F_{i+1} \to F_i \to \operatorname{im}\delta_i \to 0$. The complex $\mathbf{F}$ is minimal if and only if, for each i, the induced map

$$\bar{\delta}_{i+1} : F_{i+1}/\mathfrak{m}F_{i+1} \to F_i/\mathfrak{m}F_i$$

is zero. This holds if and only if the induced map $F_i/\mathfrak{m}F_i \to (\operatorname{im}\delta_i)/\mathfrak{m}(\operatorname{im}\delta_i)$ is an isomorphism. By Nakayama's Lemma this occurs if and only if a basis of F_i maps to a minimal set of generators of $\operatorname{im}\delta_i$. □

Considering all the choices made in the construction, it is perhaps surprising that minimal free resolutions are unique up to isomorphism:

Theorem 1.6. *Let M be a finitely generated graded S-module. If $\mathbf{F}$ and $\mathbf{G}$ are minimal graded free resolutions of M, then there is a graded isomorphism of complexes $\mathbf{F} \to \mathbf{G}$ inducing the identity map on M. Any free resolution of M contains the minimal free resolution as a direct summand.*

Proof. See [Eisenbud 1995, Theorem 20.2]. □

We can construct a minimal free resolution from any resolution, proving the second statement of Theorem 1.6 along the way. If $\mathbf{F}$ is a nonminimal complex of free modules, a matrix representing some differential of $\mathbf{F}$ must contain a nonzero element of degree 0. This corresponds to a free basis element of some F_i that maps to an element of F_{i-1} not contained in $\mathfrak{m}F_{i-1}$. By Nakayama's Lemma this element of F_{i-1} may be taken as a basis element. Thus we have found a subcomplex of $\mathbf{F}$ of the form

$$\mathbf{G}: \quad 0 \longrightarrow S(-a) \xrightarrow{\ c\ } S(-a) \longrightarrow 0$$

for a nonzero scalar c (such a thing is called a trivial complex) embedded in $\mathbf{F}$ in such a way that $\mathbf{F}/\mathbf{G}$ is again a free complex. Since $\mathbf{G}$ has no homology at all, the long exact sequence in homology corresponding to the short exact sequence of complexes $0 \to \mathbf{G} \to \mathbf{F} \to \mathbf{F}/\mathbf{G} \to 0$ shows that the homology of $\mathbf{F}/\mathbf{G}$ is the same as that of $\mathbf{F}$. In particular, if $\mathbf{F}$ is a free resolution of M, so is $\mathbf{F}/\mathbf{G}$. Continuing in this way we eventually reach a minimal complex. If $\mathbf{F}$ was a resolution of M, we have constructed the minimal free resolution.

For us the most important aspect of the uniqueness of minimal free resolutions is that, if $\mathbf{F} : \cdots \to F_1 \to F_0$ is the minimal free resolution of a finitely generated graded S-module M, the number of generators of each degree required for the free modules F_i depends only on M. The easiest way to state a precise result is to use the functor Tor; see for example [Eisenbud 1995, Section 6.2] for an introduction to this useful tool.

Proposition 1.7. *If* $\mathbf{F} : \cdots \to F_1 \to F_0$ *is the minimal free resolution of a finitely generated graded S-module M and $\mathbb{K}$ is the residue field $S/\mathfrak{m}$, then any minimal set of homogeneous generators of F_i contains exactly $\dim_{\mathbb{K}} \mathrm{Tor}_i^S(\mathbb{K}, M)_j$ generators of degree j.*

Proof. The vector space $\mathrm{Tor}_i^S(\mathbb{K}, M)_j$ is the degree j component of the graded vector space that is the i-th homology of the complex $\mathbb{K} \otimes_S \mathbf{F}$. Since $\mathbf{F}$ is minimal, the maps in $\mathbb{K} \otimes_S \mathbf{F}$ are all zero, so $\mathrm{Tor}_i^S(\mathbb{K}, M) = \mathbb{K} \otimes_S F_i$, and by Lemma 1.4 (Nakayama), $\mathrm{Tor}_i^S(\mathbb{K}, M)_j$ is the number of degree j generators that F_i requires. $\square$

Corollary 1.8. *If M is a finitely generated graded S-module then the projective dimension of M is equal to the length of the minimal free resolution.*

Proof. The projective dimension is the minimal length of a projective resolution of M, by definition. The minimal free resolution is a projective resolution, so one inequality is obvious. To show that the length of the minimal free resolution is at most the projective dimension, note that $\mathrm{Tor}_i^S(\mathbb{K}, M) = 0$ when i is greater than the projective dimension of M. By Proposition 1.7 this implies that the minimal free resolution has length less than i too. $\square$

If we allow the variables to have different degrees, $H_M(t)$ becomes, for large t, a polynomial with coefficients that are periodic in t. See Exercise 1.5 for details.

Describing Resolutions: Betti Diagrams

We have seen above that the numerical invariants associated to free resolutions suffice to describe Hilbert functions, and below we will see that the numerical invariants of minimal free resolutions contain more information. Since we will be dealing with them a lot, we will introduce a compact way to display them, called a *Betti diagram*.

To begin with an example, suppose $S = \mathbb{K}[x_0, x_1, x_2]$ is the homogeneous coordinate ring of $\mathbb{P}^2$. Theorem 3.13 and Corollary 3.10 below imply that there is a set X of 10 points in $\mathbb{P}^2$ whose homogeneous coordinate ring S_X has free resolution of the form

$$0 \longrightarrow \underset{\substack{\| \\ F_2}}{S(-6) \oplus S(-5)} \longrightarrow \underset{\substack{\| \\ F_1}}{S(-4) \oplus S(-4) \oplus S(-3)} \longrightarrow \underset{\substack{\| \\ F_0}}{S}.$$

We will represent the numbers that appear by the Betti diagram

	0	1	2
0	1	–	–
1	–	–	–
2	–	1	–
3	–	2	1
4	–	–	1

where the column labeled i describes the free module F_i.

In general, suppose that $\mathbf{F}$ is a free complex

$$\mathbf{F}: \ 0 \to F_s \to \cdots \to F_m \to \cdots \to F_0$$

where $F_i = \bigoplus_j S(-j)^{\beta_{i,j}}$; that is, F_i requires $\beta_{i,j}$ minimal generators of degree j. The Betti diagram of $\mathbf{F}$ has the form

	0	1	$\cdots$	s
i	$\beta_{0,i}$	$\beta_{1,i+1}$	$\cdots$	$\beta_{s,i+s}$
$i+1$	$\beta_{0,i+1}$	$\beta_{1,i+2}$	$\cdots$	$\beta_{s,i+s+1}$
$\cdots$	$\cdots$	$\cdots$	$\cdots$	$\cdots$
j	$\beta_{0,j}$	$\beta_{1,j+1}$	$\cdots$	$\beta_{s,j+s}$

It consists of a table with $s+1$ columns, labeled $0, 1, \ldots, s$, corresponding to the free modules $F_0, \ldots, F_s$. It has rows labeled with consecutive integers corresponding to degrees. (We sometimes omit the row and column labels when they are clear from context.) The m-th column specifies the degrees of the generators of F_m. Thus, for example, the row labels at the left of the diagram correspond to the possible degrees of a generator of F_0. For clarity we sometimes replace a 0 in the diagram by a "$-$" (as in the example given on the previous page) and an indefinite value by a "$*$".

Note that the entry in the j-th row of the i-th column is $\beta_{i,i+j}$ rather than $\beta_{i,j}$. This choice will be explained below.

If $\mathbf{F}$ is the minimal free resolution of a module M, we refer to the Betti diagram of $\mathbf{F}$ as the Betti diagram of M and the $\beta_{m,d}$ of $\mathbf{F}$ are called the *graded Betti numbers* of M, sometimes written $\beta_{m,d}(M)$. In that case the graded vector space $\mathrm{Tor}_m(M, \mathbb{K})$ is the homology of the complex $\mathbf{F} \otimes_{\mathbf{F}} \mathbb{K}$. Since $\mathbf{F}$ is minimal, the differentials in this complex are zero, so $\beta_{m,d}(M) = \dim_{\mathbb{K}} (\mathrm{Tor}_m(M, \mathbb{K})_d)$.

Properties of the Graded Betti Numbers

For example, the number $\beta_{0,j}$ is the number of elements of degree j required among the minimal generators of M. We will often consider the case where M is the homogeneous coordinate ring S_X of a (nonempty) projective variety X. As an S-module S_X is generated by the element 1, so we will have $\beta_{0,0} = 1$ and $\beta_{0,j} = 0$ for $j \neq 1$.

On the other hand, $\beta_{1,j}$ is the number of independent forms of degree j needed to generate the ideal I_X of X. If S_X is not the zero ring (that is, $X \neq \varnothing$), there are no elements of the ideal of X in degree 0, so $\beta_{1,0} = 0$. This is the case $i = d = 0$ of the following:

Proposition 1.9. *Let $\{\beta_{i,j}\}$ be the graded Betti numbers of a finitely generated S-module. If for a given i there is d such that $\beta_{i,j} = 0$ for all $j < d$, then $\beta_{i+1,j+1} = 0$ for all $j < d$.*

Proof. Suppose that the minimal free resolution is $\cdots \xrightarrow{\delta_2} F_1 \xrightarrow{\delta_1} F_0$. By minimality any generator of F_{i+1} must map to a nonzero element of the same degree in $\mathfrak{m}F_i$, the maximal homogeneous ideal times F_i. To say that $\beta_{i,j} = 0$ for all $j < d$ means that all generators — and thus all nonzero elements — of F_i have degree $\geq d$. Thus all nonzero elements of $\mathfrak{m}F_i$ have degree $\geq d+1$, so F_{i+1} can have generators only in degree $\geq d+1$ and $\beta_{i+1,j+1} = 0$ for $j < d$ as claimed. $\qquad\square$

Proposition 1.9 gives a first hint of why it is convenient to write the Betti diagram in the form we have, with $\beta_{i,i+j}$ in the j-th row of the i-th column: it says that if the i-th column of the Betti diagram has zeros above the j-th row, then the $(i+1)$-st column also has zeros above the j-th row. This allows a more compact display of Betti numbers than if we had written $\beta_{i,j}$ in the i-th column and j-th row. A deeper reason for our choice will be clear from the description of Castelnuovo–Mumford regularity in Chapter 4.

The Information in the Hilbert Function

The formula for the Hilbert function given in Corollary 1.2 has a convenient expression in terms of graded Betti numbers.

Corollary 1.10. *If $\{\beta_{i,j}\}$ are the graded Betti numbers of a finitely generated S-module M, the alternating sums $B_j = \sum_{i\geq 0}(-1)^i \beta_{i,j}$ determine the Hilbert function of M via the formula*

$$H_M(d) = \sum_j B_j \binom{r+d-j}{r}.$$

Moreover, the values of the B_j can be deduced inductively from the function $H_M(d)$ via the formula

$$B_j = H_M(j) - \sum_{k:\ k<j} B_k \binom{r+j-k}{r}.$$

Proof. The first formula is simply a rearrangement of the formula in Corollary 1.2.

Conversely, to compute the B_j from the Hilbert function $H_M(d)$ we proceed as follows. Since M is finitely generated there is a number j_0 so that $H_M(d) = 0$

for $d \leq j_0$. It follows that $\beta_{0,j} = 0$ for all $j \leq j_0$, and from Proposition 1.9 it follows that if $j \leq j_0$ then $\beta_{i,j} = 0$ for all i. Thus $B_j = 0$ for all $j \leq j_0$.

Inductively, we may assume that we know the value of B_k for $k < j$. Since $\binom{r+j-k}{r} = 0$ when $j < k$, only the values of B_k with $k \leq j$ enter into the formula for $H_M(j)$, and knowing $H_M(j)$ we can solve for B_j. Conveniently, B_j occurs with coefficient $\binom{r}{r} = 1$, and we get the displayed formula.   □

1C   Exercises

1. Suppose that f, g are polynomials (homogeneous or not) in S, neither of which divides the other, and consider the complex

$$0 \longrightarrow S \xrightarrow{\binom{g'}{-f'}} S^2 \xrightarrow{(f\ g)} S,$$

where $f' = f/h$, $g' = g/h$, and h is the greatest common divisor of f and g. Proved that this is a free resolution. In particular, the projective dimension of $S/(f, g)$ is at most 2. If f and g are homogeneous and neither divides the other, show that this is the minimal free resolution of $S/(f, g)$, so that the projective dimension of this module is exactly 2. Compute the twists necessary to make this a graded free resolution.

This exercise is a hint of the connection between syzygies and unique factorization, underlined by the famous theorem of Auslander and Buchsbaum that regular local rings (those where every module has a finite free resolution) are factorial. Indeed, refinements of the Auslander–Buchsbaum theorem by MacRae [1965] and Buchsbaum–Eisenbud [1974]) show that a local or graded ring is factorial if and only if the free resolution of any ideal generated by two elements has the form above.

In the situation of classical invariant theory, Hilbert's argument with syzygies easily gives a nice expression for the number of invariants of each degree — see [Hilbert 1993]. The situation is not quite as simple as the one studied in the text because, although the ring of invariants is graded, its generators have different degrees. Exercises 1.2–1.5 show how this can be handled. For these exercises we let $T = \mathbb{K}[z_1, \ldots, z_n]$ be a graded polynomial ring whose variables have degrees $\deg z_i = \alpha_i \in \mathbb{N}$.

2. The most obvious generalization of Corollary 1.2 is false: Compute the Hilbert function $H_T(d)$ of T in the case $n = 2, \alpha_1 = 2, \alpha_2 = 3$. Show that it is *not* eventually equal to a polynomial function of d (compare with the result of Exercise 1.5). Show that over the complex numbers this ring T is isomorphic to the ring of invariants of the cyclic group of order 6 acting on the polynomial ring $\mathbb{C}[x_0, x_1]$, where the generator acts by $x_0 \mapsto e^{2\pi i/2} x_0$, $x_1 \mapsto e^{2\pi i/3} x_1$.

Now let M be a finitely generated graded T-module. Hilbert's original argument for the Syzygy Theorem (or the modern one given in Section 2B) shows that M has a finite graded free resolution as a T-module. Let $\Psi_M(t) = \sum_d H_M(d)\, t^d$ be the generating function for the Hilbert function.

3. Two simple examples will make the possibilities clearer:
   (a) *Modules of finite length.* Show that any Laurent polynomial can be written as Ψ_M for suitable finitely generated M.
   (b) *Free modules.* Suppose $M = T$, the free module of rank 1 generated by an element of degree 0 (the unit element). Prove by induction on n that

$$\Psi_T(t) = \sum_{e=0}^{\infty} t^{e\alpha_n}\Psi_{T'}(t) = \frac{1}{1 - t^{\alpha_n}}\Psi_{T'}(t) = \frac{1}{\prod_{i=1}^n (1 - t^{\alpha_i})},$$

where $T' = \mathbb{K}[z_1, \ldots, z_{n-1}]$.
   Deduce that if $M = \sum_{i=-N}^{N} T(-i)^{\phi_i}$ then

$$\Psi_M(t) = \sum_{i=-N}^{N} \phi_i\Psi_{T(-i)}(t) = \frac{\sum_{i=-N}^{N} \phi_i t^i}{\prod_{i=1}^n (1 - t^{\alpha_i})}.$$

4. Prove:

   Theorem 1.11 (Hilbert). *Let $T = \mathbb{K}[z_1, \ldots, z_n]$, where $\deg z_i = \alpha_i$, and let M be a graded T-module with finite free resolution*

$$\cdots \longrightarrow \sum_j T(-j)^{\beta_{1,j}} \longrightarrow \sum_j T(-j)^{\beta_{0,j}}.$$

   Set $\phi_j = \sum_i (-1)^i \beta_{i,j}$ and set $\phi_M(t) = \phi_{-N}t^{-N} + \cdots + \phi_N t^N$. The Hilbert series of M is given by the formula

$$\Psi_M(t) = \frac{\phi_M(t)}{\prod_1^n (1 - t^{\alpha_i})};$$

   in particular Ψ_M is a rational function.

5. Suppose $T = \mathbb{K}[z_0, \ldots, z_r]$ is a graded polynomial ring with $\deg z_i = \alpha_i \in \mathbb{N}$. Use induction on r and the exact sequence

$$0 \to T(-\alpha_r) \xrightarrow{z_r} T \longrightarrow T/(z_r) \to 0$$

to show that the Hilbert function H_T of T is, for large d, equal to a polynomial with periodic coefficients: that is,

$$H_T(d) = h_0(d)d^r + h_1(d)d^{r-1} + \cdots$$

for some periodic functions $h_i(d)$ with values in $\mathbb{Q}$, whose periods divide the least common multiple of the α_i. Using free resolutions, state and derive a corresponding result for all finitely generated graded T-modules.

Some infinite resolutions: Let $R = S/I$ be a graded quotient of a polynomial ring $S = \mathbb{K}[x_0, \ldots, x_r]$. Minimal free resolutions exist for R, but are generally not finite. Much is known about what the resolutions look like in the case where R is a *complete intersection* — that is, I is generated by a regular sequence — and in a few other cases, but not in general. For surveys of some different areas, see [Avramov 1998, Fröberg 1999]. Here are a few sample results about resolutions of modules over a ring of the form $R = S/I$, where S is a graded polynomial ring (or a regular local ring) and I is a principal ideal. Such rings are often called *hypersurface rings.*

6. Let $S = \mathbb{K}[x_0, \ldots, x_r]$, let $I \subset S$ be a homogeneous ideal, and let $R = S/I$. Use the Auslander–Buchsbaum–Serre characterization of regular local rings (Theorem A2.19) to prove that there is a finite R-free resolution of $\mathbb{K} = R/(x_0, \ldots, x_r)R$ if and only if I is generated by linear forms.

7. Let $R = \mathbb{K}[t]/(t^n)$. Use the structure theorem for modules over the principal ideal domain $\mathbb{K}[t]$ to classify all finitely generated R-modules. Show that the minimal free resolution of the module R/t^a, for $0 < a < n$, is

$$\cdots \xrightarrow{\ t^a\ } R \xrightarrow{\ t^{n-a}\ } R \xrightarrow{\ t^a\ } \cdots \xrightarrow{\ t^a\ } R.$$

8. Let $R = S/(f)$, where f is a nonzero homogeneous form of positive degree. Suppose that A and B are two $n \times n$ matrices whose nonzero entries have positive degree in S, such that $AB = f \cdot I$, where I is an $n \times n$ identity matrix. Show that $BA = f \cdot I$ as well. Such a pair of matrices A, B is called a *matrix factorization of f*; see [Eisenbud 1980]. Let

$$\mathbf{F} : \cdots \xrightarrow{\ \bar{A}\ } R^n \xrightarrow{\ \bar{B}\ } R^n \xrightarrow{\ \bar{A}\ } \cdots \xrightarrow{\ \bar{A}\ } R^n,$$

where $\bar{A} := R \otimes_S A$ and $\bar{B} := R \otimes_S B$, denote the reductions of A and B modulo (f). Show that $\mathbf{F}$ is a minimal free resolution. (Hint: any element that goes to 0 under $\bar{A}$ lifts to an element that goes to a multiple of f over A.)

9. Suppose that M is a finitely generated R-module that has projective dimension 1 as an S-module. Show that the free resolution of M as an S-module has the form

$$0 \longrightarrow S^n \xrightarrow{\ A\ } S^n \longrightarrow M \longrightarrow 0$$

for some n and some $n \times n$ matrix A. Show that there is an $n \times n$ matrix B with $AB = f \cdot I$. Conclude that the free resolution of M as an R-module has the form given in Exercise 1.8.

10. The ring R is Cohen–Macaulay, of depth r (Example A2.40). Use part 3 of Theorem A2.14, together with the Auslander–Buchsbaum Formula A2.13, to show that if N is any finitely generated graded R-module, then the r-th syzygy of M has depth r, and thus has projective dimension 1 as an

S-module. Deduce that the free resolution of any finitely generated graded module is periodic, of period at most 2, and that the periodic part of the resolution comes from a matrix factorization.

2

First Examples of Free Resolutions

In this chapter we introduce a fundamental construction of resolutions based on simplicial complexes. This construction gives free resolutions of monomial ideals, but does not always yield minimal resolutions. It includes the Koszul complexes, which we use to establish basic bounds on syzygies of all modules, including the Hilbert Syzygy Theorem. We conclude the chapter with an example of a different kind, showing how free resolutions capture the geometry of sets of seven points in $\mathbb{P}^3$.

2A Monomial Ideals and Simplicial Complexes

We now introduce a beautiful method of writing down graded free resolutions of monomial ideals due to Bayer, Peeva and Sturmfels [Bayer et al. 1998]. So far we have used $\mathbb{Z}$-gradings only, but we can think of the polynomial ring S as $\mathbb{Z}^{r+1}$-graded, with $x_0^{a_0} \cdots x_r^{a_r}$ having degree $(a_0, \ldots, a_r) \in \mathbb{Z}^{r+1}$, and the free resolutions we write down will also be $\mathbb{Z}^{r+1}$-graded. We begin by reviewing the basics of the theory of finite simplicial complexes. For a more complete treatment, see [Bruns and Herzog 1998].

Simplicial Complexes

A *finite simplicial complex* Δ is a finite set N, called the set of *vertices* (or *nodes*) of Δ, and a collection F of subsets of N, called the *faces of* Δ, such that if $A \in F$ is a face and $B \subset A$ then B is also in F. Maximal faces are called *facets*.

A *simplex* is a simplicial complex in which every subset of N is a face. For any vertex set N we may form the *void* simplicial complex, which has no faces at all. But if Δ has any faces at all, then the empty set $\varnothing$ is necessarily a face of Δ. By contrast, we call the simplicial complex whose only face is $\varnothing$ the *irrelevant* simplicial complex on N. (The name comes from the Stanley–Reisner correspondence, which associates to any simplicial complex Δ with vertex set $N = \{x_0, \ldots, x_n\}$ the square-free monomial ideal in $S = \mathbb{K}[x_0, \ldots, x_r]$ whose elements are the monomials with support equal to a non-face of Δ. Under this correspondence the irrelevant simplicial complex corresponds to the irrelevant ideal $(x_0, \ldots, x_r)$, while the void simplicial complex corresponds to the ideal (1).)

Any simplicial complex Δ has a *geometric realization*, that is, a topological space that is a union of simplices corresponding to the faces of Δ. It may be constructed by realizing the set of vertices of Δ as a linearly independent set in a sufficiently large real vector space, and realizing each face of Δ as the convex hull of its vertex points; the realization of Δ is then the union of these faces.

An *orientation* of a simplicial complex consists of an ordering of the vertices of Δ. Thus a simplicial complex may have many orientations—this is not the same as an orientation of the underlying topological space.

Labeling by Monomials

We will say that Δ is *labeled* (by monomials of S) if there is a monomial of S associated to each vertex of Δ. We then label each face A of Δ by the least common multiple of the labels of the vertices in A. We write m_A for the monomial that is the label of A. By convention the label of the empty face is $m_\varnothing = 1$.

Let Δ be an oriented labeled simplicial complex, and write $I \subset S$ for the ideal generated by the monomials $m_j = x^{\alpha_j}$ labeling the vertices of Δ. We will associate to Δ a graded complex of free S-modules

$$\mathscr{C}(\Delta) = \mathscr{C}(\Delta; S) : \quad \cdots \longrightarrow F_i \xrightarrow{\ \delta\ } F_{i-1} \longrightarrow \cdots \xrightarrow{\ \delta\ } F_0,$$

where F_i is the free S-module whose basis consists of the set of faces of Δ having i elements, which is sometimes a resolution of S/I. The differential δ is given by the formula

$$\delta A = \sum_{n \in A} (-1)^{\mathrm{pos}(n, A)} \frac{m_A}{m_{A \setminus n}} (A \setminus n),$$

where $\mathrm{pos}(n, A)$, the *position of vertex n in A*, is the number of elements preceding n in the ordering of A, and $A \setminus n$ denotes the face obtained from A by removing n.

If Δ is not void then $F_0 = S$; the generator is the face of Δ which is the empty set. Further, the generators of F_1 correspond to the vertices of Δ, and each generator maps by δ to its labeling monomial, so

$$H_0(\mathscr{C}(\Delta)) = \mathrm{coker}\left(F_1 \xrightarrow{\ \delta\ } S\right) = S/I.$$

We set the degree of the basis element corresponding to the face A equal to the exponent vector of the monomial that is the label of A. With respect to this grading, the differential δ has degree 0, and $\mathscr{C}(\Delta)$ is a $\mathbb{Z}^{r+1}$-graded free complex.

For example we might take $S = \mathbb{K}$ and label all the vertices of Δ with $1 \in \mathbb{K}$; then $\mathscr{C}(\Delta; \mathbb{K})$ is, up to a shift in homological degree, the usual *reduced chain complex of Δ with coefficients in S*. Its homology is written $H_i(\Delta; \mathbb{K})$ and is called the *reduced homology of Δ with coefficients in S*. The shift in homological degree comes about as follows: the homological degree of a simplex in $\mathscr{C}(\Delta)$ is the number of vertices in the simplex, which is one more than the dimension of the simplex, so that $H_i(\Delta; \mathbb{K})$ is the $(i+1)$-st homology of $\mathscr{C}(\Delta; \mathbb{K})$. If $H_i(\Delta; \mathbb{K}) = 0$ for $i \geq -1$, we say that Δ is $\mathbb{K}$-*acyclic*. (Since S is a free module over $\mathbb{K}$, this is the same as saying that $H_i(\Delta; S) = 0$ for $i \geq -1$.)

The homology $H_i(\Delta; \mathbb{K})$ and the homology $H_i(\mathscr{C}(\Delta; S))$ are independent of the orientation of Δ—in fact they depend only on the homotopy type of the geometric realization of Δ and the ring $\mathbb{K}$ or S. Thus we will often ignore orientations.

Roughly speaking, we may say that the complex $\mathscr{C}(\Delta; S)$, for an arbitrary labeling, is obtained by extending scalars from $\mathbb{K}$ to S and "homogenizing" the formula for the differential of $\mathscr{C}(\Delta, \mathbb{K})$ with respect to the degrees of the generators of the F_i defined for the S-labeling of Δ.

Example 2.1. Suppose that Δ is the labeled simplicial complex

$$
\begin{array}{ccccc}
& x_0x_1x_2 & & x_0x_1x_2 & \\
\bullet\!\!\!\!-\!\!\!\!-\!\!\!\!-\!\!\!\!\bullet\!\!\!\!-\!\!\!\!-\!\!\!\!-\!\!\!\!\bullet & & & & \\
x_0x_1 & & x_0x_2 & & x_1x_2
\end{array}
$$

with the orientation obtained by ordering the vertices from left to right. The complex $\mathscr{C}(\Delta)$ is

$$
0 \longrightarrow S^2(-3) \xrightarrow{\begin{pmatrix} -x_2 & 0 \\ x_1 & -x_1 \\ 0 & x_0 \end{pmatrix}} S^3(-2) \xrightarrow{\begin{pmatrix} x_0x_1 & x_0x_2 & x_1x_2 \end{pmatrix}} S.
$$

This complex is represented by the Betti diagram

	0	1	2
0	1	–	–
1	–	3	2

As we shall soon see, the only homology of this complex is at the right-hand end, where $H_0(\mathscr{C}(\Delta)) = S/(x_0x_1, x_0x_2, x_1x_2)$, so the complex is a free resolution of this S-module.

If we took the same simplicial complex, but with the trivial labeling by 1's, we would get the complex

$$0 \longrightarrow S^2 \xrightarrow{\begin{pmatrix} -1 & 0 \\ 1 & -1 \\ 0 & 1 \end{pmatrix}} S^3 \xrightarrow{(1 \ 1 \ 1)} S,$$

represented by the Betti diagram

	0	1	2
-2	–	–	2
-1	–	3	–
0	1	–	–

which has reduced homology 0 (with any coefficients), as the reader may easily check.

We want a criterion that will tell us when $\mathscr{C}(\Delta)$ is a resolution of S/I; that is, when $H_i(\mathscr{C}(\Delta)) = 0$ for $i > 0$. To state it we need one more definition. If m is any monomial, we write Δ_m for the subcomplex consisting of those faces of Δ whose labels divide m. For example, if m is not divisible by any of the vertex labels, then Δ_m is the empty simplicial complex, with no vertices and the single face $\varnothing$. On the other hand, if m is divisible by all the labels of Δ, then $\Delta_m = \Delta$. Moreover, Δ_m is equal to $\Delta_{\mathrm{LCM}\{m_i | i \in I\}}$ for some subset Δ' of the vertex set of Δ.

A *full subcomplex* of Δ is a subcomplex of all the faces of Δ that involve a particular set of vertices. Note that all the subcomplexes Δ_m are full.

Syzygies of Monomial Ideals

Theorem 2.2 (Bayer, Peeva, and Sturmfels). *Let Δ be a simplicial complex labeled by monomials $m_1, \ldots, m_t \in S$, and let $I = (m_1, \ldots, m_t) \subset S$ be the ideal in S generated by the vertex labels. The complex $\mathscr{C}(\Delta) = \mathscr{C}(\Delta; S)$ is a free resolution of S/I if and only if the reduced simplicial homology $H_i(\Delta_m; \mathbb{K})$ vanishes for every monomial m and every $i \geq 0$. Moreover, $\mathscr{C}(\Delta)$ is a minimal complex if and only if $m_A \neq m_{A'}$ for every proper subface A' of a face A.*

By the remarks above, we can determine whether $\mathscr{C}(\Delta)$ is a resolution just by checking the vanishing condition for monomials that are least common multiples of sets of vertex labels.

Proof. Let $\mathscr{C}(\Delta)$ be the complex

$$\mathscr{C}(\Delta): \quad \cdots \longrightarrow F_i \xrightarrow{\delta} F_{i-1} \longrightarrow \cdots \xrightarrow{\delta} F_0.$$

It is clear that S/I is the cokernel of $\delta : F_1 \to F_0$. We will identify the homology of $\mathscr{C}(\Delta)$ at F_i with a direct sum of copies of the vector spaces $H_i(\Delta_m; \mathbb{K})$.

For each $\alpha \in \mathbb{Z}^{r+1}$ we will compute the homology of the complex of vector spaces

$$\mathscr{C}(\Delta)_\alpha : \quad \cdots \longrightarrow (F_i)_\alpha \xrightarrow{\ \delta\ } (F_{i-1})_\alpha \longrightarrow \cdots \xrightarrow{\ \delta\ } (F_0)_\alpha,$$

formed from the degree-α components of each free module F_i in $\mathscr{C}(\Delta)$. If any of the components of α are negative then $\mathscr{C}(\Delta)_\alpha = 0$, so of course the homology vanishes in this degree.

Thus we may suppose $\alpha \in \mathbb{N}^{r+1}$. Set $m = x^\alpha = x_0^{\alpha_0} \cdots x_r^{\alpha_r} \in S$. For each face A of Δ, the complex $\mathscr{C}(\Delta)$ has a rank-one free summand $S \cdot A$ which, as a vector space, has basis $\{n \cdot A \mid n \in S \text{ is a monomial}\}$. The degree of $n \cdot A$ is the exponent of $n m_A$, where m_A is the label of the face A. Thus for the degree α part of $S \cdot A$ we have

$$S \cdot A_\alpha = \begin{cases} \mathbb{K} \cdot (x^\alpha/m_A) \cdot A & \text{if } m_A | m, \\ 0 & \text{otherwise.} \end{cases}$$

It follows that the complex $\mathscr{C}(\Delta)_\alpha$ has a $\mathbb{K}$-basis corresponding bijectively to the faces of Δ_m. Using this correspondence we identify the terms of the complex $\mathscr{C}(\Delta)_\alpha$ with the terms of the reduced chain complex of Δ_m having coefficients in $\mathbb{K}$ (up to a shift in homological degree as for the case, described above, where the vertex labels are all 1). A moment's consideration shows that the differentials of these complexes agree.

Having identified $\mathscr{C}(\Delta)_\alpha$ with the reduced chain complex of Δ_m, we see that the complex $\mathscr{C}(\Delta)$ is a resolution of S/I if and only if $H_i(\Delta_m; \mathbb{K}) = 0$ for all $i \geq 0$, as required for the first statement.

For minimality, note that if A is an $(i+1)$-face and A' an i-face of Δ, then the component of the differential of $\mathscr{C}(\Delta)$ that maps $S \cdot A$ to $S \cdot A'$ is 0 unless $A' \subset A$, in which case it is $\pm m_A/m_{A'}$. Thus $\mathscr{C}(\Delta)$ is minimal if and only if $m_A \neq m_{A'}$ for all $A' \subset A$, as required.   □

For more information about the complexes $\mathscr{C}(\Delta)$ and about a generalization in which cell complexes replace simplicial complexes, see [Bayer et al. 1998] and [Bayer and Sturmfels 1998].

Example 2.3. We continue with the ideal $(x_0 x_1, x_0 x_2, x_1 x_2)$ as above. For the labeled simplicial complex Δ

the distinct subcomplexes Δ' of the form Δ_m are the empty complex Δ_1, the complexes $\Delta_{x_0 x_1}, \Delta_{x_0 x_2}, \Delta_{x_1 x_2}$, each of which consists of a single point, and the complex Δ itself. As each of these is contractible, they have no higher reduced homology, and we see that the complex $\mathscr{C}(\Delta)$ is the minimal free resolution of $S/(x_0 x_1, x_0 x_2, x_1 x_2)$.

Any full subcomplex of a simplex is a simplex, and since the complexes Δ_1, $\Delta_{x_0 x_1}, \Delta_{x_0 x_2}, \Delta_{x_1 x_2}$, and Δ are all contractible, they have no reduced homology (with any coefficients). This idea gives a result first proved, in a different way, by Diana Taylor [Eisenbud 1995, Exercise 17.11].

Corollary 2.4. *Let $I = (m_1, \ldots, m_n) \subset S$ be any monomial ideal, and let Δ be a simplex with n vertices, labeled $m_1, \ldots, m_n$. The complex $\mathscr{C}(\Delta)$, called the Taylor complex of $m_1, \ldots, m_n$, is a free resolution of S/I.*   □

For an interesting consequence see Exercise 2.1.

Example 2.5. The Taylor complex is rarely minimal. For instance, taking

$$(m_1, m_2, m_3) = (x_0 x_1, x_0 x_2, x_1 x_2)$$

as in the example above, the Taylor complex is a nonminimal resolution with Betti diagram

	0	1	2	3
0	1	–	–	1
1	–	3	3	–

Example 2.6. We may define the *Koszul complex* $\mathbf{K}(x_0, \ldots, x_r)$ of $x_0, \ldots, x_r$ to be the Taylor complex in the special case where the $m_i = x_i$ are variables. We have exhibited the smallest examples on page 4. By Theorem 2.2 the Koszul complex is a minimal free resolution of the residue class field $\mathbb{K} = S/(x_0, \ldots, x_r)$.

We can replace the variables $x_0, \ldots, x_r$ by any polynomials $f_0, \ldots, f_r$ to obtain a complex we will write as $\mathbf{K}(f_0, \ldots, f_r)$, the Koszul complex of the sequence $f_0, \ldots, f_r$. In fact, since the differentials have only $\mathbb{Z}$ coefficients, we could even take the f_i to be elements of an arbitrary commutative ring.

Under nice circumstances, for example when the f_i are homogeneous elements of positive degree in a graded ring, this complex is a resolution if and only if the f_i form a *regular sequence*. See Section A2F or [Eisenbud 1995, Theorem 17.6].

2B   Bounds on Betti Numbers and Proof of Hilbert's Syzygy Theorem

We can use the Koszul complex and Theorem 2.2 to prove a sharpening of Hilbert's Syzygy Theorem 1.1, which is the vanishing statement in the following proposition. We also get an alternate way to compute the graded Betti numbers.

Proposition 2.7. *Let M be a graded module over $S = \mathbb{K}[x_0, \ldots, x_r]$. The graded Betti number $\beta_{i,j}(M)$ is the dimension of the homology, at the term $M_{j-i} \otimes \bigwedge^i \mathbb{K}^{r+1}$, of the complex*

$$0 \to M_{j-(r+1)} \otimes \bigwedge\nolimits^{r+1} \mathbb{K}^{r+1} \to \cdots$$

$$\to M_{j-i-1} \otimes \bigwedge\nolimits^{i+1} \mathbb{K}^{r+1} \to M_{j-i} \otimes \bigwedge\nolimits^{i} \mathbb{K}^{r+1} \to M_{j-i+1} \otimes \bigwedge\nolimits^{i-1} \mathbb{K}^{r+1} \to$$

$$\cdots \to M_j \otimes \bigwedge\nolimits^{0} \mathbb{K}^{r+1} \to 0.$$

In particular we have $\beta_{i,j}(M) \le H_M(j-i)\binom{r+1}{i}$, *so* $\beta_{i,j}(M) = 0$ *if* $i > r+1$.

See Exercise 2.5 for the relation of this to Corollary 1.10.

Proof. To simplify the notation, let $\beta_{i,j} = \beta_{i,j}(M)$. By Proposition 1.7,

$$\beta_{i,j} = \dim_{\mathbb{K}} \mathrm{Tor}_i(M, \mathbb{K})_j.$$

Since $\mathbb{K}(x_0, \ldots, x_r)$ is a free resolution of $\mathbb{K}$, we may compute $\mathrm{Tor}_i^S(M, \mathbb{K})_j$ as the degree-j part of the homology of $M \otimes_S \mathbb{K}(x_0, \ldots, x_r)$ at the term

$$M \otimes_S \textstyle\bigwedge^i S^{r+1}(-i) = M \otimes_{\mathbb{K}} \textstyle\bigwedge^i \mathbb{K}^{r+1}(-i).$$

Decomposing M into its homogeneous components $M = \oplus M_k$, we see that the degree-j part of $M \otimes_{\mathbb{K}} \bigwedge^i \mathbb{K}^{r+1}(-i)$ is $M_{j-i} \otimes_{\mathbb{K}} \bigwedge^i \mathbb{K}^{r+1}$. The differentials of $M \otimes_S \mathbb{K}(x_0, \ldots, x_r)$ preserve degrees, so the complex decomposes as a direct sum of complexes of vector spaces of the form

$$M_{j-i-1} \otimes_{\mathbb{K}} \textstyle\bigwedge^{i+1} \mathbb{K}^{r+1} \longrightarrow M_{j-i} \otimes_{\mathbb{K}} \textstyle\bigwedge^i \mathbb{K}^{r+1} \longrightarrow M_{j-i+1} \otimes_{\mathbb{K}} \textstyle\bigwedge^{i-1} \mathbb{K}^{r+1}.$$

This proves the first statement. The inequality on $\beta_{i,j}$ follows at once.   □

The upper bound given in Proposition 2.7 is achieved when $\mathfrak{m}M = 0$ (and conversely — see Exercise 2.6). It is not hard to deduce a weak lower bound, too (Exercise 2.7), but is often a very difficult problem, to determine the actual range of possibilities, especially when the module M is supposed to come from some geometric construction.

An example will illustrate some of the possible considerations. A true geometric example, related to this one, will be given in the next section. Suppose that $r = 2$ and the Hilbert function of M has values

$$H_M(j) = \begin{cases} 0 & \text{if } j < 0, \\ 1 & \text{if } j = 0, \\ 3 & \text{if } j = 1, \\ 3 & \text{if } j = 2, \\ 0 & \text{if } j > 2. \end{cases}$$

To fit with the way we write Betti diagrams, we represent the complexes in Proposition 2.7 with maps going from right to left, and put the term $M_j \otimes \bigwedge^i \mathbb{K}^{r+1}(-i) = M_j(-i)\binom{r+1}{i}$ (the term of degree $i+j$) in row j and column i. Because the differential has degree 0, it goes diagonally down and to the left.

M	$M \otimes_{\mathbb{K}} \wedge \mathbb{K}^r(-1)$			
M_0	$\mathbb{K}^1$	$\mathbb{K}^3$	$\mathbb{K}^3$	$\mathbb{K}^1$
M_1	$\mathbb{K}^3$	$\mathbb{K}^9$	$\mathbb{K}^9$	$\mathbb{K}^3$
M_2	$\mathbb{K}^3$	$\mathbb{K}^9$	$\mathbb{K}^9$	$\mathbb{K}^3$

From this we see that the termwise maximal Betti diagram of a module with the given Hilbert function, valid if the module structure of M is trivial, is

	0	1	2	3
0	1	3	3	1
1	3	9	9	3
2	3	9	9	3

On the other hand, if the differential

$$d_{i,j} : \; M_{j-i} \otimes \bigwedge^i \mathbb{K}^3 \to M_{j-i+1} \otimes \bigwedge^{i-1} \mathbb{K}^3$$

has rank k, both $\beta_{i,j}$ and $\beta_{i-1,j}$ drop from this maximal value by k.

Other considerations come into play as well. For example, suppose that M is a cyclic module (a module requiring only one generator), generated by M_0. Equivalently, $\beta_{0,j} = 0$ for $j \neq 0$. It follows that the differentials $d_{1,1}$ and $d_{1,2}$ have rank 3, so $\beta_{1,1} = 0$ and $\beta_{1,2} \leq 6$. Since $\beta_{1,1} = 0$, Proposition 1.9 implies that $\beta_{i,i} = 0$ for all $i \geq 1$. This means that the differential $d_{2,2}$ has rank 3 and the differential $d_{3,3}$ has rank 1, so the maximal possible Betti numbers are

	0	1	2	3
0	1	–	–	–
1	–	3	8	3
2	–	9	9	3

Whatever the ranks of the remaining differentials, we see that any Betti diagram of a cyclic module with the given Hilbert function has the form

	0	1	2	3
0	1	–	–	–
1	–	3	$\beta_{2,3}$	$\beta_{3,4}$
2	–	$1+\beta_{2,3}$	$6+\beta_{3,4}$	3

for some $0 \leq \beta_{2,3} \leq 8$ and $0 \leq \beta_{3,4} \leq 3$. For example, if all the remaining differentials have maximal rank, the Betti diagram would be

	0	1	2	3
0	1	–	–	–
1	–	3	–	–
2	–	1	6	3

We will see in the next section that this diagram is realized as the Betti diagram of the homogeneous coordinate ring of a general set of 7 points in $\mathbb{P}^3$ modulo a nonzerodivisor of degree 1.

2C   Geometry from Syzygies: Seven Points in $\mathbb{P}^3$

We have seen above that if we know the graded Betti numbers of a graded S-module, then we can compute the Hilbert function. In geometric situations, the graded Betti numbers often carry information beyond that of the Hilbert function. Perhaps the most interesting current results in this direction center on *Green's Conjecture* described in Section 9B.

For a simpler example we consider the graded Betti numbers of the homogeneous coordinate ring of a set of 7 points in "linearly general position" (defined below) in $\mathbb{P}^3$. We will meet a number of the ideas that occupy the next few chapters. To save time we will allow ourselves to quote freely from material developed (independently of this discussion!) later in the text. The inexperienced reader should feel free to look at the statements and skip the proofs in the rest of this section until after having read through Chapter 6.

The Hilbert Polynomial and Function. . .

Any set X of 7 distinct points in $\mathbb{P}^3$ has Hilbert polynomial equal to the constant 7 (such things are discussed at the beginning of Chapter 4). However, not all sets of 7 points in $\mathbb{P}^3$ have the same Hilbert function. For example, if X is not contained in a plane then the Hilbert function $H = H_{S_X}(d)$ begins with the values $H(0) = 1$, $H(1) = 4$, but if X is contained in a plane then $H(1) < 4$.

To avoid such degeneracy we will restrict our attention in the rest of this section to 7-tuples of points that are in *linearly general position*. We say that a set of points $Y \subset \mathbb{P}^r$ is in linearly general position if there are no more than 2 points of Y on any line, no more than 3 points on any 2-plane, ..., no more than r points in an $r-1$ plane. Thinking of the points as coming from vectors in $\mathbb{K}^{r+1}$, this means that every subset of at most $r+1$ of the vectors is linearly independent. Of course if there are at least $r+1$ points, this is equivalent to say simply that every subset of exactly $r+1$ of the vectors is linearly independent.

The condition that a set of points is in linearly general position arises frequently. For example, the general hyperplane section of any irreducible curve over a field of characteristic 0 is a set of points in linearly general position [Harris 1980] and this is usually, though not always, true in characteristic p as well [Rathmann 1987]. See Exercises 8.17–8.20.

It is not hard to show — the reader is invited to prove a more general fact in Exercise 2.9 — that the Hilbert function of any set X of 7 points in linearly general position in $\mathbb{P}^3$ is given by the table

d	0	1	2	3	...
$H_{S_X}(d)$	1	4	7	7	...

In particular, any set X of 7 points in linearly general position lies on exactly $3 = \binom{3+2}{2} - 7$ independent quadrics. These three quadrics cannot generate the ideal: since $S = \mathbb{K}[x_0, \ldots, x_3]$ has only four linear forms, the dimension of the space of cubics in the ideal generated by the three quadrics is at most $4 \times 3 = 12$, whereas there are $\binom{3+3}{3} - 7 = 13$ independent cubics in the ideal of X. Thus the ideal of X requires at least one cubic generator in addition to the three quadrics.

One might worry that higher degree generators might be needed as well. The ideal of 7 points on a line in $\mathbb{P}^3$, for example, is minimally generated by the two linear forms that generate the ideal of the line, together with any form of degree 7 vanishing on the points but not on the line. But Theorem 4.2(c) tells us that since the 7 points of X are in linearly general position the Castelnuovo–Mumford regularity of S_X (defined in Chapter 4) is 2, or equivalently, that the Betti diagram of S_X fits into 3 rows. Moreover, the ring S_X is reduced and of dimension 1 so it has depth 1. The Auslander–Buchsbaum Formula A2.15 shows that the resolution will have length 3. Putting this together, and using Corollary 1.9 we see that the minimal free resolution of S_X must have Betti diagram of the form

	0	1	2	3
0	1	–	–	–
1	–	$\beta_{1,2}$	$\beta_{2,3}$	$\beta_{3,4}$
2	–	$\beta_{1,3}$	$\beta_{2,4}$	$\beta_{3,5}$

where the $\beta_{i,j}$ that are not shown are zero. In particular, the ideal of X is generated by quadrics and cubics.

Using Corollary 1.10 we compute successively $\beta_{1,2} = 3$, $\beta_{1,3} - \beta_{2,3} = 1$, $\beta_{2,4} - \beta_{3,4} = 6$, $\beta_{3,5} = 3$, and the Betti diagram has the form

	0	1	2	3
0	1	–	–	–
1	–	3	$\beta_{2,3}$	$\beta_{3,4}$
2	–	$1+\beta_{2,3}$	$6+\beta_{3,4}$	3

(This is the same diagram as at the end of the previous section. Here is the connection: Extending the ground field if necessary to make it infinite, we could use Lemma A2.3 and choose a linear form $x \in S$ that is a nonzerodivisor on S_X. By Lemma 3.15 the graded Betti numbers of S_X/xS_X as an S/xS-module are the same as those of S_X as an S-module. Using our knowledge of the Hilbert function of S_X and the exactness of the sequence

$$0 \longrightarrow S_X(-1) \overset{x}{\longrightarrow} S_X \longrightarrow S_X/xS_x \longrightarrow 0,$$

we see that the cyclic (S/xS)-module S_X/xS_x has Hilbert function with values $1, 3, 3$. This is what we used in Section 2B.)

... and Other Information in the Resolution

We see that even in this simple case the Hilbert function does not determine the $\beta_{i,j}$, and indeed they can take different values. It turns out that the difference reflects a fundamental geometric distinction between different sets X of 7 points in linearly general position in $\mathbb{P}^3$: whether or not X lies on a curve of degree 3.

Up to linear automorphisms of $\mathbb{P}^3$ there is only one irreducible curve of degree 3 not contained in a plane. This *twisted cubic* is one of the *rational normal curves* studied in Chapter 6. Any 6 points in linearly general position in $\mathbb{P}^3$ lie on a unique twisted cubic (see Exercise 6.5). But for a twisted cubic to pass through 7 points, the seventh must lie on the twisted cubic determined by the first 6. Thus most sets of seven points do not lie on any twisted cubic.

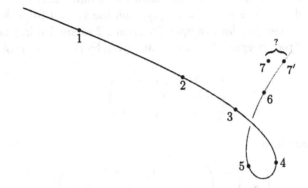

Theorem 2.8. *Let X be a set of 7 points in linearly general position in $\mathbb{P}^3$. There are just two distinct Betti diagrams possible for the homogeneous coordinate ring S_X:*

	0	1	2	3
0	1	–	–	–
1	–	3	–	–
2	–	1	6	3

and

	0	1	2	3
0	1	–	–	–
1	–	3	2	–
2	–	3	6	3

In the first case the points do not lie on any curve of degree 3. In the second case, the ideal J generated by the quadrics containing X is the ideal of the unique curve of degree 3 containing X, which is irreducible.

Proof. Let q_0, q_1, q_2 be three quadratic forms that span the degree 2 part of $I := I_X$. A *linear syzygy* of the q_i is a vector (a_0, a_1, a_2) of linear forms with $\sum_{i=0}^{2} a_i q_i = 0$. We will focus on the number of independent linear syzygies, which is $\beta_{2,3}$.

If $\beta_{2,3} = 0$, Proposition 1.9 implies that $\beta_{3,4} = 0$ and the computation of the differences of the $\beta_{i,j}$ above shows that the Betti diagram of $S_X = S/I$ is the first of the two given tables. As we shall see in Chapter 6, any irreducible curve of degree ≤ 2 lies in a plane. Since the points of X are in linearly general position, they are not contained in the union of a line and a plane, or the union of 3 lines, so any degree 3 curve containing X is irreducible. Further, if C is an irreducible degree 3 curve in $\mathbb{P}^3$, not contained in a plane, then the C is a twisted cubic, and the ideal of C is generated by three quadrics, which have 2 linear syzygies. Thus in the case where X is contained in a degree 3 curve we have $\beta_{2,3} \geq 2$.

Now suppose $\beta_{2,3} > 0$, so that there is a nonzero linear syzygy $\sum_{i=0}^{2} a_i q_i = 0$. If the a_i were linearly dependent then we could rewrite this relation as $a'_1 q'_1 + a'_2 q'_2 = 0$ for some independendent quadrics q'_1 and q'_2 in I. By unique factorization, the linear form a'_1 would divide q'_2; say $q'_2 = a'_1 b$. Thus X would be contained in the union of the planes $a_1' = 0$ and $b = 0$, and one of these planes would contain four points of X, contradicting our hypothesis. Therefore a_0, a_1, a_2 are linearly independent linear forms.

Changing coordinates on $\mathbb{P}^3$ we can harmlessly assume that $a_i = x_i$. We can then read the relation $\sum x_i q_i = 0$ as a syzygy on the x_i. But from the exactness of the Koszul complex (see for example Theorem 2.2 as applied in Example 2.6), we know that all the syzygies of x_0, x_1, x_2 are given by the columns of the matrix

$$\begin{pmatrix} 0 & x_2 & -x_1 \\ -x_2 & 0 & x_0 \\ x_1 & -x_0 & 0 \end{pmatrix},$$

and thus we must have

$$\begin{pmatrix} q_0 \\ q_1 \\ q_2 \end{pmatrix} = \begin{pmatrix} 0 & x_2 & -x_1 \\ -x_2 & 0 & x_0 \\ x_1 & -x_0 & 0 \end{pmatrix} \begin{pmatrix} b_0 \\ b_1 \\ b_2 \end{pmatrix}$$

for some linear forms b_i. Another way to express this equation is to say that q_i is $(-1)^i$ times the determinant of the 2×2 matrix formed by omitting the i-th column of the matrix

$$M = \begin{pmatrix} x_0 & x_1 & x_2 \\ b_0 & b_1 & b_2 \end{pmatrix},$$

where the columns are numbered $0, 1, 2$. The two rows of M are independent because the q_i, the minors, are nonzero. (Throughout this book we will follow the convention that a *minor* of a matrix is a subdeterminant times an appropriate sign.)

We claim that both rows of M give relations on the q_i. The vector (x_0, x_1, x_2) is a syzygy by virtue of our choice of coordinates. To see that (b_0, b_1, b_2) is also a syzygy, note that the Laplace expansion of

$$\det \begin{pmatrix} x_0 & x_1 & x_2 \\ b_0 & b_1 & b_2 \\ b_0 & b_1 & b_2 \end{pmatrix}$$

is $\sum_i b_i q_i$. However, this 3×3 matrix has a repeated row, so the determinant is 0, showing that $\sum_i b_i q_i = 0$. Since the two rows of M are linearly independent, we see that the q_i have (at least) 2 independent syzygies with linear forms as coefficients.

The ideal $(q_0, q_1, q_2) \subset I$ that is generated by the minors of M is unchanged if we replace M by a matrix PMQ, where P and Q are invertible matrices of scalars. It follows that matrices of the form PMQ cannot have any entries equal to zero. This shows that M is 1-generic in the sense of Chapter 6, and it follows from Theorem 6.4 that the ideal $J = (q_0, q_1, q_2) \subset I$ is prime and of codimension 2 — that is, J defines an irreducible curve C containing X in $\mathbb{P}^3$.

From Theorem 3.2 it follows that a free resolution of S_C may be written as

$$0 \to S^2(-3) \xrightarrow{\begin{pmatrix} x_0 & b_0 \\ x_1 & b_1 \\ x_2 & b_2 \end{pmatrix}} S^3(-2) \xrightarrow{(q_0 \quad q_1 \quad q_2)} S \longrightarrow S_C \longrightarrow 0.$$

From the resolution of S_C we can also compute its Hilbert function:

$$H_{S_C}(d) = \binom{3+d}{3} - 3\binom{3+d-2}{3} + 2\binom{3+d-3}{3}$$

$$= 3d+1 \quad \text{for } d \geq 0.$$

Thus the Hilbert polynomial of the curve is $3d+1$. It follows that C is a cubic curve — see [Hartshorne 1977, Prop. I.7.6], for example. $\qquad \square$

It may be surprising that in Theorem 2.8 the only possibilities for $\beta_{2,3}$ are 0 and 2, and that $\beta_{3,4}$ is always 0. These restrictions are removed, however, if one looks at sets of 7 points that are not in linearly general position though they have the same Hilbert function as a set of points in linearly general position; some examples are given in Exercises 2.11–2.12.

2D Exercises

1. Suppose that $m_1, \ldots, m_n$ are monomials in S. Show that the projective dimension of $S/(m_1, \ldots, m_n)$ is at most n. No such principle holds for arbitrary homogeneous polynomials; see Exercise 2.4.

2. Let $0 \leq n \leq r$. Show that if M is a graded S-module which contains a submodule isomorphic to $S/(x_0, \ldots, x_n)$ (so that $(x_0, \ldots, x_n)$ is an associated prime of M) then the projective dimension of M is at least $n+1$. If $n+1$ is equal to the number of variables in S, show that this condition is necessary as well as sufficient. (Hint: For the last statement, use the Auslander–Buchsbaum theorem, Theorem A2.15.)

3. Consider the ideal $I = (x_0, x_1) \cap (x_2, x_3)$ of two skew lines in $\mathbb{P}^3$:

Prove that $I = (x_0 x_2,\ x_0 x_3,\ x_1 x_2,\ x_1 x_3)$, and compute the minimal free resolution of S/I. In particular, show that S/I has projective dimension 3 even though its associated primes are precisely (x_0, x_1) and (x_2, x_3), which have height only 2. Thus the principle of Exercise 2.2 can't be extended to give the projective dimension in general.

4. Show that the ideal $J = (x_0 x_2 - x_1 x_3, \ x_0 x_3, x_1 x_2)$ defines the union of two (reduced) lines in $\mathbb{P}^3$, but is not equal to the saturated ideal of the two lines. Conclude that the projective dimension of S/J is 4 (you might use the Auslander–Buchsbaum formula, Theorem A2.15). In fact, three-generator ideals can have any projective dimension; see [Bruns 1976] or [Evans and Griffith 1985, Corollary 3.13].

5. Let M be a finitely generated graded S-module and let $B_j = \sum_i (-1)^i \beta_{i,j}(M)$. Show from Proposition 2.7 that

$$B_j = \sum_i (-1)^i H_M(j-i) \binom{r+1}{i}.$$

This is another form of the formula in Corollary 1.10.

6. Show that if M is a graded S module, then

$$\beta_{0,j}(M) = H_M(j) \quad \text{for all } j$$

if and only if $\mathfrak{m}M = 0$.

7. If M is a graded S-module, show that

$$\beta_{i,j}(M) \geq H_M(j-i)\binom{r+1}{i} - H_M(j-i+1)\binom{r+1}{i-1} - H_M(j-i-1)\binom{r+1}{i+1}.$$

8. Prove that the complex

$$0 \to S^2(-3) \xrightarrow{\begin{pmatrix} x_0 & x_1 \\ x_1 & x_2 \\ x_2 & x_3 \end{pmatrix}} S^3(-2) \xrightarrow{(x_1 x_3 - x_2^2 \quad -x_0 x_3 + x_1 x_2 \quad x_0 x_2 - x_1^2)} S$$

is indeed a resolution of the homogeneous coordinate ring S_C of the twisted cubic curve C, by the following steps:

(a) Identify S_C with the subring of $\mathbb{K}[s, t]$ consisting of those graded components whose degree is divisible by 3. Show in this way that $H_{S_C}(d) = 3d+1$ for $d \geq 0$.

(b) Compute the Hilbert functions of the terms S, $S^3(-2)$, and $S^2(-3)$. Show that their alternating sum $H_S - H_{S^3(-2)} + H_{S^2(-3)}$ is equal to the Hilbert function H_{S_C}.

(c) Show that the map

$$S^2(-3) \xrightarrow{\begin{pmatrix} x_0 & x_1 \\ x_1 & x_2 \\ x_2 & x_3 \end{pmatrix}} S^3(-2)$$

is a monomorphism. As a first step you might prove that it becomes a monomorphism when the polynomial ring S is replaced by its quotient field, the field of rational functions.

(d) Show that the results in parts (b) and (c) together imply that the complex exhibited above is a free resolution of S_C.

9. Let X be a set of $n \le 2r+1$ points in $\mathbb{P}^r$ in linearly general position. Show that X imposes independent conditions on quadrics: that is, show that the space of quadratic forms vanishing on X is $\binom{r+2}{2} - n$ dimensional. (It is enough to show that for each $p \in X$ there is a quadric not vanishing on p but vanishing at all the other points of X.) Use this to show that X imposes independent conditions on forms of degree ≥ 2. The same idea can be used to show that any $n \le dr+1$ points in linearly general position impose independent conditions on forms of degree d.

Deduce the correctness of the Hilbert function for 7 points in linearly general position given by the table in Section 2C.

10. The sufficient condition of Exercise 2.9 is far from necessary. One way to sharpen it is to use Edmonds' Theorem [1965], which is the following beautiful and nontrivial theorem in linear algebra (see [Graham et al. 1995, Chapter 11, Theorem 3.9] for an exposition):

Theorem 2.9. *Let $v_1, \ldots, v_{ds}$ be vectors in an s-dimensional vector space. The list $(v_1, \ldots, v_{ds})$ can be written as the union of d bases if and only if no $dk+1$ of the vectors v_i lie in a k-dimensional subspace, for every k.*   □

Now suppose that Γ is a set of at most $2r+1$ points in $\mathbb{P}^r$, and, for all $k < r$, each set of $2k+1$ points of Γ spans at least a $(k+1)$-plane. Use Edmonds' Theorem to show that Γ imposes independent conditions on quadrics in $\mathbb{P}^r$ (Hint: You can apply Edmonds' Theorem to the set obtained by counting one of the points of Γ twice.)

11. Show that if X is a set of 7 points in $\mathbb{P}^3$ with 6 points on a plane, but not on any conic curve in that plane, while the seventh point does not line in the plane, then X imposes independent conditions on forms of degree ≥ 2 and $\beta_{2,3} = 3$.

12. Let $\Lambda \subset \mathbb{P}^3$ be a plane, and let $D \subset \Lambda$ be an irreducible conic. Choose points $p_1, p_2 \notin \Lambda$ such that the line joining p_1 and p_2 does not meet D. Show that if X is a set of 7 points in $\mathbb{P}^3$ consisting of p_1, p_2 and 5 points on D, then X imposes independent conditions on forms of degree ≥ 2 and $\beta_{2,3} = 1$. (Hint: To show that $\beta_{2,3} \ge 1$, find a pair of reducible quadrics in the ideal having a common component. To show that $\beta_{2,3} \le 1$, show that the quadrics through the points are the same as the quadrics containing D and the two points. There is, up to automorphisms of $\mathbb{P}^3$, only one configuration consisting of a conic and two points in $\mathbb{P}^3$ such that the line though the two points does not meet the conic. You might produce such a configuration explicitly and compute the quadrics and their syzygies.)

13. Show that the labeled simplicial complex

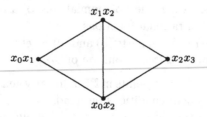

gives a nonminimal free resolution of the monomial ideal

$$(x_0x_1, x_0x_2, x_1x_2, x_2x_3).$$

Use this to prove that the Betti diagram of a minimal free resolution is

	0	1	2	3
0	1	–	–	–
1	–	4	4	1

14. Use the Betti diagram in Exercise 2.13 to show that the minimal free resolution of $(x_0x_1, x_0x_2, x_1x_2, x_2x_3)$ cannot be written as $\mathscr{C}(\Delta)$ for any labeled simplicial complex Δ. (It can be written as the free complex coming from a certain topological cell complex; for this generalization see [Bayer and Sturmfels 1998].)

15. Show the ideal

$$I = (x^3, x^2y, x^2z, y^3) \subset S = \mathbb{K}[x, y, z]$$

has minimal free resolution $\mathscr{C}(\Delta)$, where Δ is the labeled simplicial complex

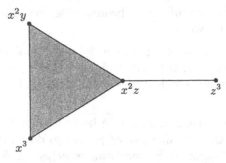

Compute the Betti diagram, the Hilbert function, and the Hilbert polynomial of S/I, and show that in this case the bound given in Corollary 1.3 is not sharp. Can you see this from the Betti diagram?

3
Points in $\mathbb{P}^2$

The first case in which the relation of syzygies and geometry becomes clear, and the one in which it is best understood, is the case where the geometric objects are finite sets of points in $\mathbb{P}^2$. We will devote this chapter to such sets. (The reader who knows about schemes, for example at the level of the first two chapters of [Eisenbud and Harris 2000], will see that exactly the same considerations apply to finite schemes in $\mathbb{P}^2$.) Of course the only intrinsic geometry of a set of points is the number of points, and we will see that this is the data present in the Hilbert polynomial. But a set of points embedded in projective space has plenty of extrinsic geometry. For example, it is interesting to ask what sorts of curves a given set of points lies on, or to ask about the geometry of the dual hyperplane arrangement (see [Orlik and Terao 1992]), or about the embedding of the "Gale transform" of the points (see [Eisenbud and Popescu 1999]). All of these things have some connections with syzygies.

Besides being a good model problem, the case of points in $\mathbb{P}^2$ arises directly in considering the plane sections of varieties of codimension 2, such as the very classical examples of curves in $\mathbb{P}^3$ and surfaces in $\mathbb{P}^4$. For example, a knowledge of the possible Hilbert functions of sets of points in "uniform position" is the key ingredient in "Castelnuovo Theory", which treats the possible genera of curves in $\mathbb{P}^3$ and related problems.

Despite this wealth of related topics, the goal of this chapter is modest: We will characterize the Betti diagrams of the possible minimal graded free resolutions of ideals of forms vanishing on sets of points in $\mathbb{P}^2$, and begin to relate these discrete invariants to geometry in simple cases.

Throughout this chapter, S will denote the graded ring $\mathbb{K}[x_0, x_1, x_2]$. All the S-modules we consider will be finitely generated and graded. Such a module admits a minimal free resolution, unique up to isomorphism. By Corollary 1.8, its length is equal to the module's projective dimension.

3A   The Ideal of a Finite Set of Points

The simplest ideals are principal ideals. As a module, such an ideal is free. The next simplest case is perhaps that of an ideal having a free resolution of length 1, and we will see that the ideal of forms vanishing on any finite set of points in $\mathbb{P}^2$ has this property.

We will write pd I for the projective dimension of I. By the *depth* of a graded ring, we mean the grade of the irrelevant ideal — that is, the length of a maximal regular sequence of homogeneous elements of positive degree. (The homogeneous case is very similar to the local case; for example, all maximal regular sequences have the same length in the homogeneous case as in the local case, and the local proofs can be modified to work in the homogeneous case. For a systematic treatement see [Goto and Watanabe 1978a; 1978b].)

Proposition 3.1. *If $I \subset S$ is the homogeneous ideal of a finite set of points in $\mathbb{P}^2$, then I has a free resolution of length 1.*

Proof. Suppose $I = I(X)$, the ideal of forms vanishing on the finite set $X \subset \mathbb{P}^2$. By the Auslander–Buchsbaum Formula (Theorem A2.15) we have

$$\text{pd } S/I = \text{depth } S - \text{depth } S/I.$$

But depth $S/I \leq \dim S/I = 1$. The ideal I is the intersection of the prime ideals of forms vanishing at the individual points of X, so the maximal homogeneous ideal $\mathfrak{m}$ of S is not associated to I. This implies that depth $S/I > 0$. Also, the depth of S is 3 (the variables form a maximal homogeneous regular sequence). Thus pd $S/I = 3 - 1 = 2$, whence pd $I = 1$, as I is the first module of syzygies in a free resolution of S/I.   □

It turns out that ideals with a free resolution of length 1 are determinantal (see Appendix A2G for some results about determinantal ideals.) This result was discovered by Hilbert in a special case and by Burch in general.

The Hilbert–Burch Theorem

In what follows, we shall work over an arbitrary Noetherian ring R. (Even more general results are possible; see for example [Northcott 1976].) For any matrix M with entries in R we write $I_t(M)$ for the ideal generated by the $t \times t$ subdeterminants of M. The length of a maximal regular sequence in an ideal I is written grade I.

Theorem 3.2 (Hilbert–Burch). *Suppose that an ideal I in a Noetherian ring R admits a free resolution of length 1:*

$$0 \longrightarrow F \overset{M}{\longrightarrow} G \longrightarrow I \longrightarrow 0.$$

If the rank of the free module F is t, then the rank of G is $t+1$, and there exists a nonzerodivisor a such that $I = aI_t(M)$. Regarding M as a matrix with respect to given bases of F and G, the generator of I that is the image of the i-th basis vector of G is $\pm a$ times the determinant of the submatrix of M formed from all except the i-th row. Moreover, the grade of $I_t(M)$ is 2.

Conversely, given a nonzerodivisor a of R and given a $(t+1) \times t$ matrix M with entries in R such that $\operatorname{grade} I_t(M) \geq 2$, the ideal $I = aI_t(M)$ admits a free resolution of length one as above. The ideal I has grade 2 if and only if the element a is a unit.

In view of the signs that appear in front of the determinants, we define the i-th *minor* of M to be $(-1)^i \det M_i'$, where M_i' is the matrix M' with the i-th row omitted. We can then say that the generator of I that is the image of the i-th basis vector of G is a times the i-th minor of M.

We postpone the proof in order to state a general result describing free resolutions. If φ is a map of free R-modules, we write $\operatorname{rank}(\varphi)$ for the rank (that is, the largest size of a nonvanishing minor) and $I(\varphi)$ for the determinantal ideal $I_{\operatorname{rank}(\varphi)}(\varphi)$. For any map φ of free modules we make the convention that $I_0(\varphi) = R$. In particular, if φ is the zero map, the rank of φ is 0, so $I(\varphi) = I_0(\varphi) = R$. We also take $\operatorname{depth}(R, R) = \infty$, so that $\operatorname{grade} I(\varphi) = \infty$ if $I(\phi) = R$.

Theorem 3.3 (Buchsbaum–Eisenbud). *A complex of free modules*

$$\mathbf{F}: 0 \longrightarrow F_m \overset{\varphi_m}{\longrightarrow} F_{m-1} \longrightarrow \cdots \longrightarrow F_1 \overset{\varphi_1}{\longrightarrow} F_0$$

over a Noetherian ring R is exact if and only if

$$\operatorname{rank} \varphi_{i+1} + \operatorname{rank} \varphi_i = \operatorname{rank} F_i \quad \text{and} \quad \operatorname{depth} I(\varphi_i) \geq i \qquad \text{for every } i.$$

For a proof see [Eisenbud 1995, Theorem 20.9]. It is crucial that the complex begin with a zero on the left; no similar result is known without such hypotheses.

In the special case where R is a polynomial ring $R = \mathbb{K}[x_0, \ldots, x_r]$ and $\mathbb{K}$ is algebraically closed, Theorem 3.3 has a simple geometric interpretation. We think of R as a ring of functions on $\mathbb{K}^{r+1}$ (in the graded case we could work with $\mathbb{P}^r$ instead). If $p \in \mathbb{K}^{r+1}$, we write $I(p)$ for the ideal of functions vanishing at p, and we write

$$\mathbf{F}(p): 0 \longrightarrow F_m(p) \overset{\varphi_m(p)}{\longrightarrow} \cdots \overset{\varphi_1(p)}{\longrightarrow} F_0(p)$$

for the result of tensoring $\mathbf{F}$ with the residue field $\kappa(p) := R/I(p)$, regarded as a complex of finite-dimensional vector spaces over $\kappa(p)$. A matrix for the map $\varphi_i(p)$ is obtained by evaluating a matrix for the map φ_i at p. Theorem 3.3 expresses the relation between the exactness of the complex of free modules $\mathbf{F}$ and the exactness of the complexes of vector spaces $\mathbf{F}(p)$.

Corollary 3.4. *Let*

$$\mathbf{F}: \quad 0 \longrightarrow F_m \xrightarrow{\varphi_m} F_{m-1} \longrightarrow \cdots \longrightarrow F_1 \xrightarrow{\varphi_1} F_0$$

be a complex of free modules over the polynomial ring $S = \mathbb{K}[x_0, \ldots, x_r]$, where $\mathbb{K}$ is an algebraically closed field. Let $X_i \subset \mathbb{K}^{r+1}$ be the set of points p such that the evaluated complex $\mathbf{F}(p)$ is not exact at $F_i(p)$. The complex $\mathbf{F}$ is exact if and only if, for every i, the set X_i is empty or $\operatorname{codim} X_i \geq i$.

Proof. Set $r_i = \operatorname{rank} F_i - \operatorname{rank} F_{i+1} + \ldots \pm \operatorname{rank} F_m$. Theorem 3.3 implies that $\mathbf{F}$ is exact if and only if $\operatorname{grade} I_{r_i}(\varphi_i) \geq i$ for each $i \geq 1$. First, if $\mathbf{F}$ is exact then by descending induction we see from condition 1 of the theorem that $\operatorname{rank} \varphi_i = r_i$ for every i, and then the condition $\operatorname{grade} I_{r_i}(\varphi_i) \geq i$ is just condition 2 of Theorem 3.3.

Conversely, suppose that $\operatorname{grade} I_{r_i}(\varphi_i) \geq i$. It follows that $\operatorname{rank} \varphi_i \geq r_i$ for each i. Tensoring with the quotient field of R we see that $\operatorname{rank} \varphi_{i+1} + \operatorname{rank} \varphi_i \leq \operatorname{rank} F_i$ in any case. Using this and the previous inequality, we see by descending induction that in fact $\operatorname{rank} \varphi_i = r_i$ for every i, so conditions 1 and 2 of Theorem 3.3 are satisfied.

Now let

$$Y_i = \{p \in \mathbb{K}^{r+1} \mid \operatorname{rank} \varphi_i(p) < r_i\}.$$

Thus Y_i is the algebraic set defined by the ideal $I_{r_i}(\varphi_i)$. Since the polynomial ring S is Cohen–Macaulay (Theorem A2.33) the grade of $I_{r_i}(\varphi_i)$ is equal to the codimension of this ideal, which is the same as the codimension of Y_i. It follows that $\mathbf{F}$ is exact if and only if the codimension of Y_i in $\mathbb{K}^{r+1}$ is at least i for each $i \geq 1$.

On the other hand, the complex of finite-dimensional $\mathbb{K}$-vector spaces $\mathbf{F}(p)$ is exact at $F_j(p)$ if and only if $\operatorname{rank} \varphi_{j+1}(p) + \operatorname{rank} \varphi_j(p) = \operatorname{rank} F_j(p)$. Since $\mathbf{F}(p)$ is a complex, this is the same as saying that $\operatorname{rank} \varphi_{j+1}(p) + \operatorname{rank} \varphi_j(p) \geq \operatorname{rank} F_j(p)$. This is true for all $j \geq i$ if and only if $\operatorname{rank} \varphi_j(p) \geq r_j$ for all $j \geq i$. Thus $\mathbf{F}(p)$ is exact at $F_j(p)$ for all $j \geq i$ if and only if $p \notin \bigcup_{j \geq i} Y_j$.

The codimension of $\bigcup_{j \geq i} Y_j$ is the minimum of the codimensions of the Y_j for $j \geq i$. Thus $\operatorname{codim} \bigcup_{j \geq i} Y_j \geq i$ for all i if and only if $\operatorname{codim} Y_i \geq i$ for all i. Thus $\mathbf{F}$ satisfies the condition of the Corollary if and only if $\mathbf{F}$ is exact. $\square$

Example 3.5. To illustrate these results, we return to the example in Exercise 2.8 and consider the complex

$$\mathbf{F}: 0 \to S^2(-3) \xrightarrow{\varphi_2 = \begin{pmatrix} x_0 & x_1 \\ x_1 & x_2 \\ x_2 & x_3 \end{pmatrix}} S^3(-2) \xrightarrow{\varphi_1 = (x_1 x_3 - x_2^2 \quad -x_0 x_3 + x_1 x_2 \quad x_0 x_2 - x_1^2)} S.$$

In the notation of the proof of Corollary 3.4 we have $r_2 = 2$, $r_1 = 1$. Further, the entries of φ_1 are the 2×2 minors of φ_2, as in Theorem 3.2 with $a = 1$. In particular $Y_1 = Y_2$ and $X_1 = X_2$. Thus Corollary 3.4 asserts that $\mathbf{F}$ is exact if and only if $\operatorname{codim} X_2 \geq 2$. But X_2 consists of the points p where φ_2 fails to be a monomorphism — that is, where $\operatorname{rank}(\varphi(p)) \leq 1$. If $p = (p_0, \dots, p_3) \in X_2$ and $p_0 = 0$ we see, by inspecting the matrix φ_2, that $p_1 = p_2 = 0$, so $p = (0,0,0,p_3)$. Such points form a set of codimension 3 in $\mathbb{K}^4$. On the other hand, if $p \in X_2$ and $p_0 \neq 0$ we see, again by inspecting the matrix φ_2, that $p_2 = (p_1/p_0)^2$ and $p_3 = (p_1/p_0)^3$. Thus p is determined by the two parameters p_0, p_1, and the set of such p has codimension at least $4 - 2 = 2$. In particular X_2, the union of these two sets, has codimension at least 2, so $\mathbf{F}$ is exact by Corollary 3.4.

In this example all the ideals are homogeneous, and the projective algebraic set X_2 is in fact the twisted cubic curve.

A consequence of Theorem 3.2 in the general case is that any ideal with a free resolution of length 1 contains a nonzerodivisor. Theorem 3.3 allows us to prove a more general result of Auslander and Buchsbaum:

Corollary 3.6 (Auslander–Buchsbaum). *An ideal I that has a finite free resolution contains a nonzerodivisor.*

In the nongraded, nonlocal case, having a finite projective resolution (finite projective dimension) would not be enough; for example, if k is a field, the ideal $k \times \{0\} \subset k \times k$ is projective but does not contain a nonzerodivisor.

Proof. In the free resolution

$$0 \longrightarrow F_n \xrightarrow{\varphi_n} \cdots \xrightarrow{\varphi_2} F_1 \xrightarrow{\varphi_1} R \longrightarrow R/I \longrightarrow 0$$

the ideal $I(\varphi_1)$ is exactly I. By Theorem 3.3 it has grade at least 1.   □

The proof of Theorem 3.2 depends on an identity:

Lemma 3.7. *If M is a $(t+1) \times t$ matrix over a commutative ring R, and $a \in R$, the composition*

$$R^t \xrightarrow{M} R^{t+1} \xrightarrow{\Delta} R$$

is zero, where the map Δ is given by the matrix $\Delta = (a\Delta_1, \dots, a\Delta_{t+1})$, the element Δ_i being the $t \times t$ minor of M omitting the i-th row (remember that by definition this minor is $(-1)^i$ times the determinant of the corresponding submatrix).

Proof. Write $a_{i,j}$ for the (i,j) entry of M. The i-th entry of the composite map ΔM is $a \sum_j \Delta_j a_{i,j}$, that is, a times the Laplace expansion of the determinant of the $(t+1) \times (t+1)$ matrix obtained from M by repeating the i-th column. Since any matrix with a repeated column has determinant zero, we get $\Delta M = 0$.   □

Proof of Theorem 3.2. We prove the last statement first: suppose that the grade of $I_t(M)$ is at least 2 and a is a nonzerodivisor. It follows that the rank of M is t, so that $I(M) = I_t(M)$, and the rank of Δ is 1. Thus $I(\Delta) = I_1(\Delta) = aI(M)$ and the grade of $I(\Delta)$ is at least 1. By Theorem 3.3,

$$0 \longrightarrow F \xrightarrow{M} G \longrightarrow I \longrightarrow 0$$

is the resolution of $I = aI(M)$, as required.

We now turn to the first part of Theorem 3.2. Using the inclusion of the ideal I in R, we see that there is a free resolution of R/I of the form

$$0 \longrightarrow F \xrightarrow{\ M\ } G \xrightarrow{\ A\ } R.$$

Since A is nonzero it has rank 1, and it follows from Theorem 3.3 that the rank of M must be t, and the rank of G must be $t+1$. The grade of $I(M) = I_t(M)$ is at least 2. Theorem A2.54 shows that the codimension of the ideal of $t \times t$ minors of a $(t+1) \times t$ matrix is at most 2. By Theorem A2.11 the codimension is an upper bound for the grade, so grade $I(M) = 2$. Write $\Delta = (\Delta_1, \dots, \Delta_{t+1})$, for the $1 \times (t+1)$ matrix whose entries Δ_i are the minors of M as in Lemma 3.7. Writing $-^*$ for $\mathrm{Hom}_R(-, R)$, it follows from Theorem 3.3 that the sequence

$$F^* \xleftarrow{\ M^*\ } G^* \xleftarrow{\ \Delta^*\ } R \longleftarrow 0,$$

which is a complex by Lemma 3.7, is exact. On the other hand, the image of the map A^* is contained in the kernel of M^*, so that there is a map $a : R \to R$ such that the diagram

$$
\begin{array}{ccccc}
F^* & \xleftarrow{\ M^*\ } & G^* & \xleftarrow{\ A^*\ } & R \\
\| & & \| & & \Big\downarrow a \\
F^* & \xleftarrow{\ M^*\ } & G^* & \xleftarrow{\ \Delta^*\ } & R
\end{array}
$$

commutes. The map a is represented by a 1×1 matrix whose entry we also call a. By Corollary 3.6, the ideal I contains a nonzerodivisor. But from the diagram above we see that $I = aI_t(M)$ is contained in (a), so a must be a nonzerodivisor.

As $I_t(M)$ has grade 2, the ideal $I = aI_t(M)$ has grade 2 if and only if a is a unit. With Theorem 3.3 this completes the proof. $\qquad\qquad\square$

Invariants of the Resolution

The Hilbert–Burch Theorem just described allows us to exhibit some interesting numerical invariants of a set X of points in $\mathbb{P}^2$. Throughout this section we will write $I = I_X \subset S$ for the homogeneous ideal of X, and $S_X = S/I_X$ for the homogeneous coordinate ring of X. By Proposition 3.1 the ideal I_X has projective dimension 1, and thus S_X has projective dimension 2. Suppose that the minimal graded free resolution of S_X has the form

$$\mathbf{F}: 0 \longrightarrow F \xrightarrow{\ M\ } G \longrightarrow S,$$

where G is a free module of rank $t+1$. By Theorem 3.2, the rank of F is t.

We can exhibit the numerical invariants of this situation either by using the degrees of the generators of the free modules or the degrees of the entries of the matrix M. We write the graded free modules G and F in the form $G = \bigoplus_1^{t+1} S(-a_i)$ and $F = \bigoplus_1^t S(-b_i)$, where, as always, $S(-a)$ denotes the free module of rank 1 with generator in degree a. The a_i are thus the degrees of the minimal generators of I. The degree of the (i,j) entry of the matrix M is then $b_j - a_i$. As we shall soon see, the degrees of the entries on the two principal diagonals of M determine all the other invariants. We write $e_i = b_i - a_i$ and $f_i = b_i - a_{i+1}$ for these degrees.

To make the data unique, we assume that the bases are ordered so that $a_1 \geq \cdots \geq a_{t+1}$ and $b_1 \geq \cdots \geq b_t$ or, equivalently, so that $f_i \geq e_i$ and $f_i \geq e_{i+1}$. Since the generators of G correspond to rows of M and the generators of F correspond to columns of M, and the e_i and f_i are degrees of entries of M, we can exhibit the data schematically as follows:

$$
\begin{array}{c}
 \\
a_1 \\
a_2 \\
\vdots \\
a_t \\
a_{t+1}
\end{array}
\begin{array}{cccc}
b_1 & b_2 & \cdots & b_t \\
\left(\begin{array}{cccc}
e_1 & * & \cdots & * \\
f_1 & e_2 & \cdots & * \\
\vdots & & \ddots & \vdots \\
* & \cdots & f_{t-1} & e_t \\
* & \cdots & * & f_t
\end{array} \right)
\end{array}
$$

The case of 8 general points in $\mathbb{P}^2$ is illustrated by the following figure. The ideal of the 8 points is generated by two cubics and a quartic (in gray); the degree matrix is

$$
\begin{array}{c}
 \\
a_1 = 4 \\
a_2 = 3 \\
a_3 = 3
\end{array}
\begin{array}{cc}
b_1 = 5 & b_2 = 5 \\
\left(\begin{array}{cc}
e_1 = 1 & \\
f_1 = 2 & e_2 = 2 \\
& f_2 = 2
\end{array} \right)
\end{array},
$$

$$e_1 f_1 + e_1 f_2 + e_2 f_2 = 8.$$

Since minimal free resolutions are unique up to isomorphism, the integers a_i, b_i, e_i, f_i are invariants of the set of points X. They are not arbitrary, however, but are determined (for example) by the e_i and f_i. The next proposition gives these relations. We shall see at the very end of this chapter that Proposition 3.8 gives all the restrictions on these invariants, so that it describes the numerical characteristics of all possible free resolutions of sets of points.

Proposition 3.8. *If*

$$\mathbf{F}: \quad 0 \longrightarrow \sum_1^t S(-b_i) \xrightarrow{\ M\ } \sum_1^{t+1} S(-a_i) \longrightarrow S,$$

is a minimal graded free resolution of S/I, with bases ordered as above, and e_i, f_i denote the degrees of the entries on the principal diagonals of M, the following statements hold for all i.

1. *$e_i \geq 1$ and $f_i \geq 1$.*
2. *$a_i = \sum_{j<i} e_j + \sum_{j \geq i} f_j$.*
3. *$b_i = a_i + e_i$ for $i = 1, \ldots, t$ and $\sum_1^t b_i = \sum_1^{t+1} a_i$.*

If the bases are ordered so that $a_1 \geq \cdots \geq a_{t+1}$ and $b_1 \geq \cdots \geq b_t$ then in addition

4. *$f_i \geq e_i,\ f_i \geq e_{i+1}$.*

This gives an upper bound on the minimal number of generators of the ideal of a set of points that are known to lie on a curve of given degree. Burch's motivation in proving her version of the Hilbert–Burch theorem was to generalize this bound, which was known independently.

Corollary 3.9. *If I is the homogeneous ideal of a set of points in $\mathbb{P}^2$ lying on a curve of degree d, then I can be generated by $d+1$ elements.*

Proof. If $t+1$ is the minimal number of generators of I then, by Proposition 3.8, the degree a_i of the i-th minimal generator is the sum of t numbers that are each at least 1, so $t \leq a_i$. Since I contains a form of degree d we must have $a_i \leq d$ for some i. $\qquad\square$

Hilbert's method for computing the Hilbert function, described in Chapter 1, allows us to compute the Hilbert function and polynomial of S_X in terms of the e_i and f_i. As we will see in Section 4A, $H_X(d)$ is constant for large d, and its value is the number of points in X, usually called the *degree of X* and written $\deg X$. If X were the complete intersection of a curve of degree e with a curve of degree f, then in the notation of Proposition 3.8 we would have $t = 1$, $e_1 = e$, $f_1 = f$, and by Bézout's Theorem the degree of X would be $ef = e_1 f_1$. The following is the generalization to arbitrary t, discovered by Ciliberto, Geramita, and Orrechia [Ciliberto et al. 1986]. For the generalization to determinantal varieties of higher codimension see [Herzog and Trung 1992, Corollary 6.5].

Corollary 3.10. *If X is a finite set of points in $\mathbb{P}^2$ then, with notation as above,*

$$\deg X = \sum_{i \leq j} e_i f_j.$$

The proof is straightforward calculation from Proposition 3.8, and we leave it and a related formula to the reader in Exercise 3.15.

Proof of Proposition 3.8. Since I has codimension 2 and S is a polynomial ring (and thus Cohen–Macaulay) I has grade 2. It follows that the nonzerodivisor a that is associated to the resolution $\mathbf{F}$ as in Theorem 3.2 is a unit. Again because S is a polynomial ring this unit must be a scalar. Thus the a_i are the degrees of the minors of M.

We may assume that the bases are ordered as in the last statement of the Proposition. We first show that the e_i (and thus also, by our ordering conventions, the f_i) are at least 1. Write $m_{i,j}$ for the (i,j) entry of M. By the minimality of $\mathbf{F}$, no $m_{i,j}$ can be a nonzero constant, so that if $e_i \leq 0$ then $m_{i,i} = 0$. Moreover if $p \leq i$ and $q \geq i$ then

$$\deg m_{p,q} = b_q - a_p \leq b_i - a_i = e_i,$$

by our ordering of the bases. If $e_i \leq 0$ then $m_{p,q} = 0$ for all (p,q) in this range, as in the following diagram, where $t = 4$ and we assume $e_3 \leq 0$:

$$M = \begin{pmatrix} * & * & 0 & 0 \\ * & * & 0 & 0 \\ * & * & 0 & 0 \\ * & * & * & * \\ * & * & * & * \end{pmatrix}.$$

We see by calculation that the determinant of the upper $t \times t$ submatrix of M vanishes. By Theorem 3.2 this determinant is a minimal generator of I, and this is a contradiction.

The identity $a_i = \sum_{j<i} e_j + \sum_{j\geq i} f_j$ again follows from Theorem 3.2, since a_i is the degree of the determinant Δ_i of the submatrix of M omitting the i-th row, and one term in the expansion of this determinant is

$$\prod_{j<i} m_{j,j} \cdot \prod_{j\geq i} m_{j+1,j}.$$

Since $e_i = b_i - a_i$. we get

$$\sum_1^t b_i = \sum_1^t a_i + \sum_1^t e_i = \sum_1^{t+1} a_i. \qquad \square$$

3B   Examples

Example 3.11 (Points on a conic). We illustrate the theory above, in particular Corollary 3.9, by discussing the possible resolutions of a set of points lying on an irreducible conic.

 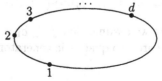

For the easy case of points on a line, and the more complicated case of points on a reducible conic, see Exercises 3.1 and 3.3–3.6 below.

Suppose that the point set $X \subset \mathbb{P}^2$ does not lie on any line, but does lie on some conic, defined by a quadratic form q. In the notation of Proposition 3.8 we have $a_{t+1} = 2$. Since $a_{t+1} = \sum_1^t e_i$, it follows from Proposition 3.8 that either $t = 1$ and $e_1 = 2$ or else $t = 2$ and $e_1 = e_2 = 1$.

1. If $t = 1$ then X is a complete intersection of the conic with a curve of degree $a_1 = d$ defined by a form g. By our formula (or Bézout's Theorem), the degree of X is $2d$. Note in particular that it is even. We have $b_1 = d + 2$, and the resolution takes the following form (see also Theorem A2.48):

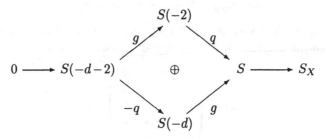

In the case $d = 2$ the Betti diagram of this resolution is

	0	1	2
0	1	–	–
1	–	2	–
2	–	–	1

while for larger d it takes the form

	0	1	2
0	1	–	–
1	–	1	–
2	–	–	–
⋮	⋮	⋮	⋮
$d-2$	–	–	–
$d-1$	–	1	–
d	–	–	1

2. The other possibility is that $t = 2$ and $e_1 = e_2 = 1$. We will treat only the case where the conic $q = 0$ is irreducible, and leave the reducible case to the reader in Exercises 3.3–3.6 at the end of the chapter. By Proposition 3.8 the resolution has the form

$$0 \longrightarrow S(-1-f_1-f_2) \oplus S(-2-f_2) \xrightarrow{M} S(-f_1-f_2) \oplus S(-1-f_2) \oplus S(-2) \longrightarrow S,$$

where we assume that $f_1 \geq e_1 = 1$, $f_1 \geq e_2 = 1$, and $f_2 \geq e_2 = 1$ as usual. If there are two quadratic generators, we further assume that the last generator is q.

By Theorem 3.2, q is (a multiple of) the determinant of the 2×2 matrix M' formed from the first two rows of M. Because q is irreducible, all four entries of the upper 2×2 submatrix of M must be nonzero. The upper right entry of M has degree $e_1 + e_2 - f_1 \leq 1$. If it were of degree 0, by the supposed minimality of the resolution it would itself be 0, contradicting the irreducibility of q. Thus $e_1 + e_2 - f_1 = 1$, so $f_1 = 1$. By our hypothesis $a_3 = 2$, and it follows from Proposition 3.8 that $a_1 = a_2 = 1 + f_2$, $b_1 = b_2 = 2 + f_2$. We deduce that the resolution has the form

$$0 \longrightarrow S(-2-f_2)^2 \xrightarrow{M} S(-1-f_2)^2 \oplus S(-2) \longrightarrow S.$$

If $f_2 = 1$ the Betti diagram is

	0	1	2
0	1	–	–
1	–	3	2

while if $f_2 > 1$ it has the form

	0	1	2
0	1	–	–
1	–	1	–
2	–	–	–
$\vdots$	$\vdots$	$\vdots$	$\vdots$
f_2-1	–	–	–
f_2	–	2	2

Applying the formula of Corollary 3.10 we get $\deg X = 2f_2 + 1$. In particular, we can distinguish this case from the complete intersection case by the fact that the number of points is odd.

Example 3.12 (Four noncolinear points). Any 5 points lie on a conic, since the quadratic forms in 3 variables form a five-dimensional vector space, and vanishing at a point is one linear condition, so there is a nonzero quadratic form vanishing at any 5 points. Thus we can use the ideas of the previous subsection to describe the possible resolutions for up to 5 points. One set of three noncolinear points in $\mathbb{P}^2$ is like another, so we treat the case of a set $X = \{p_1, \ldots, p_4\}$ of four noncolinear points, the first case where geometry enters. (For the case of 3 points see Exercise 3.2.)

Since there is a six-dimensional vector space of quadratic forms on $\mathbb{P}^2$, and the condition of vanishing at a point is a single linear condition, there must be at least two distinct conics containing X.

First suppose that no three of the points lie on a line. It follows that X is contained in the following two conics, each a union of two lines:

$$C_1 = \overline{p_1, p_2} \cup \overline{p_3, p_4} \qquad C_2 = \overline{p_1, p_3} \cup \overline{p_2, p_4}.$$

In this case, X is the complete intersection of C_1 and C_2, and we have the Betti diagram

	0	1	2
0	1	–	–
1	–	2	–
2	–	–	1

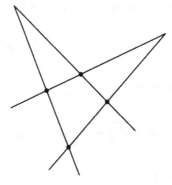

The two conics are two pairs of lines containing the four points.

On the other hand, suppose that three of the points, say p_1, p_2, p_3 lie on a line L. Let L_1 and L_2 be lines through p_4 that do not contain any of the points p_1, p_2, p_3. (See figure on the next page.) It follows that X lies on the two conics

$$C_1 = L \cup L_1 \qquad C_2 = L \cup L_2,$$

and the intersection of these two conics contains the whole line L. Thus X is not the complete intersection of these two conics containing it, so by Corollary 3.9 the ideal of X requires exactly 3 generators. From Propositions 3.8 and 3.10 it follows that

$$e_1 = e_2 = 1, \ f_1 = 2, \ f_2 = 1,$$

and the ideal I of X is generated by the quadrics defining C_1 and C_2 together with a cubic equation.

The Betti diagram will be

	0	1	2
0	1	–	–
1	–	2	1
2	–	1	1

3C   The Existence of Sets of Points with Given Invariants

This section is devoted to a proof of a converse of Proposition 3.8, which we now state.

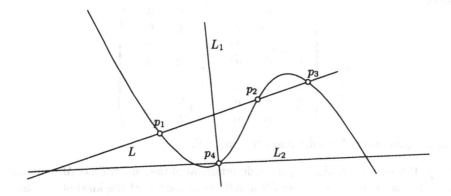

Four points, three on a line, are the intersection
of two conics (here $L \cup L_1$ and $L \cup L_2$) and a cubic.

Theorem 3.13. *If the ground field $\mathbb{K}$ is infinite and $e_i, f_i \geq 1$, for $i = 1, \ldots, t$, are integers, there is a set of points $X \subset \mathbb{P}^2$ such that S_X has a minimal free resolution whose second map has diagonal degrees e_i and f_i as in Proposition 3.8.*

The proof is in two parts. First we show that there is a *monomial ideal $J \subset \mathbb{K}[x, y]$* (that is, an ideal generated by monomials in the variables), containing a power of x and a power of y, whose free resolution has the corresponding invariants. This step is rather easy. Then, given any such monomial ideal J we will show how to produce a set of distinct points in $\mathbb{P}^2$ whose defining ideal I has free resolution with the same numerical invariants as that of J.

The second step, including Theorem 3.16, is part of a much more general theory, sometimes called the polarization of monomial ideals. We sketch its fundamentals in the exercises at the end of this chapter.

Proposition 3.14. *Let $S = \mathbb{K}[x, y]$ and let $e_1, \ldots, e_t$ and $f_1, \ldots, f_t$ be positive integers. For $i = 1, \ldots, t+1$ set*

$$m_i = \prod_{j<i} x^{e_j} \prod_{j \geq i} y^{f_j},$$

and let $I = (m_1, \ldots, m_{t+1}) \subset S$ be the monomial ideal generated by these products. Define a_i and b_i by the formulas of Proposition 3.8. The ring S/I has minimal free resolution

$$0 \longrightarrow \sum_{i=1}^{t} S(-b_i) \overset{M}{\longrightarrow} \sum_{i=1}^{t+1} S(-a_i) \longrightarrow S \longrightarrow S/I \to 0$$

where

$$M = \begin{pmatrix} x^{e_1} & 0 & 0 & \cdots & 0 & 0 \\ y^{f_1} & x^{e_2} & 0 & \cdots & 0 & 0 \\ 0 & y^{f_2} & x^{e_3} & \cdots & 0 & 0 \\ 0 & 0 & y^{f_3} & \ddots & 0 & 0 \\ \vdots & \vdots & \vdots & \ddots & \ddots & \vdots \\ 0 & 0 & 0 & \cdots & y^{f_{t-1}} & x^{e_t} \\ 0 & 0 & 0 & \cdots & 0 & y^{f_t} \end{pmatrix}$$

and the generator of $S(-a_i)$ maps to $\pm m_i \in S$.

Proof. It is easy to see that m_i is the determinant of the submatrix of M omitting the i-th row. Thus by Theorem 3.2 it suffices to show that the ideal of maximal minors of M has grade at least 2. But this ideal contains $\prod_{i=1}^{t} x^{e_i}$ and $\prod_{i=1}^{t} y^{f_i}$. $\qquad\square$

As background to the second part of the theorem's demonstration, we will prove general results that allow us to manufacture a reduced algebraic set having ideal with the same Betti diagram as any given monomial ideal, as long as the ground field $\mathbb{K}$ is infinite. This treatment follows Geramita, Gregory and Roberts [Geramita et al. 1986].

Here is the tool we will use to show that the two resolutions have the same Betti diagram:

Lemma 3.15. *Let R be a ring. If M is an R-module and $y \in R$ is a nonzerodivisor both on R and on M, then any free resolution of M over R reduces modulo (y) to a free resolution of M/yM over $R/(y)$. If, in addition, R is a graded polynomial ring, M is a graded module, and y is a linear form, then the Betti diagram of M (over R) is the same as the Betti diagram of M/yM (over the graded polynomial ring $R/(y)$.)*

Proof. Let $\mathbf{F} : \cdots \rightarrow F_1 \rightarrow F_0$ be a free resolution of M. We must show that $\mathbf{F}/y\mathbf{F} = R/(y) \otimes_R \mathbf{F}$, which is obviously a free complex of $R/(y)$-modules, is actually a free resolution — that is, its homology is trivial except at F_0, where it is M/yM. The homology of $\mathbf{F}/y\mathbf{F} = R/(y) \otimes_R \mathbf{F}$ is by definition $\operatorname{Tor}_*^R(R/(y), M)$. Because y is a nonzerodivisor on R, the complex $0 \rightarrow R \xrightarrow{\;y\;} R \rightarrow R/(y) \rightarrow 0$ is exact, and it is thus a free resolution of $R/(y)$. We can use this free resolution instead of the other to compute Tor (see [Eisenbud 1995, p. 674], for example), and we see that $\operatorname{Tor}_*^R(R/(y), M)$ is the homology of the sequence $0 \longrightarrow M \xrightarrow{\;y\;} M$. Since y is a nonzerodivisor on M, the homology is just M/yM in degree 0, as required. $\qquad\square$

We now return to the construction of sets of points. If $\mathbb{K}$ is infinite we can choose r embeddings (of sets) $\eta_i : \mathbb{N} \hookrightarrow \mathbb{K}$. (If $\mathbb{K}$ has characteristic 0, we could choose all η_i equal to the natural embedding $\eta_i(n) = n \in \mathbb{Z} \subset \mathbb{K}$, but any assignment of distinct $\eta_i(n) \in \mathbb{K}$ will do. In general we could choose all η_i to be equal, but the extra flexibility will be useful in the proof.) We use the η_i

to embed $\mathbb{N}^r$, regarded as the set of monomials of $\mathbb{K}[x_1,\ldots,x_r]$, into $\mathbb{P}_{\mathbb{K}}{}^r$: if $m = x_1^{p_1}\cdots x_r^{p_r}$ we set $\eta(m) = (1, \eta_1(p_1), \ldots, \eta_r(p_r))$, and we set

$$f_m = \prod_{i=1}^{r} \prod_{j=0}^{p_i-1}(x_i - \eta_i(j)x_0).$$

Note in particular that $f_m \equiv m \bmod (x_0)$; we maintain this notation throughout this section. We think of f_m as the result of replacing the powers of each x_i in m by products of the distinct linear forms $x_i - \eta_i(j)x_0$.

Theorem 3.16. *Let $\mathbb{K}$ be an infinite field, with an embedding $\mathbb{N}^r \subset \mathbb{P}_{\mathbb{K}}{}^r$ as above, and let J be a monomial ideal in $\mathbb{K}[x_1,\ldots,x_r]$. Let $X_J \subset \mathbb{P}^r$ be the set*

$$X_J = \{p \in \mathbb{N}^r \subset \mathbb{P}^r \mid x^p \notin J\}.$$

The ideal $I_{X_J} \subset S = \mathbb{K}[x_0,\ldots,x_r]$ has the same Betti diagram as J; in fact x_0 is a nonzerodivisor modulo I_{X_J}, and $J \equiv I_{X_J} \bmod (x_0)$. Moreover, I_{X_J} is generated by the forms f_m, where m runs over a set of monomial generators for J.

Before we give the proof, two examples will clarify the result:

Example 3.17. In the case of a monomial ideal J in $\mathbb{K}[x_1,\ldots,x_r]$ that contains a power of each variable x_i, such as the ones in $\mathbb{K}[x,y]$ described in Proposition 3.14, the set X_J is finite. Thus Theorem 3.16 and Proposition 3.14 together yield the existence of sets of points in $\mathbb{P}^2$ whose free resolution has arbitrary invariants satisfying Proposition 3.8. For example, the Betti diagram

	0	1	2
0	1	–	–
1	–	–	–
2	–	1	–
3	–	2	1
4	–	–	1

corresponds to invariants $(e_1, e_2) = (2, 1)$ and $(f_1, f_2) = (2, 2)$, and monomial ideal $J = (y^4, x^2y^2, x^3)$, where we have replaced x_1 by x and x_2 by y to simplify notation. We will also replace x_0 by z. Assuming, for simplicity, that $\mathbb{K}$ has characteristic 0 and that $\eta_i(n) = n$ for all i, the set of points X_J in the affine plane $z = 1$ looks like this:

Its ideal is generated by the polynomials

$$y(y-1)(y-2)(y-3), \quad x(x-1)y(y-1)y, \quad x(x-1)(x-2).$$

As a set of points in projective space, it has ideal $I_{X_J} \subset \mathbb{K}[z, x, y]$ generated by the homogenizations

$$f_{y^4} = y(y-z)(y-2z)(y-3z),$$
$$f_{x^2y^2} = x(x-z)y(y-z),$$
$$f_{x^4} = x(x-z)(x-2z).$$

Example 3.18. Now suppose that J does not contain any power of x ($= x_1$). There are infinitely many isolated points in X_J, corresponding to the elements $1, x, x^2, \ldots \notin J$. Thus X_J is not itself an algebraic set. Its *Zariski closure* (the smallest algebraic set containing it) is a union of planes, as we shall see. For example, if $J = (x^2y, xy^2, x^3)$, here are X_J and its Zariski closure:

For the proof of Theorem 3.16 we will use the following basic properties of the forms f_m.

Lemma 3.19. *Let* $\mathbb{K}$ *be an infinite field, and let the notation be as above.*

1. *If* $f \in S$ *is a form of degree* $\leq d$ *that vanishes on* $\eta(m) \in \mathbb{P}^r$ *for every monomial* m *with* $\deg m \leq d$, *then* $f = 0$.
2. $f_m(\eta(m)) \neq 0$.
3. $f_m(\eta(n)) = 0$ *if* $m \neq n$ *and* $\deg n \leq \deg m$.

Proof. 1. We induct on the degree $d \geq 0$ and the dimension $r \geq 1$. The cases in which $d = 0$ or $r = 1$ are easy.

For any form f of degree d we may write $f = (x_r - \eta_r(0)x_0)q + g$, where $q \in S$ is a form of degree $d - 1$ and $g \in \mathbb{K}[x_0, \ldots, x_{r-1}]$ is a form of degree $\leq d$ not involving x_r. Suppose that f vanishes on $\eta(m) = (1, \eta_1(p_1), \ldots, \eta_r(p_r))$ for every monomial $m = x_1^{p_1} \ldots x_r^{p_r}$ of degree $\leq d$. The linear form $x_r - \eta_r(0)x_0$ vanishes on $\eta(m)$ if and only if $\eta_r(p_r) = \eta_r(0)$, that is, $p_r = 0$. This means that m is not divisible by x_r. Thus g vanishes on $\eta(m)$ for all monomials m of degree $\leq d$ that are not divisible by x_r. It follows by induction on r that $g = 0$.

Since $g = 0$, the form q vanishes on $\eta(x_r n)$ for all monomials n of degree $\leq d - 1$. If we define new embeddings η_i' by the formula $\eta_i' = \eta_i$ for $i < r$ but $\eta_r'(p) = \eta_r(p+1)$, and let η' be the corresponding embedding of the set of monomials, then q vanishes on $\eta'(n)$ for all monomials n of degree at most $d-1$. By induction on d, we have $q = 0$, whence $f = 0$ as required.

2. This follows at once from the fact that $\eta : \mathbb{N} \to \mathbb{K}$ is injective.

3. Write $m = x_1^{p_1} \ldots x_r^{p_r}$ and $n = x_1^{q_1} \ldots x_r^{q_r}$. Since $\deg n \leq \deg m$ we have $q_i < p_i$ for some i. It follows that $f_m(\eta(n)) = 0$. $\square$

Proof of Theorem 3.16. Let I be the ideal generated by $\{f_m\}$ where m ranges over a set of monomial generators of J. We first prove that $I = I_{X_J}$.

For every pair of monomials $m \in J$, $n \notin J$ one of the exponents of n is strictly less than the corresponding exponent of m. It follows immediately that $I \subset I_{X_J}$.

For the other inclusion, let $f \in I_{X_J}$ be any form of degree d. Suppose that for some $e \leq d$ the form f vanishes on all the points $\eta(n)$ for $\deg n < e$, but not on some $\eta(m)$ with $\deg m = e$. By parts 2 and 3 of Lemma 3.19 we can subtract a multiple of $x_0^{d-e} f_m$ from f to get a new form of degree d vanishing on $\eta(m)$ in addition to all the points $\eta(m')$ where either $\deg m' < e$ or $\deg m' = e$ and $f(\eta(m')) = 0$. Proceeding in this way, we see that f differs from an element of I by a form g of degree d that vanishes on $\eta(m)$ for every monomial m of degree $\leq d$. By part 1 of Lemma 3.19 we have $g = 0$, so $f \in I$. This proves that $I = I_{X_J}$.

Since none of the points $\eta(m)$ lies in the hyperplane $x_0 = 0$, we see that x_0 is a nonzerodivisor modulo I_{X_J}. On the other hand it is clear from the form of the given generators that $I \cong J \mod (x_0)$. Applying Lemma 3.15 below we see that a (minimal) resolution of I over S reduces modulo x_0 to a (minimal) free resolution of J over $\mathbb{K}[x_1, \ldots, x_r]$; in particular the Betti diagrams are the same.   □

3D   Exercises

1. Let X be a set of d points on a line in $\mathbb{P}^2$. Use Corollary 3.9 to show that the ideal I_X can be generated by two elements, the linear form defining the line and one more form g, of degree $a_1 = d$. Compute the Betti diagram of S_X.

2. By a change of coordinates, any three noncolinear points can be taken to be the points $x = y = 0$, $x = z = 0$, and $y = z = 0$. Let X be this set of points. Show that X lies on a smooth conic and deduce that its ideal I_X must have 3 quadratic generators. Prove that $I_X = (yz, xz, xy)$. By Proposition 3.8 the matrix M of syzygies must have linear entries; show that it is

$$\begin{pmatrix} x & 0 \\ -y & y \\ 0 & -z \end{pmatrix}.$$

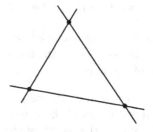

The three pairs of lines span the space of conics through the three points.

In Exercises 3.3 to 3.6 we invite the reader to treat the case where the conic in part 2 (where $t = z$) of Example 3.11 is reducible, that is, its equation is a product of linear forms. Changing coordinates, we may assume that the linear forms are x and y. The following exercises all refer to a finite set (or, in the last exercise, scheme) of points lying on the union of the lines $x = 0$ and $y = 0$, and its free resolution. We use the notation of pages 36–37. We write a for the point with coordinates $(0, 0, 1)$ where the two lines meet.

3. Show that the number of points is $f_1 + 2f_2$, which may be even or odd.

4. Suppose that $M' : F \to G_1$ is a map of homogeneous free modules over the ring $S = \mathbb{K}[x, y, z]$, and that the determinant of M' is xy. Show that with a suitable choice of the generators of F and G_1, and possibly replacing z by a linear form z' in x, y, z, the map M' can be represented by a matrix of the form

$$\begin{pmatrix} xy & 0 \\ 0 & 1 \end{pmatrix} \quad \text{or} \quad \begin{pmatrix} x & 0 \\ 0 & y \end{pmatrix} \quad \text{or} \quad \begin{pmatrix} x & 0 \\ z'^f & y \end{pmatrix}$$

for some integer $f \geq 0$.

5. Deduce that the matrix M occuring in the free resolution of the ideal of X can be reduced to the form

$$M = \begin{pmatrix} x & 0 \\ z^{f_1} & y \\ p(y, z) & q(x, z) \end{pmatrix},$$

where p and q are homogeneous forms of degrees $f_1 + f_2 - 1$ and f_2 respectively. Show that X does contains the point a if and only if $q(x, z)$ is divisible by x.

6. Supposing that X does not contain the point a, show that X contains precisely f_2 points on the line $y = 0$ and $f_1 + f_2$ points on the line $x = 0$.

7. Consider the local ring $R = k[x, y]_{(x,y)}$, and let $I \subset R$ be an ideal containing xy such that R/I is a finite-dimensional k vector space. Show that (possibly after a change of variable) $I = (xy, x^s, y^t)$ or $I = (xy, x^s + y^t)$. Show that

$$\dim_k R/(xy, x^s, y^t) = s + t - 1; \qquad \dim_k R/(xy, x^s + y^t) = s + t.$$

Regarding R as the local ring of a point in $\mathbb{P}^2$, we may think of this as giving a classfication of all the schemes lying on the union of two lines and supported at the intersection point of the lines.

8. (For those who know about schemes) In the general case of a set of points on a reducible conic, find invariants of the matrix M (after row and column transformations) that determine the length of the part of X concentrated at the point a and the parts on each of the lines $x = 0$ and $y = 0$ away from the point a.

9. Let $u \geq 1$ be an integer, and suppose that $\mathbb{K}$ is an infinite field. Show that any sufficiently general set X of $\binom{u+1}{2}$ points in $\mathbb{P}^2$ has free resolution of the form

$$0 \longrightarrow S^u(-u-1) \xrightarrow{M} S^{u+1}(-u) \longrightarrow S;$$

that is, the equations of X are the minors of a $(u+1) \times u$ matrix of linear forms.

Exercises 3.10–3.14 explain when we can apply the technique of Theorem 3.16 to the set $X \subset \mathbb{P}^r$. As an application, we produce very special sets of points with the same Betti diagrams as general sets of points.

10. (Monomial ideals and partitions.) Suppose that $J \subset \mathbb{K}[x_1, x_2]$ is a monomial ideal such that $\dim_\mathbb{K} \mathbb{K}[x_1, x_2]/J = d < \infty$. For each i, let σ_i be the number of monomials of the form $x_1^i x_2^j$ not in J. Similarly, for each j let τ_j be the number of monomials of the form $x_1^i x_2^j$ not in J.

(a) Show that σ is a *partition* of d, in the sense that

$$\sigma_0 \geq \sigma_1 \geq \cdots \geq 0 \qquad \text{and} \qquad \sum_w \sigma_w = d.$$

(b) Show that the function $J \mapsto \sigma$ is a one-to-one correspondence between monomial ideals J with $\dim_\mathbb{K} \mathbb{K}[x_1, x_2]/J = d$ and partitions of d.

(c) Show that σ and τ are *dual partitions* in the sense that σ_w is the number of integers v such that $\tau_v = w$, and conversely.

11. Now suppose that U, V are finite subsets of $\mathbb{K}$, and identify $U \times V$ with its image in $\mathbb{K}^2 \subset \mathbb{P}^2_\mathbb{K}$. Let $X \subset \mathbb{P}^2$ be a finite subset of $U \times V$. For $i \in U$ and $j \in V$, let g_i be the number of points of X in $\{i\} \times V$, and let h_j be the number of points of X in $U \times \{j\}$.

Write $\sigma = (\sigma_0, \sigma_1, \ldots)$ with $\sigma_0 \geq \sigma_1 \geq \cdots$ for the sequence of nonzero numbers g_i, written in decreasing order, and similarly for τ and h_j. Show that if σ and τ are dual partitions, and J is the monomial ideal corresponding to σ as in Exercise 3.10, then the Betti diagram of S_X is the same as that of $\mathbb{K}[x_1, x_2]/J$. (Hint: You can use Theorem 3.16.)

12. (Gaeta.) Suppose $d \geq 0$ is an integer, and let t, s be the coordinates of d in the diagram

Algebraically speaking, s, t are the unique nonnegative integers such that

$$d = \binom{s+t+1}{2} + s, \quad \text{or equivalently } d = \binom{s+t+2}{2} - t - 1.$$

(a) Use Theorem 3.16 to show that there is a set of d points $X \subset \mathbb{P}^2$ with Betti diagram

	0	1	2
0	1	–	–
1	–	–	–
$\vdots$	$\vdots$	$\vdots$	$\vdots$
$s+t-1$	–	$t+1$	$t-s$
$s+t$	–	–	s

or

	0	1	2
0	1	–	–
1	–	–	–
$\vdots$	$\vdots$	$\vdots$	$\vdots$
$s+t-1$	–	$t+1$	–
$s+t$	–	$s-t$	s

according as $s \leq t$ or $t \leq s$. (This was proved by Gaeta [1951] using the technique of linkage; see [Eisenbud 1995, Section 21.10] for the definition of linkage and modern references.)

(b) Let M_X be the presentation matrix for the ideal of a set of points X as above. Show that if $s \leq t$ then M_X has $t+1$ rows, with s columns of quadrics followed by $t-s$ columns of linear forms; while if $t \leq s$ then M_X has $s+1$ rows, with $s-t$ rows of linear forms followed by $t+1$ rows of quadrics.

(c) (The Gaeta set.) Suppose that $\mathbb{K}$ has characteristic 0. Define the *Gaeta set* of d points to be the set of points in the affine plane with labels $1, 2, \ldots, d$ in the picture above, regarded as a set of points in $\mathbb{P}^2$. Show that if X is the Gaeta set of d points, then the Betti diagram of S_X has the form given in part (a). (Hint: Theorem 3.16 can still be used.)

(d) Find the smallest d such that the d points $0, 1, \ldots, d-1$ have a different Hilbert function than the d points $1, 2, \ldots, d$ in the diagram above.

13. Although the Gaeta set X (part (c) of preceding exercise) is quite special — for example, it is usually not in linearly general position — show that the free resolution of S_X as above has the same Betti diagram as that of the generic set of d points. (Saying that "the generic set of d points in $\mathbb{P}^2$" has a certain property is saying that this property is shared by all d-tuples of points in some open dense set of $(\mathbb{P}^2)^d$.) One way to prove this is to follow these steps. Let Y be the generic set of d points.

(a) Show that the generic set of points Y has Hilbert function $H_{S_Y}(n) = \min\{H_S(n), d\}$, and that this is the same as for the Gaeta set.

(b) Deduce that with s, t defined as above, the ideal I_Y of Y does not contain any form of degree $< s+t$, and contains exactly $t+1$ independent forms of degree $s+t$; and that I_Y requires at least $(s-t)_+$ generators of degree $s+t+1$, where $(s-t)_+$ denotes $\max\{0, s-t\}$, the "positive part" of $s-t$.

(c) Show that the fact that the ideal of the Gaeta set requires only $(s-t)_+$ generators of degree $s+t+1$, and none of higher degree, implies that the same is true for an open (and thus dense) set of sets of points with d elements, and thus is true for Y.

(d) Conclude that the resolution of S_X has the same Betti diagram as that of S_Y.

Despite quite a lot of work we do not know how to describe the free resolution of a general set of d points in $\mathbb{P}^r$. It would be natural to conjecture that the resolution is the "simplest possible, compatible with the Hilbert function", as in the case above, and this is known to be true for $r \leq 4$. On the other hand it fails in general; the simplest case, discovered by Schreyer, is for 11 points in $\mathbb{P}^6$, and many other examples are known. See Eisenbud, Popescu, Schreyer and Walter [Eisenbud et al. 2002] for a recent account.

14. (Geramita, Gregory, and Roberts [Geramita et al. 1986]) Suppose that $J \subset \mathbb{K}[x_1,\ldots,x_r]$ is a monomial ideal, and that the cardinality of $\mathbb{K}$ is q. Suppose further that no variable x_i appears to a power higher than q in a monomial minimal generator of J. Show that there is a radical ideal $I \subset S = \mathbb{K}[x_0,\ldots,x_r]$ such that x_0 is a nonzerodivisor modulo I and $J \cong I \bmod (x_0)$. (Hint: Although X_J may not make any sense over $\mathbb{K}$, the generators of I_{X_J} defined in Theorem 3.16 can be defined in S. Show that they generate a radical ideal.)

15. (Degree formulas.) We will continue to assume that $X \subset \mathbb{P}^2$ is a finite set of points, and to use the notation for the free resolution of S_X developed in Proposition 3.8.

Show that

$$H_X(d) = H_S(d) - \sum_{i=1}^{t+1} H_S(d-a_i) + \sum_{i=1}^{t} H_S(d-b_i)$$
$$= \binom{d+2}{2} - \sum_{i=1}^{t+1}\binom{d-a_i+2}{2} + \sum_{i=1}^{t}\binom{d-b_i+2}{2}$$

and

$$P_X(d) = \frac{(d+2)(d+1)}{2} - \sum_{i=1}^{t+1}\frac{(d-a_i+2)(d-a_i+1)}{2} + \sum_{i=1}^{t}\frac{(d-b_i+2)(d-b_i+1)}{2}.$$

Deduce that in $P_X(d)$ the terms of degree ≥ 1 in d all cancel. (This can also be deduced from the fact that the degree of P_X is the dimension of X.) Prove that

$$P_X(0) = \frac{1}{2}\left(\sum_{i=1}^{t} b_i^2 - \sum_{i=1}^{t+1} a_i^2\right) = \sum_{i \leq j} e_i f_j.$$

In Chapter 4 we will see that $P_X(0) = \deg X$.

16. (Sturmfels.) Those who know about Gröbner bases (see [Eisenbud 1995, Chapter 15], for instance) may show that, with respect to a suitable term order, the ideal I_{X_J} constructed in Theorem 3.16 has initial ideal J.

Monomial ideals. This beautiful theory is one of the main links between commutative algebra and combinatorics, and has been strongly developed in recent years. We invite the reader to work out some of this theory in Exercises 3.17–3.24. These results only scratch the surface. For more information see [Eisenbud 1995, Section 15.1 and Exercises 15.1–15.6] and [Miller and Sturmfels 2004].

17. (Ideal membership for monomial ideals.) Show that if $J = (m_1, \ldots, m_g) \subset T = \mathbb{K}[x_1, \ldots, x_n]$ is the ideal generated by monomials $m_1, \ldots, m_g$ then a polynomial p belongs to J if and only if each term of p is divisible by one of the m_i.

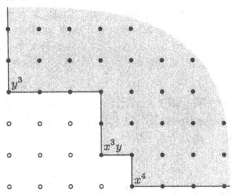

18. (Intersections and quotients of monomial ideals.) Let $I = (m_1, \ldots, m_s)$, $J = (n_1, \ldots, n_t)$ be two monomial ideals. Show that

$$I \cap J = (\{\mathrm{LCM}(m_i, n_j) \mid i = 1 \ldots s, \; j = 1, \ldots, t\})$$

and

$$(I : J) = \bigcap_{j=1 \ldots t} (\{m_i : n_j \mid i = 1 \ldots s\}),$$

where we write $m : n$ for the "quotient" monomial $p = \mathrm{LCM}(m, n)/n$, so that $(m) : (n) = (p)$.

19. (Decomposing a monomial ideal.) Let $J = (m_1, \ldots, m_t) \subset T = \mathbb{K}[x_1, \ldots, x_n]$ be a monomial ideal. If $m_t = ab$ where a and b are monomials with no common divisor, show that

$$(m_1, \ldots, m_t) = (m_1, \ldots, m_{t-1}, a) \cap (m_1, \ldots, m_{t-1}, b).$$

20. Use the preceding exercise to decompose the ideal (x^2, xy, y^3) into the simplest pieces you can.

21. The only monomial ideals that cannot be decomposed by the technique of Exercise 3.19 are those generated by powers of the variables. Let

$$J_\alpha = (x_1^{\alpha_1}, x_2^{\alpha_2}, \ldots, x_n^{\alpha_n})$$

where we allow some of the α_i to be ∞ and make the convention that $x_i^\infty = 0$. Set $N_* = \mathbb{Z} \cup \{\infty\}$. Now show that any monomial ideal J may be written as $J = \bigcap_{\alpha \in A} J_\alpha$ for some finite set $A \subset N_*^n$. The ideal J_α is $P_\alpha := (\{x_i \mid \alpha_i \neq \infty\})$-primary. Deduce that the variety corresponding to any associated prime of a monomial ideal J is a plane of some dimension.

22. If P is a prime ideal, show that the intersection of finitely many P-primary ideals is P-primary. Use the preceding exercise to find an irredundant primary decomposition, and the associated primes, of $I = (x^2, yz, xz, y^2z, yz^2, z^4)$. (Note that the decomposition produced by applying Exercise 3.19 may produce redundant components, and also may produce several irreducible components corresponding to the same associated prime.)

23. We say that an ideal J is *reduced* if it is equal to its own radical; that is, if $p^n \in J$ implies $p \in J$ for any ring element p. An obvious necessary condition for a monomial ideal to be reduced is that it is *square-free* in the sense that none of its minimal generators is divisible by the square of a variable. Prove that this condition is also sufficient.

24. (Polarization and Hartshorne's proof of Theorem 3.16.) An older method of proving Theorem 3.16, found in [Hartshorne 1966], uses a process called *polarization*. If $m = x_1^{a_1} x_2^{a_2} \ldots x_n^{a_n}$ is a monomial, the polarization of m is a monomial (in a larger polynomial ring) obtained by replacing each power $x_i^{a_i}$ by a product of a_i distinct new variables $P(x_i^{a_i}) = x_{i,1} \ldots x_{i,a_i}$. Thus

$$P(m) = \prod_i \prod_j x_{ij} \in \mathbb{K}[x_{1,1}, \ldots, x_{n,a_n}].$$

Similarly, if $J = (m_1, \ldots, m_t) \subset T = \mathbb{K}[x_1, \ldots, x_n]$ is a monomial ideal, then we define the polarization $P(J)$ as the ideal generated by $P(m_1), \ldots, P(m_t)$ in a polynomial ring $T = \mathbb{K}[x_{1,1}, \ldots]$ large enough to form all the $P(m_i)$. For example, if $J = (x_1^2, x_1 x_2^2) \subset \mathbb{K}[x_1, x_2]$ then

$$P(J) = (x_{1,1} x_{1,2}, \; x_{1,1} x_{2,1} x_{2,2}) \subset \mathbb{K}[x_{1,1}, x_{1,2}, x_{2,1}, x_{2,2}].$$

(a) Show that $P(J)$ is square-free and that if $x_{i,j+1}$ divides a polarized monomial $P(m)$, then $x_{i,j}$ divides $P(m)$. Let L be the ideal of T generated by all the differences $x_{i,j} - x_{i,k}$. Note that $T/L \simeq S$, and that the image of $P(J)$ in S is J. Prove that a minimal set of generators for L is a regular sequence modulo $P(J)$. Conclude that the Betti diagram of $P(J)$ is equal to the Betti diagram of J.

(b) Suppose the ground field $\mathbb{K}$ is infinite and J is a monomial ideal in $\mathbb{K}[x_1, x_2]$ containing a power of each variable, with polarization $P(J) \subset \mathbb{K}[x_0, \ldots, x_r]$. Show that for a general set of $r-2$ linear forms $y_3, \ldots, y_r$ in the x_i the ideal $P(J) + (y_3, \ldots, y_r)/(y_3, \ldots, y_r)$ is reduced and defines a set of points in the 2-plane defined by $y_3 = 0, \ldots, y_r = 0$. Show that the ideal I of this set of points has the same Betti diagram as J.

4

Castelnuovo–Mumford Regularity

4A   Definition and First Applications

The *Castelnuovo–Mumford regularity*, or simply *regularity*, of an ideal in S is an important measure of how complicated the ideal is. A first approximation is the maximum degree of a generator the ideal requires; the actual definition involves the syzygies as well. Regularity is actually a property of a complex, defined as follows.

Let $S = \mathbb{K}[x_0, \ldots, x_r]$ and let

$$\mathbf{F} : \cdots \longrightarrow F_i \longrightarrow F_{i-1} \longrightarrow \cdots$$

be a graded complex of free S-modules, with $F_i = \sum_j S(-a_{i,j})$. The *regularity* of $\mathbf{F}$ is the supremum of the numbers $a_{i,j} - i$. The regularity of a finitely generated graded S-module M is the regularity of a minimal graded free resolution of M. We will write reg M for this number. The most important special case gets its own terminology: if $X \subset \mathbb{P}_\mathbb{K}^r$ is a projective variety or scheme and I_X is its ideal, then reg I is called the *regularity of X*, or reg X.

For example, if M is free, the regularity of M is the supremum of the degrees of a set of homogeneous minimal generators of M. In general, the regularity of M is an upper bound for the largest degree of a minimal generator of M, which is the supremum of the numbers $a_{0,j} - 0$. Assuming that M is generated by elements of degree 0, the regularity of M is the index of the last nonzero row in the Betti diagram of M. Thus, in Example 3.17, the regularity of the homogeneous coordinates ring of the points is 4.

The power of the notion of regularity comes from an alternate description, in terms of cohomology, which might seem to have little to do with free resolutions. Historically, the cohomological interpretation came first. David Mumford defined the regularity of a coherent sheaf on projective space in order to generalize a classic argument of Castelnuovo. Mumford's definition is given in terms of sheaf cohomology; see Section 4D below. The definition for modules, which extends that for sheaves, and the equivalence with the condition on the resolution used as a definition above, come from [Eisenbud and Goto 1984]. In most cases the regularity of a sheaf, in the sense of Mumford, is equal to the regularity of the graded module of its twisted global sections.

To give the reader a sense of how regularity is used, we postpone the technical treatment to describe two applications.

The Interpolation Problem

We begin with a classic problem. It is not hard to show that if X is a finite set of points in $\mathbb{A}^r = \mathbb{A}_{\mathbb{K}}^r$, all functions from X to $\mathbb{K}$ are induced by polynomials. Indeed, if X has n points, polynomials of degree at most $n-1$ suffice. To see this, let $X = \{p_1, \ldots, p_n\}$ and assume for simplicity that the field $\mathbb{K}$ is infinite (we will soon see that this is unnecessary). Using this assumption we can choose an affine hyperplane passing through p_i but not through any of the other p_j. Let ℓ_i be a linear function vanishing on this hyperplane: that is, a linear function on $\mathbb{A}^r$ such that $\ell_i(p_i) = 0$ but $\ell_i(p_j) \neq 0$ for all $j \neq i$. If we set $Q_i = \prod_{j \neq i} \ell_j$, the polynomial

$$\sum_{i=1}^{n} \frac{a_i}{Q_i(p_i)} Q_i$$

takes the value a_i at the point p_i for any desired values $a_i \in \mathbb{K}$.

The polynomials Q_i have degree $n-1$. Can we find polynomials of strictly lower degree that give the same functions on X? Generally not: a polynomial of degree less than $n-1$ that vanishes at $n-1$ points on a line vanishes on the entire line, so if all the points of X lie on a line, the lowest possible degree is $n-1$. On the other hand, if we consider the set of three noncolinear points $\{(0,0), (0,1), (1,0)\}$ in the plane with coordinates x, y, the linear function $a(1-x-y) + by + cx$ takes arbitrary values a, b, c at the three points, showing that degree 1 polynomials suffice in this case although $1 < n-1 = 2$. This suggests the following problem.

Interpolation Problem. Given a finite set of points X in $\mathbb{A}^r$, what is the smallest degree d such that every function $X \to \mathbb{K}$ can be realized as the restriction from $\mathbb{A}^r$ of a polynomial function of degree at most d?

The problem has nothing to do with free resolutions; but its solution lies in the regularity.

Theorem 4.1. *Let $X \subset \mathbb{A}^r \subset \mathbb{P}^r$ be a finite collection of points, and let S_X be the homogeneous coordinate ring of X as a subset of $\mathbb{P}^r$. The interpolation degree of X is equal to $\operatorname{reg} S_X$, the regularity of S_X.*

The proof comes in Section 4B. As we shall see there, the interpolation problem is related to the question of when the Hilbert function of a module becomes equal to the Hilbert polynomial.

When Does the Hilbert Function Become a Polynomial?

As a second illustration of how the regularity is used, we consider the Hilbert polynomial. Recall that $H_M(d) = \dim_{\mathbb{K}} M_d$ is the Hilbert function of M, and that it is equal to a polynomial function $P_M(d)$ for large d. How large does d have to be for $H_M(d) = P_M(d)$? We will show that the regularity of M provides a bound, which is sharp in the case of a Cohen–Macaulay module.

Recall that a graded S-module is said to be *Cohen–Macaulay* if its depth is equal to its dimension. For any finite set of points $X \subset \mathbb{P}^r$ we have depth $S_X = 1 = \dim S_X$, so S_X is a Cohen–Macaulay module.

Theorem 4.2. *Let M be a finitely generated graded module over the polynomial ring $S = \mathbb{K}[x_0, \ldots, x_r]$.*

1. *The Hilbert function $H_M(d)$ agrees with the Hilbert polynomial $P_M(d)$ for $d \geq \operatorname{reg} M + 1$.*
2. *More precisely, if M is a module of projective dimension δ, then $H_M(d) = P_M(d)$ for $d \geq \operatorname{reg} M + \delta - r$.*
3. *If $X \subset \mathbb{P}^r$ is a nonempty set of points and $M = S_X$, then $H_M(d) = P_M(d)$ if and only if $d \geq \operatorname{reg} M$. In general, if M is a Cohen–Macaulay module the bound in part 2 is sharp.*

Proof. Part 1 follows at once from part 2 and the Hilbert Syzygy Theorem (Theorem 1.1), which shows that $\delta \leq r + 1$. To prove part 2, consider the minimal graded free resolution of M. By assumption, it has the form

$$0 \longrightarrow \sum_j S(-a_{\delta,j}) \longrightarrow \cdots \longrightarrow \sum_j S(-a_{0,j}) \longrightarrow M \longrightarrow 0,$$

and in these terms $\operatorname{reg} M = \max_{i,j}(a_{i,j} - i)$.

We can compute the Hilbert function or polynomial of M by taking the alternating sum of the Hilbert functions or polynomials of the free modules in the resolution of M. In this way we obtain the expressions

$$H_M(d) = \sum_{i,\,j} (-1)^i \binom{d - a_{i,j} + r}{r},$$

$$P_M(d) = \sum_{i,\,j} (-1)^i \frac{(d - a_{i,j} + r)(d - a_{i,j} + r - 1) \cdots (d - a_{i,j} + 1)}{r!}$$

where i runs from 0 to δ. This expansion for P_M is the expression for H_M with each binomial coefficient replaced by the polynomial to which it is eventually equal. In fact the binomial coefficient $\binom{d-a+r}{r}$ has the same value as the polynomial

$$(d-a+r)(d-a+r-1)\cdots(d-a+1)/r!$$

for all $d \geq a-r$. Thus from $d \geq \operatorname{reg} M + \delta - r$ we get $d \geq a_{i,j} - i + \delta - r \geq a_{i,j} - r$ for each $a_{i,j}$ with $i \leq \delta$. For such d, each term in the expression of the Hilbert function is equal to the corresponding term in the expression of the Hilbert polynomial, proving part 2.

Half of part 3 follows from part 2: The ideal defining X is reduced, and thus S_X is of depth ≥ 1 so, by the Auslander–Buchsbaum formula (Theorem A2.15), the projective dimension of S_X is r. Thus by part 2, the Hilbert function and polynomial coincide for $d \geq \operatorname{reg} S_X$. The converse, and the more general fact about Cohen–Macaulay modules, is more delicate. Again, we will complete the proof in Section 4B, after developing some general theory. A different, more direct proof is sketched in Exercises 4.8–4.10.   □

4B   Characterizations of Regularity: Cohomology

Perhaps the most important characterization of the regularity is cohomological. One way to state it is that the regularity of a module M can be determined from the homology of the complex $\operatorname{Hom}(\mathbf{F}, S)$, where $\mathbf{F}$ is a free resolution of M. This homology is actually dual to the local cohomology of M. We will formulate the results in terms of local cohomology. The reader not already familiar with this notion which, in the case we will require, is a simple extension of the notion of the (Zariski) cohomology of sheaves, should probably take time out to browse at least the first sections of Appendix 1 (through Corollary A1.12). The explicit use of local cohomology can be eliminated — by local duality, many statements about local cohomology can be turned into statements about Ext modules. For a treatment with this flavor see [Eisenbud 1995, Section 20.5]. We follow the convention that the maximum of the empty set is $-\infty$.

Theorem 4.3. *Let M be a finitely generated graded S-module and let d be an integer. The following conditions are equivalent:*

1. $d \geq \operatorname{reg} M$.
2. $d \geq \max\{e \mid \operatorname{H}^i_{\mathfrak{m}}(M)_e \neq 0\} + i$ for all $i \geq 0$.
3. $d \geq \max\{e \mid \operatorname{H}^0_{\mathfrak{m}}(M)_e \neq 0\}$; and $\operatorname{H}^i_{\mathfrak{m}}(M)_{d-i+1} = 0$ for all $i > 0$.

The proof of this result will occupy most of this section. Before beginning it, we illustrate with four corollaries.

Corollary 4.4. *If M is a graded S-module of finite length, then*

$$\operatorname{reg} M = \max\{d \mid M_d \neq 0\}.$$

Proof. $\operatorname{H}^0_{\mathfrak{m}}(M) = M$ and all the higher cohomology of M vanishes by by Corollary A1.5.   □

Corollary 4.4 suggests a convenient reformulation of the definition and of a (slightly weaker) formulation of Theorem 4.3. We first extend the result of the Corollary with a definition: *If $M = \bigoplus M_d$ is an Artinian graded S-module*, then

$$\operatorname{reg} M := \max\{d \mid M_d \neq 0\}.$$

This does not conflict with our previous definition because an Artinian module that is finitely generated over a Noetherian ring is of finite length. The local cohomology modules of any finitely generated graded module are graded Artinian modules by local duality, Theorem A1.9. Thus the following formulas make sense:

Corollary 4.5. *With the preceding notation,*

$$\operatorname{reg} M = \max_i \operatorname{reg} \operatorname{Tor}_i(M, \mathbb{K}) - i = \max_j \operatorname{reg} \operatorname{H}_{\mathfrak{m}}^j(M) + j.$$

There is also a term-by-term comparison,

$$\operatorname{reg} \operatorname{H}_{\mathfrak{m}}^j(M) + j \leq \operatorname{reg} \operatorname{Tor}_{r+1-j}(M, \mathbb{K}) - (r+1-j) \quad \text{for each } j,$$

as we invite the reader to prove in Exercise 4.12.

Proof. The formula $\operatorname{reg} M = \max_j \operatorname{reg} \operatorname{H}_{\mathfrak{m}}^j(M) + j$ is part of Theorem 4.3. For the rest, let $\mathbf{F} : \cdots \to F_i \to \cdots$ be the minimal free resolution of M, with $F_i = \sum_j S(-a_{ij})$. The module $\operatorname{Tor}_i(M, \mathbb{K}) = F_i/\mathfrak{m}F_i$ is a finitely generated graded vector space, thus a module of finite length. By Nakayama's Lemma, the numbers $a_{i,j}$, which are the degrees of the generators of F_i, are also the degrees of the nonzero elements of $\operatorname{Tor}_i(M, \mathbb{K})$. Thus $\operatorname{reg} \operatorname{Tor}_i(M, \mathbb{K}) - i = \max_j\{a_{i,j}\} - i$ and the first equality follows. □

It follows from Corollary 4.4 that the regularity of a module M of finite length is a property that has nothing to do with the S-module structure of M — it would be the same if we replaced S by $\mathbb{K}$. Theorem 4.3 allows us to prove a similar independence for any finitely generated module. To express the result, we write $\operatorname{reg}_S M$ to denote the regularity of M considered as an S-module.

Corollary 4.6. *Let M be a finitely generated graded S-module, and let $S' \to S$ be a homomorphism of graded rings generated by degree 1 elements. If M is also a finitely generated S'-module, then $\operatorname{reg}_S M = \operatorname{reg}_{S'} M$.*

Proof of Corollary 4.6. The statement of finite generation is equivalent to the statement that the maximal ideal of S is nilpotent modulo the ideal generated by the maximal ideal of S' and the annihilator of M. By Corollary A1.8 the local cohomology of M with respect to the maximal ideal of S' is thus the same as that with respect to the maximal ideal of S, so Theorem 4.3 gives the same value for the regularity in either case. □

Theorem 4.3 allows us to complete the proof of part 3 of Theorem 4.2. We first do the case of points.

Corollary 4.7. *If X is a set of n points in $\mathbb{P}^r$ then the regularity of S_X is the smallest integer d such that the space of forms vanishing on the points X has codimension n in the space of forms of degree d.*

Proof. The ring S_X has depth at least 1 because it is reduced, so $H^0_m(S_X) = 0$. Further, since $\dim S_X = 1$ we have $H^i_m(S_X) = 0$ for $i > 1$ by Proposition A1.16. Using Theorem 4.3, we see that the regularity is the smallest integer d such that $H^1_m(S_X)_d = 0$. On the other hand, by Proposition A1.11, there is an exact sequence

$$0 \longrightarrow H^0_m(S_X) \longrightarrow S_X \longrightarrow \bigoplus_d H^0(\mathcal{O}_X(d)) \longrightarrow H^1_m S_X \longrightarrow 0.$$

Since X is just a finite set of points, it is isomorphic to an affine variety, and every line bundle on X is trivial. Thus for every d the sheaf $\mathcal{O}_X(d) \cong \mathcal{O}_X$, a sheaf whose sections are the locally polynomial functions on X. This is just $\mathbb{K}^X$, a vector space of dimension n. Thus $(H^1_m S_X)_d = 0$ if and only if $(S_X)_d = (S/I_X)_d$ has dimension n as a vector space, or equivalently, the space of forms $(I_X)_d$ of degree d that vanish on X has codimension n. $\qquad\square$

Corollary 4.8. *Let M be a finitely generated graded Cohen–Macaulay S-module. If s is the smallest number such that $H_M(d) = P_M(d)$ for all $d \geq s$, then $s = 1 - \operatorname{depth} M + \operatorname{reg} M$.*

Proof. Since M is Cohen–Macaulay we have $\dim M = \operatorname{depth} M$, so Proposition A1.16 shows that the only local cohomology module of M that is nonzero is $H^{\operatorname{depth} M}_m M$. Given this, there can be no cancellation in the formula of Corollary A1.15. Thus s is the smallest number such that $H^{\operatorname{depth} M}(M)_d = 0$ for all $d \geq s$, and Corollary 4.8 follows by Theorem 4.3. $\qquad\square$

See Exercise 4.6 for an example showing that the Cohen–Macaulay hypothesis is necessary, and Exercise 4.9 for a proof that gives some additional information.

It will be convenient to introduce a temporary definition. We call a module *weakly d-regular* if $H^i_m(M)_{d-i+1} = 0$ for every $i > 0$, and *d-regular* if in addition $d \geq \operatorname{reg} H^0_m(M)$. In this language, Theorem 4.3 asserts that M is d-regular if and only if $\operatorname{reg} M \leq d$.

Proof of Theorem 4.3. For the implication $1 \Rightarrow 2$ we do induction on the projective dimension of M. If $M = \bigoplus S(-a_j)$ is a graded free module, this is easy: $\operatorname{reg} M = \max_j a_j$ by definition, and the computation of local cohomology in Corollary A1.6 shows that M is d-regular if and only if $a_i \leq d$ for all i.

Next suppose that $d \geq \operatorname{reg} M$ and the minimal free resolution of M begins

$$\cdots \longrightarrow L_1 \xrightarrow{\varphi_1} L_0 \longrightarrow M \longrightarrow 0.$$

Let $M' = \operatorname{im} \varphi_1$ be the first syzygy module of M. By the definition of regularity, $\operatorname{reg} M' \leq 1 + \operatorname{reg} M$. By induction on projective dimension, we may assume that

M' is $(d+1)$-regular; in fact, since $e \geq \operatorname{reg} M$ for every $e \geq d$ we may assume that M' is $e+1$-regular for every $e \geq d$. The long exact sequence in local cohomology

$$\cdots \longrightarrow \operatorname{H}^i_{\mathfrak{m}}(L_0) \longrightarrow \operatorname{H}^i_{\mathfrak{m}}(M) \longrightarrow \operatorname{H}^{i+1}_{\mathfrak{m}}(M') \longrightarrow \cdots$$

yields exact sequences in each degree, and shows that M is e-regular for every $e \geq d$. This is condition 2.

The implication $2 \Rightarrow 3$ is obvious, but $3 \Rightarrow 1$ requires some preparation. For $x \in R$ we set

$$(0 :_M x) = \{m \in M \mid xm = 0\} = \ker(M \xrightarrow{\ x\ } M).$$

This is a submodule of M that is zero when x is a nonzerodivisor (that is, a *regular element*) on M. When $(0 :_M x)$ has finite length, we say that x is *almost regular* on M.

Lemma 4.9. *Let M be a finitely generated graded S-module, and suppose that $\mathbb{K}$ is infinite. If x is a sufficiently general linear form, then x is almost regular on M.*

The meaning of the conclusion is that the set of linear forms x for which $(0 :_M x)$ is of finite length contains the complement of some proper algebraic subset of the space S_1. The same argument would work for forms of any degree d.

Proof. The module $(0 :_M x)$ has finite length if the radical of the annihilator of $(0 :_M x)$ is the maximal homogeneous ideal $\mathfrak{m}$, or equivalently, if the annihilator of $(0 :_M x)$ is not contained in any prime ideal $P \neq \mathfrak{m}$. This is equivalent to the condition that for all primes $P \neq \mathfrak{m}$, the localization $(0 :_M x)_P = 0$ or equivalently that x is a nonzerodivisor on the localized module M_P. For this it suffices that x not be contained in any associated prime ideal of M except possibly $\mathfrak{m}$.

Each prime ideal P of S other than $\mathfrak{m}$ intersects S_1 in a proper subspace, since otherwise $P \supset \mathfrak{m}$, whence $\mathfrak{m} = P$. Since there are only finitely many associated prime ideals of M, an element $x \in S_1$ has the desired property if it is outside a certain finite union of proper subspaces. $\square$

Proposition 4.10. *Suppose that M is a finitely generated graded S-module, and suppose that x is almost regular on M.*

1. *If M is weakly d-regular, M/xM is weakly d-regular.*
2. *If M is (weakly) d-regular, M is (weakly) $(d+1)$-regular.*
3. *M is d-regular if and only if M/xM is d-regular and $\operatorname{H}^0_{\mathfrak{m}}(M)$ is d-regular.*

Here is a consequence that seems surprising from the free resolution point of view.

Corollary 4.11. *If a linear form x is almost regular on M, then*

$$\operatorname{reg} M = \max\{\operatorname{reg} \operatorname{H}^0_{\mathfrak{m}}(M), \operatorname{reg} M/xM\} = \max\{\operatorname{reg}(0 :_M x), \operatorname{reg} M/xM\}. \quad \square$$

Proof. The first equality follows at once from Theorem 4.3 and part 3 of the proposition. For the second, note that $(0 :_M x) \subset H^0_m(M)$, so $\mathrm{reg}(0 :_M x) \leq \mathrm{reg}\, H^0_m(M)$. On the other hand, if $f \in H^0_m(M)$ is a nonzero element of (maximal) degree $\mathrm{reg}\, H^0_m(M)$, then $xf = 0$, giving the opposite inequality.   □

Proof of Proposition 4.10. *Part* 1. We set $\overline{M} = M/(0 :_M x)$. Using Corollary A1.5 and the long exact sequence of local cohomology we obtain $H^i_m(M) = H^i_m(\overline{M})$ for every $i > 0$.

Consider the exact sequence

$$0 \longrightarrow (\overline{M})(-1) \stackrel{x}{\longrightarrow} M \longrightarrow M/xM \longrightarrow 0, \qquad (*)$$

where the left-hand map is induced by multiplication with x. The associated long exact sequence in local cohomology contains the sequence

$$H^i_m(M)_{d+1-i} \to H^i_m(M/xM)_{d+1-i} \to H^{i+1}_m(\overline{M}(-1))_{d+1-i}.$$

By definition $H^{i+1}_m((\overline{M})(-1))_{d+1-i} \simeq H^{i+1}_m(\overline{M})_{d-i}$. If M is weakly d-regular then the modules on the left and right vanish for every $i \geq 1$. Thus the module in the middle vanishes too, proving that M/xM is weakly d-regular.

Part 2. Suppose M is weakly d-regular. To prove that M is weakly $d+1$-regular we do induction on $\dim M$. If $\dim M = 0$, then $H^i_m(M) = 0$ for all $i \geq 1$ by Corollary A1.5, so M is weakly e-regular for all e and there is nothing to prove.

Now suppose that $\dim M > 0$. Since

$$(0 :_M x) = \ker M \stackrel{x}{\longrightarrow} M$$

has finite length, the Hilbert polynomial of M/xM is the first difference of the Hilbert polynomial of M. From Theorem A2.11 we deduce $\dim M/xM = \dim M - 1$. We know from part 1 that M/xM is weakly d-regular. It follows from our inductive hypothesis that M/xM is weakly $d+1$-regular.

From the exact sequence $(*)$ we get an exact sequence

$$H^i_m(\overline{M}(-1))_{(d+1)-i+1} \longrightarrow H^i_m(M)_{(d+1)-i+1} \longrightarrow H^i_m(M/xM)_{(d+1)-i+1}.$$

For $i \geq 1$, we have $H^i_m(\overline{M}(-1)) = H^i_m(M)$, and since M is weakly d-regular the left-hand term vanishes. The right-hand term is zero because M/xM is weakly $d+1$-regular. Thus M is weakly $d+1$-regular as asserted.

If M is d-regular then as before M is weakly $(d+1)$-regular; and since the extra condition on $H^0_m(M)$ for $(d+1)$-regularity is included in the corresponding condition for d-regularity, we see that M is actually $(d+1)$ regular as well.

Part 3. Suppose first that M is d-regular. The condition that $H^0_m(M)_e = 0$ for all $e > d$ is part of the definition of d-regularity, so it suffices to show that M/xM is d-regular. Since we already know that M/xM is weakly d-regular, it remains to show that if $e > d$ then $H^0_m(M/xM)_e = 0$. Using the sequence $(*)$ once more we get the exact sequence

$$H^0_m(M)_e \to H^0_m(M/xM)_e \to H^1_m(\overline{M}(-1))_e.$$

The left-hand term is 0 by hypothesis. The right-hand term equals $H^1_m(M)_{e-1}$. From part 2 we see that M is weakly e-regular, so the right-hand term is 0. Thus $H^0_m(M/xM)_e = 0$ as required.

Suppose conversely that $H^0_m(M)_e = 0$ for $e > d$ and that M/xM is d-regular. To show that M is d-regular, it suffices to show that $H^i_m(M)_{d-i+1} = 0$ for $i \geq 1$. From the exact sequence (∗) we derive, for each e, an exact sequence

$$H^{i-1}_m(M/xM)_{e+1} \longrightarrow H^i_m(\overline{M})_e \xrightarrow{\alpha_e} H^i_m(M)_{e+1}.$$

Since M/xM is d-regular, part 2 shows it is e-regular for $e \geq d$. Thus the left-hand term vanishes for $e \geq d-i+1$, so α_e is a monomorphism. From $H^i_m(\overline{M}) \cong H^i_m(M)$ we thus get an infinite sequence of monomorphisms

$$H^i_m(M)_{d-i+1} \longrightarrow H^i_m(M)_{d-i+2} \longrightarrow H^i_m(M)_{d-i+3} \longrightarrow \cdots,$$

induced by multiplication by x on $H^i_m(M)$. But by Proposition A1.1 every element of $H^i_m(M)$ is annihilated by some power of x, so $H^i_m(M)_{d-i+1}$ itself is 0, as required.   □

We now complete the proof of Theorem 4.3. Assuming that M is d-regular, it remains to show that $d \geq \text{reg } M$. Since extension of our base field commutes with the formation of local cohomology, these conditions are independent of such an extension, and we may assume for the proof that $\mathbb{K}$ is infinite.

Suppose that the minimal free resolution of M has the form

$$\cdots \longrightarrow L_1 \xrightarrow{\varphi_1} L_0 \longrightarrow M \longrightarrow 0.$$

To show that the generators of the free module L_0 are all of degree $\leq d$ we must show that M is generated by elements of degrees $\leq d$. For this purpose we induct on $\dim M$. If $\dim M = 0$ the result is easy: M has finite length, so by d-regularity $M_e = H^0_m(M)_e = 0$ for $e > d$.

Set $\overline{M} := M/H^0_m(M)$. From the short exact sequence

$$0 \longrightarrow H^0_m(M) \longrightarrow M \longrightarrow \overline{M} \longrightarrow 0,$$

we see that it suffices to prove that both $H^0_m(M)$ and $\overline{M}$ are generated in degrees at most d. For $H^0_m(M)$ this is easy, since $H^0_m(M)_e = 0$ for $e > d$.

By Lemma 4.9 we may choose a linear form x that is a nonzerodivisor on $\overline{M}$. By Proposition 4.10 we see that $\overline{M}/x\overline{M}$ is d-regular. As $\dim \overline{M}/x\overline{M} < \dim \overline{M}$, the induction shows that $\overline{M}/x\overline{M}$, and thus $\overline{M}/m\overline{M}$, are generated by elements of degrees $\leq d$. Nakayama's Lemma allows us to conclude that $\overline{M}$ is also generated by elements of degrees $\leq d$.

If M is free, this concludes the argument. Otherwise, we induct on the projective dimension of M. Let $M' = \text{im } \varphi_1$ be the first syzygy module of M. The long exact sequence in local cohomology coming from the exact sequence

$$0 \longrightarrow M' \longrightarrow L_0 \longrightarrow M \longrightarrow 0$$

shows that M' is $d+1$-regular. By induction $\operatorname{reg} M' \leq d+1$; that is, the part of the resolution of M that starts from L_1 satisfies exactly the conditions that make $\operatorname{reg} M \leq d$. □

Solution of the Interpolation Problem

The first step in solving the interpolation problem is to reformulate the question solely in terms of projective geometry. To do this we first have to get away from the language of functions. A homogeneous form $F \in S$ does not define a function with a value at a point $p = (p_0, \ldots, p_r) \in \mathbb{P}^r$: for we could also write $p = (\lambda p_0, \ldots, \lambda p_r)$ for any nonzero λ, but if $\deg F = d$ then $F(\lambda p_0, \ldots, \lambda p_r) = \lambda^d F(p_0, \ldots, p_r)$ which may not be equal to $F(p_0, \ldots, p_q)$. But the trouble disappears if $F(p_0, \ldots, p_r) = 0$, so it does make sense to speak of a homogeneous form vanishing at a point. This is a linear condition on the coefficients of the form (Reason: choose homogeneous coordinates for the point and substitute them into the monomials in the form, to get a value for each monomial. The linear combination of the coefficients given by these values is zero if and only if the form vanishes at the point.) We will say that X imposes *independent conditions* on the forms of degree d if the linear conditions associated to the distinct points of X are independent, or equivalently if we can find a form vanishing at any one of the points without vanishing at the others. In this language, Corollary 4.7 asserts that the regularity of S_X is equal to the smallest degree d such that X imposes independent conditions on forms of degree d. The following result completes the proof of Theorem 4.1.

Proposition 4.12. *A finite set of points $X \subset \mathbb{A}^r \subset \mathbb{P}^r$ imposes independent conditions on forms of degree d in $\mathbb{P}^r$ if and only if every function on the points is the restriction of a polynomial of degree $\leq d$ on $\mathbb{A}^r$.*

Proof. We think of $\mathbb{A}^r \subset \mathbb{P}^r$ as the complement of the hyperplane $x_0 = 0$. If the points impose independent conditions on forms of degree d then we can find a form $F_i(x_0, \ldots, x_r)$ of degree d vanishing on p_j for exactly those $j \neq i$. The polynomial $f_i(x_1, \ldots, x_r) = F_i(1, x_1, \ldots, x_r)$ has degree $\leq d$ and the same vanishing/nonvanishing property, so the function $\sum_i (a_i / f_i(p_i)) f_i$ takes values a_i on p_i for any desired a_i.

Conversely, if every function on X is induced by a polynomial of degree $\leq d$ on $\mathbb{A}^r$, then for each i there is a function f_i of degree $\leq d$ that vanishes at p_j for $j \neq i$ but does not vanish at p_i. The degree d homogenization $F_i(x_0, \ldots, x_r) = x_0^d f_i(x_1/x_0, \ldots, x_r/x_0)$ has corresponding vanishing properties. The existence of the F_i shows that the points p_i impose independent conditions on forms of degree d. □

The maximal number of independent linear equations in a certain set of linear equations — the rank of the system of equations — does not change when we extend the field, so Proposition 4.12 shows that the interpolation degree is independent of field extensions.

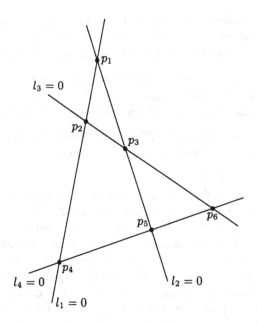

$l_3 = 0$

p_1

p_2

p_3

p_5

p_6

p_4

$l_4 = 0$

$l_2 = 0$

$l_1 = 0$

Every function $\{p_1, \ldots, p_6\} \to \mathbb{R}$ is the restriction of a quadratic polynomial.

4C   The Regularity of a Cohen–Macaulay Module

In the special case of Proposition 4.10 where x is a regular element, $H^0_{\mathfrak{m}}(M)$ must vanish, so part 3 of the proposition together with Theorem 4.3 says that $\operatorname{reg} M/xM = \operatorname{reg} M$. This special case admits a simple proof without cohomology.

Corollary 4.13. *Suppose that M is a finitely generated graded S-module. If x is a linear form in S that is a nonzerodivisor on M then $\operatorname{reg} M = \operatorname{reg} M/xM$.*

Proof. Let $\mathbf{F}$ be the minimal free resolution of M. From the free resolution

$$\mathbf{G}: \quad 0 \longrightarrow S(-1) \overset{x}{\longrightarrow} S$$

of $S/(x)$, we can compute $\operatorname{Tor}_*(M, S/(x))$. Since x is a nonzerodivisor on M, the sequence $0 \to M(-1) \to M$ obtained by tensoring M with $\mathbf{G}$ has homology

$$\operatorname{Tor}_0(M, S/(x)) = M/xM; \quad \operatorname{Tor}_i(M, S/(x)) = 0$$

for $i > 0$. We can also compute Tor as the homology of the free complex $\mathbf{F} \otimes \mathbf{G}$, so we see that $\mathbf{F} \otimes \mathbf{G}$ is the minimal free resolution of M/xM. The i-th free module in $\mathbf{F} \otimes \mathbf{G}$ is $F_i \oplus F_{i-1}(-1)$, so $\operatorname{reg} M/xM = \operatorname{reg} M$. □

We can apply this to get another means of computing the regularity in the Cohen–Macaulay case.

Proposition 4.14. *Let M be a finitely generated Cohen–Macaulay graded S-module, and let $y_1, \ldots, y_t$ be a maximal M-regular sequence of linear forms. The regularity of M is the largest d such that $(M/(y_1, \ldots, y_t)M)_d \neq 0$.*

Proof. Because M is Cohen–Macaulay we have $\dim M/(y_1, \ldots, y_t)M = 0$. With Corollary 4.13, this allows us to reduce the statement to the zero-dimensional case. But if $\dim M = 0$ the result follows at once from Theorem 4.3.   □

As a consequence, we can give a general inequality on the regularity of the homogeneous coordinate ring of an algebraic set X that strengthens the computation done at the beginning of Section 4A — so long as S_X is Cohen–Macaulay.

Corollary 4.15. *Suppose that $X \subset \mathbb{P}^r$ is not contained in any hyperplane. If S_X is Cohen–Macaulay, then* $\operatorname{reg} S_X \leq \deg X - \operatorname{codim} X$.

Proof. Let $t = \dim X$, so that the dimension of S_X as a module is $t+1$. We may harmlessly extend the ground field and assume that it is algebraically closed, and in particular infinite. Thus we may assume that a sufficiently general sequence of linear forms $y_0, \ldots, y_t$ is a regular sequence in any order on S_X. Set $\bar{S}_X = S_X/(y_0, \ldots, y_t)$. Since X is not contained in a hyperplane, we have $\dim_{\mathbb{K}}(S_X)_1 = r+1$, and thus $\dim_{\mathbb{K}}(\bar{S}_X)_1 = r-t = \operatorname{codim} X$. If the regularity of S_X is d, then by Proposition 4.14 we have $H_{\bar{S}_X}(d) \neq 0$. Because $\bar{S}_X$ is generated as an S-module in degree 0, this implies that $H_{\bar{S}_X}(e) \neq 0$ for all $0 \leq e \leq d$. On the other hand, $\deg X$ is the number of points in which X meets the linear space defined by $y_1 = \cdots = y_t = 0$. By induction, using the exact sequence

$$0 \to S_X/(y_1, \ldots, y_t)(-1) \xrightarrow{y_0} S_X/(y_1, \ldots, y_t) \longrightarrow \bar{S}_X \to 0,$$

we see that $H_{S_X/(y_1,\ldots,y_t)}(d) = \sum_{e=0}^{d} H_{\bar{S}_X}(e)$. For large d the polynomials of degree d induce all possible functions on the set $X \cap L$, so $\deg X = H_{S_X/(y_1,\ldots,y_t)}(l)$. It follows that for large d

$$\deg X = \sum_{e=0}^{d} H_{\bar{S}_X}(e) \geq 1 + (\operatorname{codim} X) + (\operatorname{reg} X - 1)$$

since there are at least $\operatorname{reg} X - 1$ more nonzero values of $H_{\bar{S}_X}(e) \neq 0$ for $e = 2, \ldots, d$. This gives $\operatorname{reg} X \leq \deg X - \operatorname{codim} X$ as required.   □

In the most general case, the regularity can be very large. Consider the case of a module of the form $M = S/I$. Gröbner basis methods give a general bound for the regularity of M in terms of the degrees of generators of I and the number of variables, but these bounds are doubly exponential in the number of variables. Moreover, there are examples of ideals I such that the regularity of S/I really is doubly exponential in r (see [Bayer and Stillman 1988] and [Koh 1998]). (Notwithstanding, I know few examples in small numbers of variables of ideals I where $\operatorname{reg} S/I$ is much bigger than the sum of the degree of the generators of I. Perhaps the best is due to Caviglia [2004], who has proved that if $S = \mathbb{K}[s, t, u, v]$ and $d > 1$ then

$$I = (s^d, t^d, su^{d-1} - tv^{d-1}) \subset \mathbb{K}[s, t, u, v]$$

has $\operatorname{reg} S/I = d^2 - 2$. It would be interesting to have more and stronger examples with high regularity.)

In contrast with the situation of general ideals, prime ideals seem to behave very well. For example, in Chapter 5.1 we will prove a theorem of Gruson, Lazarsfeld, and Peskine to the effect that if $\mathbb{K}$ is algebraically closed and X is an irreducible (reduced) curve in projective space, not contained in a hyperplane then again $\operatorname{reg} S_X \leq \deg X - \operatorname{codim} X$, even if S_X is not Cohen–Macaulay, and we will discuss some conjectural extensions of this result.

4D   The Regularity of a Coherent Sheaf

Mumford originally defined a coherent sheaf $\mathscr{F}$ on $\mathbb{P}^r$ to be d-regular if

$$H^i \mathscr{F}(d-i) = 0 \quad \text{for every } i \geq 1$$

(see [Mumford 1966, Lecture 14].) When $\mathscr{F}$ is a sheaf, we will write $\operatorname{reg} \mathscr{F}$ for the least number d such that $\mathscr{F}$ is d-regular (or $-\infty$ if $\mathscr{F}$ is d-regular for every d.) The connection with our previous notion is the following:

Proposition 4.16. *Let M be a finitely generated graded S-module, and let $\widetilde{M}$ be the coherent sheaf on $\mathbb{P}^r_{\mathbb{K}}$ that it defines. The module M is d-regular if and only if*

1. *$\widetilde{M}$ is d-regular;*
2. *$H^0_{\mathfrak{m}}(M)_e = 0$ for every $e > d$; and*
3. *the canonical map $M_d \to H^0(\widetilde{M}(d))$ is surjective.*

In particular, $\operatorname{reg} M \geq \operatorname{reg} \widetilde{M}$ always, with equality if $M = \bigoplus_d H^0(\mathscr{F}(d))$.

Proof. By Proposition A1.12, $H^i_{\mathfrak{m}}(M)_e = H^{i-1}(\widetilde{M}(e))$ for all $i \geq 2$. Thus M is d-regular if and only if it fulfills conditions 1, 2, and

3'. $H^1_{\mathfrak{m}}(M)_e = 0$ for all $e \geq d$.

The exact sequence of Proposition A1.12 shows that conditions 3' and 3 are equivalent.   □

We can give a corresponding result starting with the sheaf. Suppose $\mathscr{F}$ is a nonzero coherent sheaf on $\mathbb{P}^r_{\mathbb{K}}$. The S-module $\Gamma_*(\mathscr{F}) := \bigoplus_{e \in \mathbb{Z}} H^0(\mathscr{F}(e))$ is not necessarily finitely generated (the problem comes about if $\mathscr{F}$ has 0-dimensional associated points); but for every e_0 its truncation

$$\Gamma_{\geq e_0}(\mathscr{F}) := \bigoplus_{e \geq e_0} H^0(\mathscr{F}(e))$$

is a finitely generated S-module. We can compare its regularity with that of $\mathscr{F}$.

Corollary 4.17. *For a coherent sheaf $\mathscr{F}$ on $\mathbb{P}^r_{\mathbb{K}}$,*

$$\operatorname{reg} \Gamma_{\geq e_0}(\mathscr{F}) = \max(\operatorname{reg} \mathscr{F}, e_0).$$

Proof. Suppose first that $M := \Gamma_{\geq e_0}(\mathscr{F})$ is d-regular. The sheaf associated to M is $\mathscr{F}$. Proposition A1.12 shows that $\mathscr{F}$ is d-regular. Since M is d-regular it is generated in degrees $\leq d$. If $d < e_0$ then $M = 0$, contradicting our hypothesis $\mathscr{F} \neq 0$. Thus $d \geq e_0$.

It remains to show that if $\mathscr{F}$ is d-regular and $d \geq e_0$, then M is d-regular. We again want to apply Proposition A1.12. Conditions 1 and 3 are clearly satisfied, while 2 follows from Proposition A1.12. □

It is now easy to give the analogue for sheaves of Proposition 4.10. The first statement is one of the key results in the theory.

Corollary 4.18. *If $\mathscr{F}$ is a d-regular coherent sheaf on $\mathbb{P}^r$ then $\mathscr{F}(d)$ is generated by global sections. Moreover, $\mathscr{F}$ is e-regular for every $e \geq d$.*

Proof. The module $M = \Gamma_{\geq d}(\mathscr{F})$ is d-regular by Corollary 4.17, and thus it is generated by its elements of degree d, that is to say, by $H^0 \mathscr{F}(d)$. Since $\tilde{M} = \mathscr{F}$, the first conclusion follows.

By Proposition 4.10 M is e-regular for $e \geq d$. Using Corollary 4.17 again we see that $\mathscr{F}$ is e-regular. □

4E   Exercises

1. For a set of points X in $\mathbb{P}^2$, with notation e_i, f_i as in Proposition 3.8, show that $\operatorname{reg} S_X = e_1 + \sum_i f_i - 2$. Use this to compute the possible regularities of all sets of 10 points in $\mathbb{P}^2$.

2. Suppose that
$$0 \to M' \to M \to M'' \to 0$$
is an exact sequence of finitely generated graded S-modules. Show that
   (a) $\operatorname{reg} M' \leq \max\{\operatorname{reg} M,\ \operatorname{reg} M'' - 1\}$;
   (b) $\operatorname{reg} M \leq \max\{\operatorname{reg} M',\ \operatorname{reg} M''\}$;
   (c) $\operatorname{reg} M'' \leq \max\{\operatorname{reg} M,\ \operatorname{reg} M' + 1\}$.

3. Suppose $X \subset \mathbb{P}^r$ is a projective scheme and $r > 0$. Show that $\operatorname{reg} I_X = \operatorname{reg} \mathscr{I}_X = 1 + \operatorname{reg} \mathscr{O}_X = 1 + \operatorname{reg} S_X$.

4. We say that a variety in a projective space is *nondegenerate* if it is not contained in any hyperplane. Correspondingly, we say that a homogeneous ideal is nondegenerate if it does not contain a linear form. Most questions about the free resolutions of ideals can be reduced to the nondegenerate case, just as can most questions about varieties in projective space. Here is the basic idea:
   (a) Show that if $I \subset S$ is a homogeneous ideal in a polynomial ring containing linearly independent linear forms $\ell_0, \ldots, \ell_t$, there are linear forms $\ell_{t+1}, \ldots, \ell_r$ such that $\{\ell_0, \ldots, \ell_t, \ell_{t+1}, \ldots, \ell_r\}$ is a basis for S_1, and such that I may be written in the form $I = JS + (\ell_1, \ldots, \ell_t)$ where J is a homogeneous ideal in the smaller polynomial ring $R = \mathbb{K}[\ell_{t+1}, \ldots, \ell_r]$.

(b) Show that the minimal S-free resolution of JS is obtained from the minimal R-free resolution of J by tensoring with S. Thus they have the same graded Betti numbers.

(c) Show that the minimal S-free resolution of S/I is obtained from the minimal S-free resolution of S/JS by tensoring with the Koszul complex on $\ell_0, \ldots, \ell_t$. Deduce that the regularity of S/I is the same as that of R/J.

5. Suppose that M is a finitely generated graded Cohen–Macaulay S-module, with minimal free resolution

$$0 \to F_c \to \cdots \to F_1 \to F_0,$$

and write $F_i = \bigoplus S(-j)^{\beta_{i,j}}$ as usual. Show that

$$\operatorname{reg} M = \max\{j \mid \beta_{c,j} \neq 0\} - c;$$

that is, the regularity of M is measured "at the end of the resolution" in the Cohen–Macaulay case. Find an example of a module for which the regularity cannot be measured just "at the end of the resolution."

6. Find an example showing that Corollary 4.8 may fail if we do not assume that M is Cohen–Macaulay. (If this is too easy, find an example with $M = S/I$ for some ideal I.)

7. Show that if X consists of d distinct point in $\mathbb{P}^r$ then the regularity of S_X is bounded below by the smallest integer s such that $d \leq \binom{r+s}{r}$. Show that this bound is attained by the general set of d points.

8. Recall that the generating function of the Hilbert function of a (finitely generated graded) module M is $\Psi_M(t) = \sum_{-\infty}^{\infty} H_M(d)t^d$, and that by Theorem 1.11 (with all x_i of degree 1) it can be written as a rational $\phi_M(t)/(1-t)^{r+1}$. Show that if $\dim M < r+1$ then $1-t$ divides the numerator; more precisely, we can write

$$\Psi_M(t) = \frac{\phi'_M(t)}{(1-t)^{\dim M}}.$$

for some Laurent polynomial ϕ'_M, and this numerator and denominator are relatively prime.

9. With notation as in the previous exercise, suppose M is a Cohen–Macaulay S-module, and let $y_0, \ldots, y_s$ be a maximal M-regular sequence of linear forms, so that $M' = M/(y_0, \ldots, y_s)$ has finite length. Let $\Psi_{M'} = \sum H_{M'}(d)t^d$ be the generating function of the Hilbert function of M', so that $\Psi_{M'}$ is a polynomial with positive coefficients in t and t^{-1}. Show that

$$\Psi_M(t) = \frac{\Psi_{M'}(t)}{(1-t)^{\dim M}}.$$

In the notation of Exercise 4.8 $\phi'_M = \Psi_{M'}$. Deduce that

$$H_M(d) = \sum_{e \le d} \binom{\dim M + d}{\dim M} H_{M'}(d-e).$$

10. Use Proposition 4.14 and the result of Exercise 4.9 to give a direct proof of Theorem 4.3.2.

11. Find an example of a finitely generated graded S-module M such that $\phi'_M(t)$ does not have positive coefficients.

12. Use local duality to refine Corollary 4.5 by showing that for each j we have

$$\text{reg } H^j_{\mathfrak{m}}(M) + j \le \text{reg Tor}_{r+1-j}(M, \mathbb{K}) - (r+1-j).$$

13. (The basepoint-free pencil trick.) Here is the idea of Castelnuovo that led Mumford to define what we call Castelnuovo–Mumford regularity: Suppose that $\mathscr{L}$ is a line bundle on a curve $X \subset \mathbb{P}^r$ over an infinite field, and suppose and that $\mathscr{L}$ is basepoint-free. Show that we may choose 2 sections σ_1, σ_2 of $\mathscr{L}$ which together form a basepoint-free pencil — that is, $V := \langle \sigma_1, \sigma_2 \rangle$ is a two-dimensional subspace of $H^0(\mathscr{L})$ which generates $\mathscr{L}$ locally everywhere. Show that the Koszul complex of σ_1, σ_2

$$\mathbf{K}: \quad 0 \to \mathscr{L}^{-2} \to \mathscr{L}^{-1} \oplus \mathscr{L}^{-1} \to \mathscr{L} \to 0$$

is exact, and remains exact when tensored with any sheaf.

   Now let $\mathscr{F}$ be a coherent sheaf on X with $H^1\mathscr{F} = 0$ (or, as we might say now, such that the Castelnuovo–Mumford regularity of $\mathscr{F}$ is at most -1.) Use the sequence $\mathbf{K}$ above to show that the multiplication map map $V \otimes \mathscr{F} \to \mathscr{L} \otimes \mathscr{F}$ induces a surjection $V \otimes H^0 \mathscr{F} \to H^0(\mathscr{L} \otimes \mathscr{F})$.

   Suppose that X is embedded in $\mathbb{P}^r$ as a curve of degree $d \ge 2g+1$, where g is the genus of X. Use the argument above to show that

$$H^0(\mathscr{O}_X(1)) \otimes H^0(\mathscr{O}_X(n)) \to H^0(\mathscr{O}_X(n+1))$$

is surjective for $n \ge 1$. This result is a special case of what is proved in Theorem 8.1.

14. Suppose that $X \subset \mathbb{P}^r$ is a projective scheme. We say that X is d-normal if the restriction map

$$H^0(\mathscr{O}_{\mathbb{P}^r}(d)) \to H^0(\mathscr{O}_X(d))$$

is surjective. We say that X is d-regular if $\mathscr{I}_X$ is d-regular.

   (a) Show that X is d-normal if and only if $H^1(\mathscr{I}_X(d)) = 0$.

   (b) Let $x \in S_1$ be a linear form that is almost regular on S_X, and let $H \subset \mathbb{P}^r$ be the hyperplane defined by the vanishing of x. Show that X is d-regular if and only if $H \cap X$ is d-regular and X is $(d-1)$-normal.

15. Surprisingly few general bounds on the regularity of ideals are known. As we have seen, if X is the union of n points on a line, then $\operatorname{reg} S_X = n-1$. The following result [Derksen and Sidman 2002] shows (in the case $I_0 = (0)$) that this is in some sense the worst case: no matter what the dimensions, the ideal of the union of n planes in $\mathbb{P}^r$ has regularity at most n. Here is the algebraic form of the result. The extra generality is used for an induction.

Theorem 4.19. *Suppose $I_0, \ldots, I_n$ are ideals generated by spaces of linear forms in S. Then, for $n \geq 1$, the regularity of $I = I_0 + \bigcap_1^n I_j$ is at most n.*

Prove this result by induction on $\dim S/I_0$, the case $\dim S/I_0 = 0$ being trivial.

(a) Show that it is equivalent to prove that $\operatorname{reg} S/I = n-1$.
(b) Reduce to the case where $I_0 + I_1 + \cdots + I_n = \mathfrak{m}$ and the ground field is infinite.
(c) Use Corollary 4.11 to reduce the problem to proving $\operatorname{reg} H^0_{\mathfrak{m}}(S/I) \leq n-1$; that is, reduce to showing that if f is an element of degree at least n in $H^0_{\mathfrak{m}}(S/I)$ then $f = 0$.
(d) Let x be a general linear form in S. Show that $f = xf'$ for some f' of degree $n-1$ in S/I. Use the fact that x is general to show that the image of f' is in $H^0_{\mathfrak{m}}(S/(I_0 + \bigcap_{j \neq i} I_j))$ for $i = 1, \ldots, n$. Conclude by induction on n that the image of f' is zero in
$$S/(I_0 + \bigcap_{j \neq i} I_j).$$

(e) Use part (b) to write $x = \sum x_i$ for linear forms $x_i \in I_i$. Now show that $f = xf' \in I$.

Putting Theorem 4.19 together with Conjecture 5.2, that the regularity is bounded (roughly) by the degree in the irreducible case, one might be tempted to guess that the regularity of an algebraic set would be bounded by the sum of the degrees of its components. This is false: Daniel Giaimo [2004] has given a series of examples of algebraic sets $X_r = L_r \cup Y_r \subset \mathbb{P}^r$, where L_r is a linear subspace and Y_r is a reduced, irreducible complete intersection of the same dimension as L_r, but $\operatorname{reg} X_r$ is exponential in the degree of X_r (and doubly exponential in r).

5

The Regularity of Projective Curves

This chapter is devoted to a theorem of Gruson, Lazarsfeld and Peskine [Gruson et al. 1983] giving an optimal upper bound for the regularity of a projective curve in terms of its degree. The result had been proven for smooth curves in $\mathbb{P}^3$ by Castelnuovo [1893].

Theorem 5.1 (Gruson–Lazarsfeld–Peskine). *Suppose $\mathbb{K}$ is an algebraically closed field. If $X \subset \mathbb{P}^r_{\mathbb{K}}$ is a reduced and irreducible curve, not contained in a hyperplane, then $\operatorname{reg} S_X \leq \deg X - r + 1$, and thus $\operatorname{reg} I_X \leq \deg X - r + 2$.*

In particular, Theorem 5.1 implies that the degrees of the polynomials needed to generate I_X are bounded by $\deg X - r + 2$. If the field $\mathbb{K}$ is the complex numbers, the degree of X may be thought of as the homology class of X in $H_2(\mathbb{P}^r; \mathbb{K}) = \mathbb{Z}$, so the bound given depends only on the topology of the embedding of X.

5A A General Regularity Conjecture

We have seen in Corollary 4.15 that if $X \subset \mathbb{P}^r$ is arithmetically Cohen–Macaulay (that is, if S_X is a Cohen–Macaulay ring) and nondegenerate (that is, not contained in a hyperplane), then $\operatorname{reg} S_X \leq \deg X - \operatorname{codim} X$, which gives $\deg X - r + 1$ in the case of curves. This suggests that some version of Theorem 5.1 could hold much more generally. However, this bound can fail for schemes that are not arithmetically Cohen–Macaulay, even in the case of curves; the simplest example is where X is the union of two disjoint lines in $\mathbb{P}^3$ (see Exercise 5.2). The result can also fail when X is not reduced or the ground field is not algebraically closed

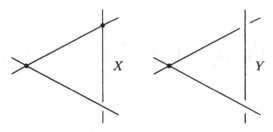

X is 2-regular but Y is not.

(see Exercises 5.3 and 5.4). And it is not enough to assume that the scheme is reduced and connected, since the cone over a disconnected set is connected and has the same codimension and regularity.

A possible way around these examples is to insist that X be reduced, and *connected in codimension 1*, meaning that X is pure-dimensional and cannot be disconnected by removing any algebraic subset of codimension 2.

Conjecture 5.2. [Eisenbud and Goto 1984] *If* $\mathbb{K}$ *is algebraically closed and* $X \subset \mathbb{P}_\mathbb{K}^r$ *be a nondegenerate algebraic set that is connected in codimension 1, then*

$$\operatorname{reg}(S_X) \le \deg X - \operatorname{codim} X.$$

For example, in dimension 1 the conjecture just says that the bound should hold for connected reduced curves. This was recently proved in [Giaimo 2003]. In addition to the Cohen–Macaulay and one-dimensional cases, the conjecture is known to hold for smooth surfaces in characteristic 0 [Lazarsfeld 1987], arithmetically Buchsbaum surfaces [Stückrad and Vogel 1987] and toric varieties of low codimension [Peeva and Sturmfels 1998]. Somewhat weaker results are known more generally; see [Kwak 1998] and [Kwak 2000] for the best current results and [Bayer and Mumford 1993] for a survey.

Of course for the conjecture to have a chance, the number $\deg X - \operatorname{codim} X$ must at least be nonnegative. The next proposition establishes this inequality. The examples in Exercises 5.2–5.4 show that the hypotheses are necessary.

Proposition 5.3. *Suppose that* $\mathbb{K}$ *is algebraically closed. If* X *is a nondegenerate algebraic set in* $\mathbb{P}^r = \mathbb{P}_\mathbb{K}^r$ *and* X *is connected in codimension 1, then* $\deg X \ge 1 + \operatorname{codim} X$.

To understand the bound, set $c = \operatorname{codim} X$ and let $p_1, \dots, p_{c+1}$ be general points on X. Since X is nondegenerate, these points can be chosen to span a plane L of dimension c. The degree of X is the number of points in which X meets a general c-plane, and it is clear that L meets X in at least c points. The problem with this argument is that L might, a priori, meet X in a set of positive dimension, and this can indeed happen without some extra hypothesis, such as "connected in codimension 1".

As you may see using the ideas of Corollary 4.15, the conclusion of Proposition 5.3 also holds for any scheme $X \subset \mathbb{P}^r$ such that S_X is Cohen–Macaulay.

Proof. We do induction on the dimension of X. If $\dim X = 0$, then X cannot span $\mathbb{P}^r$ unless it contains at least $r+1$ points; that is, $\deg X \geq 1+r = 1+\operatorname{codim} X$. If $\dim X > 0$ we consider a general hyperplane section $Y = H \cap X \subset H = \mathbb{P}^{r-1}$. The degree and codimension of Y agree with those for X. Further, since H was general, Bertini's Theorem [Hartshorne 1977, p. 179] tells us that Y is reduced. It remains to show that Y is connected in codimension 1 and nondegenerate.

The condition that X is pure-dimensional and connected in codimension 1 can be reinterpreted as saying that the irreducible components of X can be ordered, say $X_1, X_2, \ldots$ in such a way that if $i > 1$ then X_i meets some X_j, with $j < i$, in a set of codimension 1 in each. This condition is inherited by $X \cap H$ so long as the H does not contain any of the X_i or $X_i \cap X_j$.

For nondegeneracy we need only the condition that X is connected. Lemma 5.4 completes the proof. $\qquad\square$

Lemma 5.4. *If $\mathbb{K}$ is algebraically closed and X is a connected algebraic set in $\mathbb{P}^r = \mathbb{P}^r_{\mathbb{K}}$, not contained in any hyperplane, then for every hyperplane in $\mathbb{P}^r$ the scheme $X \cap H$ is nondegenerate in H.*

For those who prefer not to deal with schemes: the general hyperplane section of any algebraic set is reduced, and thus can be again considered an algebraic set. So scheme theory can be avoided at the expense of taking general hyperplane sections.

Proof. Let x be the linear form defining H. There is a commutative diagram with exact rows

$$
\begin{array}{ccccccccc}
0 & \longrightarrow & H^0(\mathscr{O}_{\mathbb{P}^r}) & \xrightarrow{\ x\ } & H^0(\mathscr{O}_{\mathbb{P}^r}(1)) & \longrightarrow & H^0(\mathscr{O}_H(1)) & \longrightarrow & H^1(\mathscr{O}_{\mathbb{P}^r}) \\
& & \downarrow & & \downarrow & & \downarrow & & \\
0 & \longrightarrow & H^0(\mathscr{O}_X) & \xrightarrow{\ x\ } & H^0(\mathscr{O}_X(1)) & \longrightarrow & H^0(\mathscr{O}_{X \cap H}(1)) & \longrightarrow & \cdots
\end{array}
$$

The hypotheses that X is connected and projective, together with the hypothesis that $\mathbb{K}$ is algebraically closed, imply that the only regular functions defined everywhere on X are constant; that is, $H^0(\mathscr{O}_X) = \mathbb{K}$, so the left-hand vertical map is surjective (in fact, an isomorphism). The statement that X is nondegenerate means that the middle vertical map is injective. Using the fact that $H^1(\mathscr{O}_{\mathbb{P}^r}) = 0$, the Snake Lemma shows that the right-hand vertical map is injective, so $X \cap H$ is nondegenerate. $\qquad\square$

5B Proof of the Gruson–Lazarsfeld–Peskine Theorem

Here is a summary of the proof: We will find a complex that is almost a resolution of an ideal that is almost the ideal I_X of X. Miraculously, this will establish the regularity of I_X.

More explicitly, we will find a module F over S_X which is similar to S_X but admits a free presentation by a matrix of linear forms ψ, and such that the Eagon–Northcott complex associated with the ideal of maximal minors of ψ is nearly a resolution of I_X. We will then prove that the regularity of this Eagon–Northcott complex is a bound for the regularity of I_X. The module F will come from a line bundle on the normalization of the curve X. From the cohomological properties of the line bundle we will be able to control the properties of the module.

Still more explicitly, let $\pi : C \to X \subset \mathbb{P}^r_{\mathbf{K}}$ be the normalization of X. Let $\mathscr{A}$ be an invertible sheaf on C and let $\mathscr{F} = \pi_* \mathscr{A}$. The sheaf $\mathscr{F}$ is locally isomorphic to $\mathscr{O}_X$ except at the finitely many points where π fails to be an isomorphism. Let $F = \bigoplus_{n \geq 0} \mathrm{H}^0 \mathscr{F}(n)$, and let

$$L_1 \xrightarrow{\psi} L_0 \longrightarrow F$$

be a minimal free presentation of F. We write $I(\psi)$ for the ideal generated by the rank L_0-sized minors (subdeterminants) of a matrix representing ψ; this is the *0-th Fitting ideal of F*. We will use three facts about Fitting ideals presented in Section A2G (page 220): they do not depend on the free presentations used to define them; they commute with localization; and the 0-th Fitting ideal of a module is contained in the annihilator of the module. Write $\mathscr{I}(\psi)$ for the sheafification of the Fitting ideal (which is also the sheaf of Fitting ideals of the sheaf $\mathscr{A}$, by our remark on localization). This sheaf is useful to us because of the last statement of the following result.

Proposition 5.5. *With notation above, $\mathscr{I}(\psi) \subseteq \mathscr{I}_X$. The quotient $\mathscr{I}_X / \mathscr{I}(\psi)$ is supported on a finite set of points in $\mathbb{P}^r$, and $\operatorname{reg} I(\psi) \geq \operatorname{reg} I_X$.*

Proof. The 0-th Fitting ideal of a module is quite generally contained in the annihilator of the module. The construction of the Fitting ideal commutes with localization (see [Eisenbud 1995, Corollary 20.5] or Section A2G.) At any point $p \in \mathbb{P}^r$ such that π is an isomorphism we have $(\pi_* A)_p \cong (\mathscr{O}_X)_p$, where the subscript denotes the stalk at the point p. Since the Fitting ideal of S_X is I_X, we see that $(\mathscr{I}_X)_p = \mathscr{I}(\psi)_p$. Since X is reduced and one-dimensional, the map π is an isomorphism except at finitely many points.

Consider the exact sequence

$$0 \longrightarrow \mathscr{I}(\psi) \longrightarrow \mathscr{I}_X \longrightarrow \mathscr{I}_X / \mathscr{I}(\psi) \longrightarrow 0.$$

Since $\mathscr{I}_X / \mathscr{I}(\psi)$ is supported on a finite set, we have $\mathrm{H}^1(\mathscr{I}_X(d) / \mathscr{I}(\psi)(d)) = 0$ for every d. From the long exact sequence in cohomology we see that $\mathrm{H}^1(\mathscr{I}_X(d))$ is a quotient of $\mathrm{H}^1(\mathscr{I}(\psi)(d))$, while $\mathrm{H}^i(\mathscr{I}_X(d)) = \mathrm{H}^i(\mathscr{I}(\psi)(d))$ for $i > 1$. In particular, $\operatorname{reg} \mathscr{I}(\psi) \geq \operatorname{reg} \mathscr{I}_X$. Since I_X is saturated, we obtain $\operatorname{reg} I(\psi) \geq \operatorname{reg} I_X$ as well. $\square$

Thus it suffices to find a line bundle $\mathscr{A}$ on C such that the regularity of $\mathscr{I}(\psi)$ is low enough. We will show that when $\mathscr{F}$ has a linear presentation, as defined below, the regularity of this Fitting ideal is bounded by the dimension of $H^0(\mathcal{F})$. We begin with a criterion for linear presentation.

Linear Presentations

The main results in this section were proved by Green [1984a; 1984b; 1989] in his exploration of Koszul cohomology.

If F is any finitely generated graded S module, we say that F has a *linear presentation* if, in the minimal free resolution

$$\cdots \longrightarrow L_1 \xrightarrow{\varphi_1} L_0 \longrightarrow F \longrightarrow 0,$$

we have $L_i = \bigoplus S(-i)$ for $i = 0, 1$. This signifies that F is generated by elements of degree 0 and the map φ_1 can be represented by a matrix of linear forms.

The condition of having a linear presentation implies that the homogeneous components F_d are 0 for $d < 0$. Note that if F is any module with $F_d = 0$ for $d < 0$, and $L_1 \to L_0$ is a minimal free presentation, then the free module L_0 is generated in degrees ≥ 0. By Nakayama's lemma the kernel of $L_0 \to F$ is contained in the homogeneous maximal ideal times L_0 so it is generated in degrees ≥ 1, and it follows from minimality that L_1 is generated in degrees ≥ 1. Thus a module F generated in degrees ≥ 0 has a linear presentation if and only if L_i requires no generators of degree $> i$ for $i = 0, 1$ — we do not have to worry about generators of too low degree.

In the following results we will make use of the *tautological rank-r subbundle* $\mathcal{M}$ on $\mathbb{P} := \mathbb{P}_K^r$. It is defined as the subsheaf of $\mathcal{O}_\mathbb{P}^{r+1}$ that fits into the exact sequence

$$0 \longrightarrow \mathcal{M} \longrightarrow \mathcal{O}_\mathbb{P}^{r+1} \xrightarrow{(x_0 \ \cdots \ x_r)} \mathcal{O}_\mathbb{P}(1) \longrightarrow 0,$$

where $x_0, \ldots, x_r$ generate the linear forms on $\mathbb{P}$. This subsheaf is a subbundle (that is, locally a direct summand, and in particular locally free) because, locally at each point of $\mathbb{P}^r$, at least one of the x_i is a unit. (The bundle $\mathcal{M}$ may be identified with the twist $\Omega_\mathbb{P}(1)$ of the cotangent sheaf $\Omega = \Omega_\mathbb{P}$; see for example [Eisenbud 1995, Section 17.5]. We will not need this fact.)

The result that we need for the proof of Theorem 5.1 is:

Theorem 5.6. *Let $\mathcal{F}$ be a coherent sheaf on $\mathbb{P} = \mathbb{P}_K^r$ with $r \geq 2$ and let $\mathcal{M}$ be the tautological rank-r subbundle on $\mathbb{P}$. If the support of $\mathcal{F}$ has dimension ≤ 1 and*

$$H^1\left(\bigwedge\nolimits^2 \mathcal{M} \otimes \mathcal{F}\right) = 0,$$

the graded S-module $F := \bigoplus_{n \geq 0} H^0 \mathcal{F}(n)$ has a linear free presentation.

Before giving the proof we explain how the exterior powers of $\mathcal{M}$ arise in the context of syzygies. Set $S = \mathbb{K}[x_0, \ldots, x_r]$ and let

$$\mathbf{K}: \ 0 \longrightarrow K_{r+1} \longrightarrow \cdots \longrightarrow K_0$$

be the minimal free resolution of the residue field $\mathbb{K} = S/(x_0, \ldots, x_r)$ as an S-module. By Theorem A2.50 we may identify $\mathbb{K}$ with the dual of the Koszul complex of $x = (x_0, \ldots, x_r) \in (S^{r+1})^*$ (as ungraded modules). To make the grading correct, so that the copy of $\mathbb{K}$ that is resolved is concentrated in degree 0, we must set $K_i = \bigwedge^i (S^{r+1}(-1)) = (\bigwedge^i S^{r+1})(-i)$, so that the complex begins with the terms

$$\mathbb{K}: \quad \cdots \xrightarrow{\varphi_3} (\textstyle\bigwedge^2 S^{r+1})(-2) \xrightarrow{\varphi_2} S^{r+1}(-1) \xrightarrow{\varphi_1 = (x_0 \ \cdots \ x_r)} S.$$

Let $M_i = (\ker \varphi_i)(i)$, that is, the module $\ker \varphi_i$ shifted so that it is a submodule of the free module $\bigwedge^{i-1} S^{r+1}$ generated in degree 0. For example, the tautological subbundle $\mathcal{M} \subset \mathcal{O}_{\mathbb{P}^r}^{r+1}$ on projective space is the sheafification of M_1. We need the following generalization of this remark.

Proposition 5.7. *With notation as above, the i-th exterior power $\bigwedge^i \mathcal{M}$ of the tautological subbundle on $\mathbb{P}^r$ is the sheafification of M_i.*

This result is only true at the sheaf level: $\bigwedge^i M_1$ is not isomorphic to M_i.

Proof. Locally at any point of $\mathbb{P}^r$ at least one of the x_i is invertible, so the sheafification of the Koszul complex is split exact. Thus the sheafifications of all the M_i are vector bundles, and it suffices to show that $(\widetilde{M_i})^* \cong (\bigwedge^i \mathcal{M})^*$. Since Hom is left exact, the module M_i is the dual of the module $N_i = (\operatorname{coker} \varphi_i^*)(-i)$. Being a vector bundle, $\tilde{N}_i$ is reflexive, so $\tilde{M}_i^* = \tilde{N}_i$. Thus it suffices to show that $N_i \cong \bigwedge^i N_1$ (it would be enough to prove this for the associated sheaves, but in this case it is true for the modules themselves.)

As described above, the complex $\mathbb{K}$ is the dual of the Koszul complex of the element $x = (x_0, \ldots, x_r) \in (S^{r+1})^*(1)$. By the description in Section A2F (page 217), the map $\varphi_i^* : \bigwedge^{i-1}((S^{r+1})^*(1)) \to \bigwedge^i((S^{r+1})^*(1))$ is given by exterior multiplication with x. But the exterior algebra functor is right exact. Thus from

$$N_1 = \frac{(S^{r+1})^*(1)}{Sx}$$

we deduce that

$$\textstyle\bigwedge N_1 = \frac{\bigwedge((S^{r+1})^*(1))}{x \wedge (\bigwedge S^{r+1})^*(1)}$$

as graded algebras. In particular

$$\textstyle\bigwedge^i N_1 = \frac{\bigwedge^i((S^{r+1})^*(1))}{x \wedge (\bigwedge^{i-1}(S^{r+1})^*(1))} = \operatorname{coker}(\varphi_i)^*$$

as required.   □

With this preamble, we can state the general connection between syzygies and the sort of cohomology groups that appear in Theorem 5.6:

Theorem 5.8. *Let $\mathcal{F}$ be a coherent sheaf on $\mathbb{P}^r_{\mathbb{K}}$, and set $F = \bigoplus_{n\geq 0} H^0\mathcal{F}(n)$. Let $\mathcal{M}$ be the tautological rank-r subbundle on $\mathbb{P}$. If $d \geq i+1$ then there is an exact sequence*

$$0 \longrightarrow \operatorname{Tor}^S_i(F,\mathbb{K})_d \longrightarrow H^1\big(\textstyle\bigwedge^{i+1}\mathcal{M}\otimes\mathcal{F}(d-i-1)\big) \xrightarrow{\ \alpha\ }$$
$$\xrightarrow{\ \alpha\ } H^1\big(\textstyle\bigwedge^{i+1}\mathcal{O}^{r+1}_{\mathbb{P}}\otimes\mathcal{F}(d-i-1)\big),$$

where the map α is induced by the inclusion $\mathcal{M}\subset\mathcal{O}^{r+1}_{\mathbb{P}}$.

Proof. The vector space $\operatorname{Tor}^S_i(F,\mathbb{K})$ can be computed as the homology of the sequence obtained by tensoring the Koszul complex, which is a free resolution of $\mathbb{K}$, with F. In particular, $\operatorname{Tor}^S_i(F,\mathbb{K})_d$ is the homology of of the sequence

$$\big(\textstyle\bigwedge^{i+1} S^{r+1}(-i-1)\otimes F\big)_d \to \big(\textstyle\bigwedge^i S^{r+1}(-i)\otimes F\big)_d \to \big(\textstyle\bigwedge^{i-1} S^{r+1}(-i+1)\otimes F\big)_d.$$

For any t the module $\bigwedge^t S^{r+1}(-t)\otimes F$ is just a sum of copies of $F(-t)$, and thus if $d \geq t$, so that $F_{d-t} = H^0\mathcal{F}(d-t)$, then

$$\big(\textstyle\bigwedge^t S^{r+1}(-t)\otimes F\big)_d = \big(\textstyle\bigwedge^t S^{r+1}\otimes F\big)_{d-t} = H^0\big(\textstyle\bigwedge^t \mathcal{O}^{r+1}_{\mathbb{P}}\otimes\mathcal{F}(d-t)\big).$$

For this reason we can compute Tor through sheaf cohomology. Since the sheafification of $\mathbb{K}$ is locally split, it remains exact when tensored by any sheaf, for example $\mathcal{F}(d)$. With notation as in Proposition 5.7 we get short exact sequences

$$0 \to \textstyle\bigwedge^t \mathcal{M}\otimes\mathcal{F}(d-t) \to \textstyle\bigwedge^t \mathcal{O}^{r+1}_{\mathbb{P}}\otimes\mathcal{F}(d-t) \to \textstyle\bigwedge^{t-1}\mathcal{M}\otimes\mathcal{F}(d-t+1) \to 0 \quad (5.1)$$

that fit into a diagram

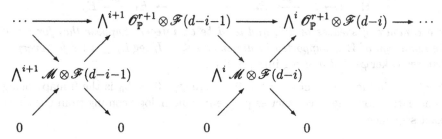

where every left-to-right path is exact. It follows that $\operatorname{Tor}^S_i(F,\mathbb{K})_d$ is the cokernel of the diagonal map

$$H^0\big(\textstyle\bigwedge^{i+1}\mathcal{O}^{r+1}_{\mathbb{P}}\otimes\mathcal{F}(d-i-1)\big) \longrightarrow H^0\big(\textstyle\bigwedge^i\mathcal{M}\otimes\mathcal{F}(d-i)\big).$$

The long exact sequence in cohomology associated to the sequence 5.1 now gives the desired result. $\qquad\square$

Proof of Theorem 5.6. Let

$$\mathbf{L}: \quad \cdots \longrightarrow L_1 \xrightarrow{\ \varphi_1\ } L_0 \longrightarrow F \longrightarrow 0$$

be the minimal free resolution of F. By the definition of F the free module L_0 has no generators of degrees ≤ 0. As we saw at the beginning of this section, this implies that L_1 has no generators of degrees < 1.

Since $H^1(\bigwedge^2 \mathcal{M} \otimes \mathcal{F}) = 0$ and $\bigwedge^2 \mathcal{M} \otimes \mathcal{F}$ has support of dimension at most 1, it has no higher cohomology and is thus a 1-regular sheaf. It follows that this sheaf is s-regular for all $s \geq 2$ as well, so that

$$H^1\left(\bigwedge^2 \mathcal{M} \otimes \mathcal{F}(t)\right) = 0$$

for all $t \geq 0$. By Theorem 5.8 we have $\mathrm{Tor}_1^S(F, \mathbb{K})_d = 0$ for all $d \geq 2$. We can compute this Tor as the homology of the complex $\mathbf{L} \otimes \mathbb{K}$. As $\mathbf{L}$ is minimal, the complex $\mathbf{L} \otimes \mathbb{K}$ has differentials equal to 0, so $\mathrm{Tor}_i^S(F, \mathbb{K}) = L_i \otimes \mathbb{K}$. In particular, L_1 has no generators of degrees ≥ 2.

Being a torsion S-module, F has no free summands, so for any summand L_0' of L_0 the composite map $L_1 \to L_0 \to L_0'$ is nonzero. From this and the fact that L_1 is generated in degree 1 it follows that L_0 can have no generator of degree ≥ 1. By construction, F is generated in degrees ≥ 0 so L_0 is actually generated in degree 0, completing the proof. □

Regularity and the Eagon–Northcott Complex

To bound the regularity of the Fitting ideal of the sheaf $\pi_* \mathcal{A}$ that will occur in the proof of Theorem 5.1 we will use the following generalization of the argument at the beginning of the proof of Theorem 4.3.

Lemma 5.9. *Let*

$$\mathbf{E}: \quad 0 \to E_t \xrightarrow{\varphi_t} E_{t-1} \longrightarrow \cdots \longrightarrow E_1 \xrightarrow{\varphi_1} E_0$$

be a complex of sheaves on $\mathbb{P}^r$, and let d be an integer. Suppose that for $i > 0$ the homology of $\mathbf{E}$ is supported in dimension ≤ 1. If $\mathrm{reg}\, E_s \leq d+s$ for every s, then $\mathrm{reg}\, \mathrm{coker}\, \varphi_1 \leq d$ and $\mathrm{reg}\, \mathrm{im}\, \varphi_1 \leq d+1$.

Proof. We induct on t, the case $t = 0$ (where $\varphi_1 : 0 \to E_0$ is the 0 map) being immediate. From the long exact sequence in cohomology coming from the short exact sequence

$$0 \longrightarrow \mathrm{im}\, \varphi_1 \longrightarrow E_0 \longrightarrow \mathrm{coker}\, \varphi_1 \longrightarrow 0$$

we see that the regularity bound for $\mathrm{im}\, \varphi_1$ implies the one for $\mathrm{coker}\, \varphi_1$.

Since the homology $H_1(\mathbf{E})$ is supported in dimension at most 1, we have $H^i(H_1(\mathbf{E})(s)) = 0$ for all $i > 1$. Thus the long exact sequence in cohomology coming from the short exact sequence

$$0 \longrightarrow H_1(\mathbf{E}) \longrightarrow \mathrm{coker}\, \varphi_2 \longrightarrow \mathrm{im}\, \varphi_1 \longrightarrow 0$$

gives surjections $H^i(\mathrm{coker}\, \phi_2(s)) \to H^i(\mathrm{im}\, \phi_i(s))$ for all $i > 0$ and all s, showing that $\mathrm{reg}\, \mathrm{im}\, \varphi_1 \leq \mathrm{reg}\, \mathrm{coker}\, \varphi_2$. By induction, we have $\mathrm{reg}\, \mathrm{coker}\, \phi_2 \leq d+1$, and we are done. □

From Lemma 5.9 we derive a general bound on the regularity of Fitting ideals:

Corollary 5.10. *Suppose* $\varphi : \mathscr{F}_1 \to \mathscr{F}_0$ *is a map of vector bundles on* $\mathbb{P}^r$ *with* $\mathscr{F}_1 = \bigoplus_{i=1}^{n} \mathscr{O}_{\mathbb{P}^r}(-1)$ *and* $\mathscr{F}_0 = \bigoplus_{i=1}^{h} \mathscr{O}_{\mathbb{P}^r}$. *If the ideal sheaf* $\mathscr{I}_h(\varphi)$ *generated by the* $h \times h$ *minors of* φ *defines a scheme of dimension* ≤ 1, *then*

$$\operatorname{reg} \mathscr{I}_h(\varphi) \leq h.$$

Proof. We apply Lemma 5.9 to the Eagon–Northcott complex $\mathbf{E} = \mathbf{EN}(\varphi)$ of φ (see Section A2H). The zeroth term of the complex is isomorphic to $\mathscr{O}_{\mathbb{P}^r}$, while for $s > 0$ the s-th term is isomorphic to

$$E_s = (\operatorname{Sym}_{s-1} \mathscr{F}_0)^* \otimes \textstyle\bigwedge^{h+s-1} \mathscr{F}_1 \otimes \bigwedge^h \mathscr{F}_0^*.$$

This sheaf is a direct sum of copies of $\mathscr{O}_{\mathbb{P}^r}(-h-s+1)$. Thus it has regularity $h+s-1$, so we may take $d = h-1$ in Lemma 5.9 and the result follows. $\square$

The following Theorem, a combination of Corollary 5.10 with Theorem 5.6, summarizes our progress. For any sheaf $\mathscr{F}$, we set $\mathrm{h}^0(\mathscr{F}) = \dim_\mathbb{K} \mathrm{H}^0(\mathscr{F})$.

Theorem 5.11. *Let* $X \subset \mathbb{P}^r_\mathbb{K}$ *be a reduced irreducible curve with* $r \geq 3$. *Let* $\mathscr{F}$ *be a coherent sheaf on* X *which is locally free of rank 1 except at finitely many points of* X, *and let* $\mathscr{M}$ *be the tautological rank-r subbundle on* $\mathbb{P}^r_\mathbb{K}$. *If*

$$\mathrm{H}^1\big(\textstyle\bigwedge^2 \mathscr{M} \otimes \mathscr{F}\big) = 0$$

then $\operatorname{reg} \mathscr{I}_X \leq \mathrm{h}^0 \mathscr{F}$.

Proof. By Theorem 5.6 the module $F = \bigoplus_{n \geq 0} \mathrm{H}^0(\mathscr{F}(n))$ has a linear presentation matrix; in particular, $\mathscr{F}$ is the cokernel of a matrix $\varphi : \mathscr{O}_{\mathbb{P}^r}^n(-1) \to \mathscr{O}_{\mathbb{P}^r}^h$. Applying Corollary 5.10 we see that $\operatorname{reg} \mathscr{I}_h(\varphi) \leq \mathrm{h}^0 \mathscr{F}$. But by Proposition 5.5 we have $\operatorname{reg} \mathscr{I}_X \leq \operatorname{reg} \mathscr{I}_h(\varphi)$. $\square$

Even without further machinery, Theorem 5.11 is quite powerful. See Exercise 5.7 for a combinatorial statement proved by S. Lvovsky using it, for which I don't know a combinatorial proof.

Filtering the Restricted Tautological Bundle

With this reduction of the problem in hand, we can find the solution by working on the normalization $\pi : C \to X$ of X. If $\mathscr{A}$ is a line bundle on C then $\mathscr{F} = \pi_* \mathscr{A}$ is locally free except at the finitely many points where X is singular, and

$$\mathrm{H}^1\big(\textstyle\bigwedge^2 \mathscr{M} \otimes \pi_* \mathscr{A}\big) = \mathrm{H}^1\big(\pi^* \bigwedge^2 \mathscr{M} \otimes \mathscr{A}\big) = \mathrm{H}^1\big(\bigwedge^2 \pi^* \mathscr{M} \otimes \mathscr{A}\big).$$

On the other hand, since π is a finite map we have $\mathrm{h}^0 \pi_* \mathscr{A} = \mathrm{h}^0 \mathscr{A}$. It thus suffices to investigate the bundle $\pi^* \mathscr{M}$ and to find a line bundle $\mathscr{A}$ on C such that the cohomology above vanishes and $\mathrm{h}^0 \mathscr{A}$ is small enough.

We need three facts about $\pi^* \mathscr{M}$. This is where we use the hypotheses on the curve X in Theorem 5.1.

Proposition 5.12. *Let $\mathbb{K}$ be an algebraically closed field, and let $X \subset \mathbb{P}^r_{\mathbb{K}}$ be a nondegenerate, reduced and irreducible curve. Suppose that $\pi : C \to \mathbb{P}^r$ is a map from a reduced and irreducible curve C onto X, and that $\pi : C \to X$ is birational. If $\mathscr{M}$ denotes the tautological subbundle on $\mathbb{P}^r$, then*

1. $\pi^* \mathscr{M}$ *is contained in a direct sum of copies of $\mathscr{O}_C$;*
2. $H^0(\pi^* \mathscr{M}) = 0$; *and*
3. $\deg \pi^* M = -\deg X$.

Proof. Since any exact sequence of vector bundles is locally split, we can pull back the defining sequence

$$0 \to \mathscr{M} \to \mathscr{O}_{\mathbb{P}^r}^{r+1} \to \mathscr{O}_{\mathbb{P}^r}(1) \to 0$$

to get an exact sequence

$$0 \to \pi^* \mathscr{M} \to \mathscr{O}_C^{r+1} \to \mathscr{L} \to 0,$$

where we have written $\mathscr{L}$ for the line bundle $\pi^* \mathscr{O}_{\mathbb{P}^r}(1)$. This proves part 1.

Using the sequence above, it suffices in order to prove part 2 to show that the map on cohomology

$$H^0(\mathscr{O}_C^{r+1}) \to H^0(\mathscr{L})$$

is a monomorphism. Since π is finite, we can compute the cohomology after pushing forward to X. Since X is reduced and irreducible and $\mathbb{K}$ is algebraically closed we have $H^0 \mathscr{O}_X = \mathbb{K}$, generated by the constant section 1. For the same reason $\mathbb{K} = H^0 \mathscr{O}_C = H^0(\pi_* \mathscr{O}_C)$ is also generated by 1. The map $\mathscr{O}_X(1) \to \pi_* \mathscr{L} = \pi_* \pi^* \mathscr{O}_X(1)$ looks locally like the injection of $\mathscr{O}_X$ into $\mathscr{O}_C$, so it is a monomorphism. Thus the induced map $H^0 \mathscr{O}_X(1) \to H^0 \mathscr{L}$ is a monomorphism, and it suffices to show that the map on cohomology

$$H^0(\mathscr{O}_X^{r+1}) \to H^0(\mathscr{O}_X(1))$$

coming from the embedding of X in $\mathbb{P}^r$ is a monomorphism. This is the restriction to X of the map

$$H^0(\mathscr{O}_{\mathbb{P}^r}^{r+1}) \to H^0(\mathscr{O}_{\mathbb{P}^r}(1))$$

sending the generators of $\mathscr{O}_{\mathbb{P}^r}^{r+1}$ to linear forms on $\mathbb{P}^r$. Since X is nondegenerate, no nonzero linear form vanishes on X, so the displayed maps are all monomorphisms.

Finally, we must prove that $\deg \pi^* M = -\deg X$. The bundle $\mathscr{M}$ has rank r, and so does its pullback $\pi^* \mathscr{M}$. The degree of the latter is, by definition, the degree of its highest nonvanishing exterior power, $\bigwedge^r \pi^* \mathscr{M} = \pi^* \bigwedge^r \mathscr{M}$. From the exact sequence defining $\mathscr{M}$ we see that $\bigwedge^r \mathscr{M} \cong \mathscr{O}_{\mathbb{P}^r}(-1)$, and it follows that $\pi^* \bigwedge^r \mathscr{M} = \pi^* \mathscr{O}_X(-1)$ has degree $-\deg X$.   □

Any vector bundle on a smooth curve can be filtered by a sequence of subbundles in such a way that the successive quotients are line bundles (Exercise 5.8). Using Proposition 5.12 we can find a special filtration.

Proposition 5.13. *Let $\mathscr{N}$ be a vector bundle on a smooth curve C over an algebraically closed field $\mathbb{K}$. If $\mathscr{N}$ is contained in a direct sum of copies of $\mathscr{O}_C$ and $h^0 \mathscr{N} = 0$ then $\mathscr{N}$ has a filtration*

$$\mathscr{N} = \mathscr{N}_1 \supset \cdots \supset \mathscr{N}_{r+1} = 0,$$

such that $\mathscr{L}_i := \mathscr{N}_i/\mathscr{N}_{i+1}$ is a line bundle of strictly negative degree.

Proof. We will find an epimorphism $\mathscr{N} \to \mathscr{L}_1$ from $\mathscr{N}$ to a line bundle $\mathscr{L}_1$ of negative degree. Given such a map, the kernel $\mathscr{N}' \subset \mathscr{N}$ automatically satisfies the hypotheses of the proposition, and thus by induction $\mathscr{N}$ has a filtration of the desired type.

By hypothesis there is an embedding $\mathscr{N} \hookrightarrow \mathscr{O}_C^n$ for some n. We claim that we can take $n = \operatorname{rank} \mathscr{N}$. For simplicity, set $r = \operatorname{rank} \mathscr{N}$. Tensoring the given inclusion with the field K of rational functions on C, we get a map of K-vector spaces $K^r \cong K \otimes \mathscr{N} \to K \otimes \mathscr{O}_C^n = K^n$. Since this map is a monomorphism, one of its $r \times r$ minors must be nonzero. Thus we can factor out a subset of $n-r$ of the given basis elements of K^n and get a monomorphism $K^r \cong K \otimes \mathscr{N} \to K \otimes \mathscr{O}_C^r = K^r$. Since $\mathscr{N}$ is torsion free, the corresponding projection of $\mathscr{O}_C^n \to \mathscr{O}_C^r$ gives a composite monomorphism $\alpha : \mathscr{N} \hookrightarrow \mathscr{O}_C^r$ as claimed.

Since $\mathscr{N}$ has no global sections, the map α cannot be an isomorphism. Since the rank of $\mathscr{N}$ is r, the cokernel of α is torsion; that is, it has finite support. Let p be a point of its support. Since we have assumed that $\mathbb{K}$ is algebraically closed, the residue class field $\kappa(p)$ is $\mathbb{K}$. We may choose an epimorphism from $\mathscr{O}_C^r/\mathscr{N} \to \mathscr{O}_p$, the skyscraper sheaf at p. Since $\mathscr{O}_C^r$ is generated by its global sections, the image of the global sections of $\mathscr{O}_C^r$ generate the sheaf $\mathscr{O}_p$, and thus the map $\mathbb{K}^r = H^0(\mathscr{O}_C^r) \to H^0(\mathscr{O}_p) = \mathbb{K}$ is onto, and its kernel has dimension $r-1$. Any subspace of $H^0(\mathscr{O}_C^r)$ generates a direct summand, so we get a summand $\mathscr{O}_C^{r-1}$ of $\mathscr{O}_C^r$ which maps to a proper subsheaf of $\mathscr{O}_C^r/\mathscr{N}$. The map $\mathscr{O}_C^r \to \mathscr{O}_p$ factors through the quotient $\mathscr{O}_C^r/\mathscr{O}_C^{r-1} = \mathscr{O}_C$, as in the diagram

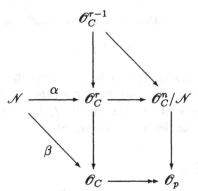

The composite map $\mathscr{N} \to \mathscr{O}_p$ is zero, so $\beta : \mathscr{N} \to \mathscr{O}_C^r \to \mathscr{O}_C$ is not an epimorphism. Thus the ideal sheaf $\mathscr{L}_1 = \beta(\mathscr{N})$ is properly contained in $\mathscr{O}_C$. It defines a nonempty finite subscheme Y of C, so $\deg \mathscr{L}_1 = -\deg Y < 0$. Since C is smooth, $\mathscr{L}_1$ is a line bundle, and we are done. $\qquad\square$

Multilinear algebra gives us a corresponding filtration for the exterior square.

Proposition 5.14. *If $\mathscr{N}$ is a vector bundle on a variety V which has a filtration*

$$\mathscr{N} = \mathscr{N}_1 \supset \cdots \supset \mathscr{N}_r \supset \mathscr{N}_{r+1} = 0,$$

such that the successive quotients $\mathscr{L}_i := \mathscr{N}_i/\mathscr{N}_{i+1}$ are line bundles, then $\bigwedge^2 \mathscr{N}$ has a similar filtration whose successive quotients are the line bundles $\mathscr{L}_i \otimes \mathscr{L}_j$ with $1 \le i < j \le r$.

Proof. We induct on r, the rank of $\mathscr{N}$. If $r = 1$ then $\bigwedge^2 \mathscr{N} = 0$, and we are done. From the exact sequence

$$0 \to \mathscr{N}_r \to \mathscr{N} \to \mathscr{N}/\mathscr{N}_r \to 0,$$

and the right exactness of the exterior algebra functor we deduce that

$$\bigwedge(\mathscr{N}/\mathscr{N}_r) = \bigwedge \mathscr{N}/(\mathscr{N}_r \wedge \bigwedge \mathscr{N})$$

as graded algebras. In degree 2 this gives a right exact sequence

$$\mathscr{N} \otimes \mathscr{N}_r \to \bigwedge^2 \mathscr{N} \to \bigwedge^2 (\mathscr{N}/\mathscr{N}_r) \to 0.$$

Since $\mathscr{N}_r$ is a line bundle, the left-hand map kills $\mathscr{N}_r \otimes \mathscr{N}_r$, and thus factors through $\mathscr{N}/\mathscr{N}_r \otimes \mathscr{N}_r$. The induced map $\mathscr{N}/\mathscr{N}_r \otimes \mathscr{N}_r \to \bigwedge^2 \mathscr{N}$ is a monomorphism because

$$\mathrm{rank}(\mathscr{N}/(\mathscr{N}_r \otimes \mathscr{N}_r)) = (r-1) \cdot 1 = r - 1$$

is the same as the difference of the ranks of the right-hand bundles,

$$r - 1 = \binom{r}{2} - \binom{r-1}{2}.$$

Thus we can construct a filtration of $\bigwedge^2 \mathscr{N}$ by combining a filtration of

$$(\mathscr{N}/\mathscr{N}_r) \otimes \mathscr{N}_r$$

with a filtration of $\bigwedge^2(N/N_r)$. The subbundles $(\mathscr{N}_i/\mathscr{N}_r) \otimes \mathscr{N}_r \subset (\mathscr{N}/\mathscr{N}_r) \otimes \mathscr{N}_r$ give a filtration of $\mathscr{N}/\mathscr{N}_r$ with successive quotients $\mathscr{L}_i \otimes \mathscr{L}_r = \mathscr{N}_r$ for $i < r$. By induction on the rank of $\mathscr{N}$, the bundle $\bigwedge^2(N/N_r)$ has a filtration with subquotients $\mathscr{L}_i \otimes \mathscr{L}_j$, completing the argument. $\square$

General Line Bundles

To complete the proof of Theorem 5.1 we will use a general result about line bundles on curves:

Proposition 5.15. *Let C be a smooth curve of genus g over an algebraically closed field. If $\mathscr{B}$ is a general line bundle of degree $\ge g-1$ then $\mathrm{h}^1 \mathscr{B} = 0$.*

To understand the statement, the reader needs to know that the set $\text{Pic}_d(C)$ of isomorphism classes of line bundles of degree d on C form an irreducible variety, called the Picard variety. The statement of the proposition is shorthand for the statement that the set of line bundles $\mathscr{B}$ of degree $g - 1$ that have vanishing cohomology is an open dense subset of this variety.

We will need this Proposition and related results in Chapter 8, Lemma 8.5 and we postpone the proof until then.

Proof of Theorem 5.1. Set $d = \deg X$. By Propositions 5.12–5.14, the bundle $\bigwedge^2 \pi^* \mathscr{M}$ can be filtered in such a way that the successive quotients are the tensor products $\mathscr{L}_i \otimes \mathscr{L}_j$ of two line bundles, each of strictly negative degree.

To achieve the vanishing of $H^1(\bigwedge^2 M \otimes \mathscr{A})$ it suffices to choose $\mathscr{A}$ such that $h^1(\mathscr{L}_i \otimes \mathscr{L}_j \otimes \mathscr{A}) = 0$ for all i, j. By Proposition 5.15, it is enough to choose $\mathscr{A}$ general and of degree e such that $\deg(\mathscr{L}_i \otimes \mathscr{L}_j \otimes \mathscr{A}) = \deg \mathscr{L}_i + \deg \mathscr{L}_j + e \geq g-1$ for every i and j.

Again by Proposition 5.12 we have $-d = \deg \pi^* \mathscr{M} = \sum_i \deg \mathscr{L}_i$. Since the $\deg \mathscr{L}_i$ are negative integers,

$$\deg \mathscr{L}_i + \deg \mathscr{L}_j = -d - \sum_{k \neq i,j} \deg \mathscr{L}_k \geq -d - r + 2,$$

and it suffices to take $e = g - 1 + d - r + 2$. In sum, we have shown that if $\mathscr{A}$ is general of degree $g - 1 + d - r + 2$ then $\text{reg } \mathscr{I}_X \leq h^0 \mathscr{A}$. By the Riemann–Roch theorem we have $h^0 \mathscr{A} = h^1 \mathscr{A} + d - r + 2$. By Proposition 5.3, $d \geq r$, so $\deg \mathscr{A} \geq g+1$, and Proposition 5.15 implies that $h^1 \mathscr{A} = 0$. Thus $\text{reg } \mathscr{I}_X \leq h^0 \mathscr{A} = d - r + 2$, completing the proof. □

As we shall see in the next chapter, the bound we have obtained is sometimes optimal. But the examples that we know in which this happens are very special — rational and elliptic curves. Are there better bounds if we take into account more about the curve? At any rate, we shall see in Corollary 8.2 that there are much better bounds for curves embedded by complete series of high degree. (Exercise 8.5 gives a weak form of this for varieties, even schemes, of any dimension.)

5C   Exercises

1. Show that if the curve $X \subset \mathbb{P}^r$ has an n-secant line (that is, a line that meets the curve in n points) then $\text{reg } \mathscr{I}_X \geq n$. Deduce that that there are nondegenerate smooth rational curves X in $\mathbb{P}^3$ of any degree $d \geq 3$ with $\text{reg } S_X = \deg X - \text{codim } X$. (Hint: consider curves on quadric surfaces.)

2. Show that if X is the union of 2 disjoint lines in $\mathbb{P}^3$, or a conic contained in a plane in $\mathbb{P}^3$, then $2 = \text{reg } \mathscr{I}_X > \deg X - \text{codim } X + 1$.

3. Let X_d be the scheme in $\mathbb{P}^3$ given by the equations

$$x_0^2, \; x_0 x_1, \; x_1^2, \; x_0 x_2^d - x_1 x_3^d.$$

Show that X_d is one-dimensional, irreducible, and not contained in a hyperplane. Show that the degree of X_d is 2 but the regularity of S_{X_d} is $\geq d$. In case $\mathbb{K}$ is the field of complex numbers, the scheme X_d can be visualized as follows: It lies in the first infinitesimal neighborhood, defined by the ideal $(x_0^2, x_0 x_1, x_1^2)$, of the line X defined by $x_0 = x_1 = 0$, which has affine coordinate x_2/x_3. In this sense X_d can be thought of as a subscheme of the normal bundle of X in $\mathbb{P}^3$. Identifying the normal bundle with $X \times \mathbb{K}^2$ the scheme X_d meets each $p \times \mathbb{K}^2 = \mathbb{K}^2$ as a line through the origin of $\mathbb{K}^2$, and is identified by its slope $x_0/x_1 = (x_2/x_3)^d$. Thus for example if we restrict to values of x_2/x_3 in the unit circle, we see that X_d is a ribbon with d twists:

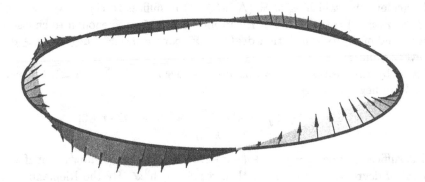

4. Consider the reduced irreducible one-dimensional subscheme X of the real projective space $\mathbb{P}_{\mathbb{R}}^3$ defined by the equations

$$x_0^2 - x_1^2, \ x_2^2 - x_3^2, \ x_3 x_0 - x_1 x_2, \ x_0 x_2 - x_1 x_3.$$

Show that $\deg X = 2$ and $\operatorname{reg} S_X > \deg X - \operatorname{codim} X$, so the conclusion of Theorem 5.1 does not hold for X. Show that after a ground field extension X becomes the union of two disjoint lines. Hint: consider the rows of the matrix

$$\begin{pmatrix} x_0 + ix_1 & x_2 - ix_3 \\ x_2 + ix_3 & x_0 - ix_1 \end{pmatrix}.$$

5. Show that Proposition 5.7 is only true on the sheaf level; the i-th syzygy module of $\mathbb{K}$ itself is not isomorphic to a twist of the i-th exterior power of the first one. (Hint: Consider the number of generators of each module, which can be deduced from Nakayama's Lemma and the right exactness of the exterior algebra functor; see [Eisenbud 1995, Proposition A2.1].)

6. Generalizing Corollary 5.10, suppose $\varphi : \mathscr{F}_1 \to \mathscr{F}_0$ is a map of vector bundles on $\mathbb{P}^r$ with $F_1 = \bigoplus_{i=1}^n \mathcal{O}_{\mathbb{P}^r}(-b_i)$ and $F_0 = \bigoplus_{i=1}^h \mathcal{O}_{\mathbb{P}^r}(-a_i)$. Suppose that $\min a_j < \min b_j$ (as would be the case if φ were a minimal presentation of a coherent sheaf.) Show that if the ideal sheaf $\mathscr{I}_h(\varphi)$ generated by the $h \times h$ minors of φ defines a scheme of dimension ≤ 1, then

$$\operatorname{reg} \mathscr{I}_h \leq \sum b_i - \sum a_i - (n-h)(1 + \min_i a_i)$$

7. The *monomial curve in* $\mathbb{P}^r$ *with exponents* $a_1 \leq a_2 \leq \cdots \leq a_{r-1}$ is the curve $X \subset \mathbb{P}^r$ of degree $d = a_r$ parametrized by

$$\phi : \mathbb{P}^1 \ni (s,t) \mapsto (s^d, s^{d-a_1}t^{a_1}, \ldots, s^{d-a_{r-1}}t^{a_{r-1}}, t^d).$$

Set $a_0 = 0$, $a_r = d$, and for $i = 1, \ldots, r$ set $\alpha_i = a_i - a_{i-1}$. With notation as in Theorem 5.11, show that

$$\phi^*(\mathcal{M}) = \bigoplus_{i \neq j} \mathcal{O}_{\mathbb{P}^1}(-\alpha_i - \alpha_j).$$

Now use Theorem 5.11 to show that the regularity of I_X is at most $\max_{i \neq j} \alpha_i + \alpha_j$. This exercise is taken from [L'vovsky 1996].

8. Let C be a smooth curve, and let $\mathcal{E}$ be a vector bundle. Let $\mathcal{F}$ be any coherent subsheaf of $\mathcal{E}$, and denote by $\mathcal{F}' \subset \mathcal{E}$ the preimage of the torsion subsheaf of $\mathcal{E}/\mathcal{F}$. Show that $\mathcal{F}'$ is a subbundle of $\mathcal{E}$; that is, both $\mathcal{F}'$ and $\mathcal{E}/\mathcal{F}'$ are vector bundles. Show that rank $\mathcal{F}' = $ rank $\mathcal{F}$. Show that any bundle has subsheaves of rank 1. Conclude that $\mathcal{E}$ has a filtration by subbundles.

6

Linear Series and 1-Generic Matrices

In this chapter we introduce two techniques useful for describing embeddings of curves and other varieties: linear series and the 1-generic matrices they give rise to. By way of illustration we treat in detail the free resolutions of ideals of curves of genus 0 and 1 in their "nicest" embeddings.

In the case of genus-0 curves we are looking at embeddings of degree at least 1; in the case of genus-1 curves we are looking at embeddings of degree at least 3. It turns out that the technique of this chapter gives very explicit information about the resolutions of ideal of any hyperelliptic curves of any genus g embedded by complete linear series of degree at least $2g + 1$. We will see in Chapter 8 that some qualitative aspects extend to all curves in such "high-degree" embeddings.

The specific constructions for elliptic curves made in the last part of this chapter are rather complex, and involve the theory of divisors on a ruled surface, which we will not need later in this book. The qualitative properties of these curves that we deduce from the resolutions we construct, such as the Castelnuovo–Mumford regularity, can be seen with much less work from the general theory to be developed later. I felt that it was worth including the most explicit treatment I could for these resolutions, but the reader may skim the material from Proposition 6.19 to Theorem 6.26, and return to detailed reading in Chapter 7 without missing anything needed in the rest of the book.

For simplicity we suppose throughout this chapter that $\mathbb{K}$ is an algebraically closed field and we work with projective varieties, that is, irreducible algebraic subsets of a projective space $\mathbb{P}^r$.

6A   Rational Normal Curves

Consider first the plane conics. One such conic — we will call it the *standard conic* in $\mathbb{P}^2$ with respect to coordinates x_0, x_1, x_2 — is the curve with equation $x_0 x_2 - x_1^2 = 0$. It is the image of the map $\mathbb{P}^1 \to \mathbb{P}^2$ defined by

$$(s,t) \mapsto (s^2, st, t^2).$$

Any irreducible conic is obtained from this one by an automorphism — that is, a linear change of coordinates — of $\mathbb{P}^2$.

Analogously, we consider the curve $X \in \mathbb{P}^r$ that is the image of the map $\mathbb{P}^1 \xrightarrow{\nu_r} \mathbb{P}^r$ defined by

$$(s,t) \mapsto (s^r, s^{r-1}t, \ldots, st^{r-1}, t^r).$$

We call X the *standard rational normal curve* in $\mathbb{P}^r$. By a *rational normal curve* in $\mathbb{P}^r$ we will mean any curve obtained from this standard one by an automorphism — a linear change of coordinates — of $\mathbb{P}^r$. Being an image of $\mathbb{P}^1$, a rational normal curve is irreducible. In fact, the map ν_r is an embedding, so $X \cong \mathbb{P}^1$ is a smooth rational (genus 0) curve. Because the monomials $s^r, s^{r-1}t, \ldots, t^r$ are linearly independent, it is *nondegenerate*, that is, not contained in a hyperplane. The intersection of X with the hyperplane $\sum a_i x_i = 0$ is the set of nontrivial solutions of the homogeneous equation

$$\sum a_i s^{r-i} t^i.$$

Up to scalars there are (with multiplicity) r such solutions, so that X has degree r. We will soon see (Theorem 6.8) that any irreducible, nondegenerate curve of degree r in $\mathbb{P}^r$ is a rational normal curve in $\mathbb{P}^r$.

In algebraic terms, the standard rational normal curve X is the variety whose ideal is the kernel of the ring homomorphism

$$\alpha : S = \mathbb{K}[x_0, \ldots, x_r] \to \mathbb{K}[s,t]$$

sending x_i to $s^{r-i}t^i$. Since $\mathbb{K}[s,t]$ is a domain, this ideal is prime. Since $\mathbb{K}[s,t]$ is generated as a module over the ring $\alpha(S) \subset \mathbb{K}[s,t]$ by the the finitely many monomials in $\mathbb{K}[s,t]$ of degree $< r$, we see that $\dim \alpha(S) = 2$. This is the algebraic counterpart of the statement that X is an irreducible curve.

The defining equation $x_0 x_2 - x_1^2$ of the standard conic can be written in a simple way as a determinant:

$$x_0 x_2 - x_1^2 = \det \begin{pmatrix} x_0 & x_1 \\ x_1 & x_2 \end{pmatrix}.$$

This whole chapter concerns the systematic understanding and exploitation of such determinants!

6A.1   Where'd That Matrix Come From?

If we replace the variables x_0, x_1, x_2 in the matrix above by their images s^2, st, t^2 under ν_2 we get the interior of the "multiplication table"

	s	t
s	s^2	st
t	st	t^2

The determinant of M goes to zero under the homomorphism α because $(s^2)(t^2)$ equals $(st)(st)$, by associativity and commutativity.

To generalize this to the rational normal curve of degree r we may take any d with $1 \le d < r$ and write the multiplication table

	s^{r-d}	$s^{r-d-1}t$	$\cdots$	t^{r-d}
s^d	s^r	$s^{r-1}t$	$\cdots$	$s^d t^{r-d}$
$s^{d-1}t$	$s^{r-1}t$	$s^{r-2}t^2$	$\cdots$	$s^{d-1}t^{r-d+1}$
$\vdots$	$\vdots$	$\vdots$	$\vdots$	$\vdots$
t^d	$s^{r-d}t^d$	$s^{r-d-1}t^{d+1}$	$\cdots$	t^r

Substituting x_i for $s^{r-i}t^i$ we see that the 2×2 minors of the $(d+1) \times (r-d+1)$ matrix

$$M_{r,d} = \begin{pmatrix} x_0 & x_1 & \cdots & x_{r-d} \\ x_1 & x_2 & \cdots & x_{r-d+1} \\ \vdots & \vdots & \ddots & \vdots \\ x_d & x_{d+1} & \cdots & x_r \end{pmatrix}$$

vanish on X. Arthur Cayley called the matrices $M_{r,d}$ *catalecticant* matrices (see Exercises 6.2 and 6.3 for the explanation), and we will follow this terminology. They are also called *generic Hankel matrices*; a Hankel matrix is any matrix whose antidiagonals are constant.

Generalizing the result that the quadratic form $q = \det M_{2,1}$ generates the ideal of the conic in the case $r = 2$, we now prove:

Proposition 6.1. *The ideal $I \subset S = \mathbb{K}[x_0, \ldots, x_r]$ of the rational normal curve $X \subset \mathbb{P}^r$ of degree r is generated by the 2×2 minors of the matrix*

$$M_{r,1} = \begin{pmatrix} x_0 & \cdots & x_{r-1} \\ x_1 & \cdots & x_r \end{pmatrix}.$$

Proof. Consider the homogeneous coordinate ring $S_X = S/I$ which is the image of the homomorphism

$$\alpha : S \to \mathbb{K}[s, t]; \quad x_i \mapsto s^{r-i}t^i.$$

The homogeneous component $(S/I)_d$ is equal to $\mathbb{K}[s, t]_{rd}$, which has dimension $rd+1$.

On the other hand, let $J \subset I$ be the ideal of 2×2 minors of $M_{r,1}$, so S/I is a homomorphic image of S/J. To prove $I = J$ it thus suffices to show that $\dim(S/J)_d \leq rd+1$ for all d.

We have $x_i x_j \equiv x_{i-1} x_{j+1} \mod (J)$ as long as $i-1 \geq 0$ and $j+1 \leq r$. Thus, modulo J, any monomial of degree d is congruent either to $x_0^a x_r^{d-a}$, with $0 \leq a \leq d$, or to $x_0^a x_i x_r^{d-1-a}$ with $0 \leq a \leq d-1$ and $1 \leq i \leq r-1$. There are $d+1$ monomials of the first kind and $d(r-1)$ of the second, so $\dim(S/J)_d \leq (d+1)+d(r-1) = rd+1$ as required. $\square$

By using the (much harder!) Theorem 5.1 we could have simplified the proof a little: Since the degree of the rational normal curve is r, Theorem 5.1 shows that $\text{reg}\, I \leq 2$, and in particular I is generated by quadratic forms. Thus it suffices to show that comparing the degree-2 part of J and I we have $\dim J_2 \geq \dim(I)_2$. This reduces the proof to showing that the minors of $M_{1,r}$ are linearly independent; one could do this as in the proof above, or using the result of Exercise 6.7.

Corollary 6.2. *The minimal free resolution of the homogeneous coordinate ring S_X of the rational normal curve X of degree r in $\mathbb{P}^r$ is given by the Eagon–Northcott complex of the matrix $M_{r,1}$,*

$$\mathbf{EN}(M_{r,1}): \ 0 \longrightarrow (\text{Sym}_{r-2} S^2)^* \otimes \bigwedge^r S^r \longrightarrow \cdots$$

$$\longrightarrow (S^2)^* \otimes \bigwedge^3 S^r \longrightarrow \bigwedge^2 S^r \xrightarrow{\bigwedge^2 M_{r,1}} \bigwedge^2 S^2$$

(see Theorem A2.60). It has Betti diagram

	0	1	2	$\cdots$	$r-1$
0	1	–	–	$\cdots$	–
1	–	$\binom{r}{2}$	$2\binom{r}{3}$	$\cdots$	$(r-1)\binom{r}{r} = r-1$

In particular, S_X is a Cohen–Macaulay ring.

Proof. The codimension of $X \subset \mathbb{P}^r$, and thus of $I \subset S$, is $r-1$, which is equal to the codimension of the ideal of 2×2 minors of a generic $2 \times r$ matrix. Thus by Theorem A2.60 the Eagon–Northcott complex is exact. The entries of $M_{r,1}$ are of degree 1. From the construction of the Eagon–Northcott complex given in Section A2H we see that the Eagon–Northcott complex is minimal. In particular, the Betti diagram is as claimed. The length of $\mathbf{EN}(M_{r,1})$ is $r-1$, the codimension of X, so S_X is Cohen–Macaulay by the Auslander–Buchsbaum Theorem (A2.15). $\square$

6B   1-Generic Matrices

To describe some of what is special about the matrices $M_{r,d}$ we introduce some terminology: If M is a matrix of linear forms with rows $\ell_i = (\ell_{i,1}, \dots, \ell_{i,n})$ then a *generalized row* of M is by definition a row

$$\sum_i \lambda_i \ell_i = \left(\sum_i \lambda_i \ell_{i,1}, \ \ldots, \ \sum_i \lambda_i \ell_{i,n} \right),$$

that is, a scalar linear combination of the rows of M, with coefficients $\lambda_i \in \mathbb{K}$ that are not all zero. We similarly define *generalized columns* of M. In the same spirit, a *generalized entry* of M is a nonzero linear combination of the entries of some generalized row of M or, equivalently, a nonzero linear combination of the entries of some generalized column of M. We will say that M is *1-generic* if every generalized entry of M is nonzero. This is the same as saying that every generalized row (or column) of M consists of linearly independent linear forms.

Proposition 6.3. *For each $0 < d < r$ the matrix $M_{r,d}$ is 1-generic.*

Proof. A nonzero linear combination of the columns of the multiplication table corresponds to a nonzero form of degree $r - d$ in s and t, and, similarly, a nonzero linear combination of the rows corresponds to a nonzero form of degree d. A generalized entry of $M_{r,d}$ corresponds to a product of such nonzero forms, and so is nonzero. $\qquad\Box$

The same argument would work for a matrix made from part of the multiplication table of any graded domain; we shall further generalize and apply this idea later.

Determinantal ideals of 1-generic matrices have many remarkable properties. See [Room 1938] for a classical account and [Eisenbud 1988] for a modern treatment. In particular, they satisfy a generalization of Proposition 6.1 and Corollary 6.2.

Theorem 6.4. *If M is a 1-generic matrix of linear forms in $S = \mathbb{K}[x_0, \ldots, x_r]$, of size $p \times q$ with $p \le q$, over an algebraically closed field $\mathbb{K}$, then the ideal $I_p(M)$ generated by the maximal minors of M is prime of codimension $q - p + 1$; in particular, its free resolution is given by an Eagon–Northcott complex, and $S/I_p(M)$ is a Cohen–Macaulay domain.*

Note that $q - p + 1$ is the codimension of the ideal of $p \times p$ minors of the generic matrix (Theorem A2.54).

Proof. Set $I = I_p(M)$. We first show that $\operatorname{codim} I = q - p + 1$; equivalently, if X is the projective algebraic set defined by I, we will show that the dimension of X is $r - (q - p + 1)$. By Theorem A2.54 the codimension of I cannot be any greater than $q - p + 1$, so it suffices to show that $\dim X \le r - (q - p + 1)$.

Let $a \in \mathbb{P}^r$ be a point with homogeneous coordinates $a_0, \ldots, a_r$. The point a lies in X if and only if the rows of M become linearly dependent when evaluated at a. This is equivalent to saying that some generalized row vanishes at a, so X is the union of the zero loci of the generalized rows of M. As M is 1-generic, each generalized row has zero locus equal to a linear subspace of $\mathbb{P}^r$ of dimension precisely $r - q$. A generalized row is determined by an element of the vector space $\mathbb{K}^p$ of linear combinations of rows. Two generalized rows have the same zero locus if they differ by a scalar, so X is the union of a family of linear spaces of dimension $r - q$, parametrized by a projective space $\mathbb{P}^{p-1}$. Thus $\dim X \leq (r-q) + (p-1) = r - (q-p+1)$. More formally, we could define

$$X' = \{(y, a) \in \mathbb{P}^{p-1} \times \mathbb{P}^r \mid R_y \text{ vanishes at } a\},$$

where R_y denotes the generalized row corresponding to the parameter value y. The set X' fibers over $\mathbb{P}^{p-1}$ with fibers isomorphic to $\mathbb{P}^{r-q}$ so

$$\dim X' = (r - q) + (p - 1) = r - (q - p + 1).$$

Also, the projection of X' to $\mathbb{P}^r$ carries X' onto X, so $\dim X \leq \dim X'$.

A projective algebraic set, such as X', which is fibered over an irreducible base with irreducible equidimensional fibers is irreducible; see [Eisenbud 1995, Exercise 14.3]. It follows that the image X is also irreducible. This proves that the radical of $I_p(M)$ is prime.

From the codimension statement, and the Cohen–Macaulay property of S, it follows that the Eagon–Northcott complex associated to M is a free resolution of S/I, and we see that the projective dimension of S/I is $q - p + 1$. By the Auslander–Buchsbaum Formula (Theorem A2.15) the ring S/I is Cohen–Macaulay.

It remains to show that I itself is prime. From the fact that S/I is Cohen–Macaulay, it follows that all the associated primes of I are minimal, and have codimension precisely $q - p + 1$. Since the radical of I is prime, we see that in fact I is a primary ideal.

The submatrix M_1 of M consisting of the first $p - 1$ rows is also 1-generic so, by what we have already proved, the ideal $I_{p-1}(M_1)$ has codimension $q - p$. Thus some $(p - 1) \times (p - 1)$ minor Δ of M_1 does not vanish identically on X. Since X is the union of the zero loci of the generalized rows of M, there is even a generalized row whose elements generate an ideal that does not contain Δ. This generalized row cannot be in the span of the first $p - 1$ rows alone, so we may replace the last row of M by this row without changing the ideal of minors of M, and we may assume that $\Delta \notin Q := (x_{p,1}, \ldots, x_{p,q})$. On the other hand, since we can expand any $p \times p$ minor of M along its last row, we see that I is contained in Q.

Since the ideal Q is generated by a sequence of linear forms, it is prime. Since we have seen that I is primary, it suffices to show that IS_Q is prime, where S_Q denotes the local ring of S at Q. Since Δ becomes a unit in S_Q we may make an S_Q-linear invertible transformation of the columns of M to bring M into the form

$$M' = \begin{pmatrix} 1 & 0 & \cdots & 0 & 0 & \cdots & 0 \\ 0 & 1 & \cdots & 0 & 0 & \cdots & 0 \\ \cdots & \cdots & \cdots & \cdots & \cdots & \cdots & \cdots \\ 0 & 0 & \cdots & 1 & 0 & \cdots & 0 \\ x'_{p,1} & x'_{p,2} & \cdots & x'_{p,p-1} & x'_{p,p} & \cdots & x'_{p,q} \end{pmatrix},$$

where $x'_{p,1}, \ldots, x'_{p,q}$ is the result of applying an invertible S_Q-linear transformation to $x_{p,1}, \ldots, x_{p,q}$, and the $(p-1) \times (p-1)$ matrix in the upper left-hand corner is the identity. It follows that $IS_Q = (x'_{p,p}, \ldots, x'_{p,q})S_Q$.

Since $x_{p,1}, \ldots, x_{p,q}$ are linearly independent modulo $Q^2 S_Q$, so are $x'_{p,1}, \ldots, x'_{p,q}$. It follows that $S_Q/(x'_{p,p}, \ldots, x'_{p,q}) = S_Q/IS_Q$ is a regular local ring and thus a domain (see [Eisenbud 1995, Corollary 10.14]). This shows that IS_Q is prime. $\square$

Theorem 6.4 gives another proof of Proposition 6.1; see Exercise 6.4.

6C   Linear Series

We can extend these ideas to give a description of certain embeddings of genus-1 curves. At least over the complex numbers, this could be done very explicitly, replacing monomials by doubly periodic functions. Instead, we approach the problem algebraically, using the general notion of linear series. For simplicity, we continue to suppose that the curves and other algebraic sets we consider are irreducible, and that the ground field $\mathbb{K}$ is algebraically closed.

A *linear series* $(\mathscr{L}, V, \alpha)$ on a variety X over $\mathbb{K}$ consists of a line bundle $\mathscr{L}$ on X, a finite dimensional $\mathbb{K}$-vector space V and a nonzero homomorphism $\alpha : V \to \mathrm{H}^0 \mathscr{L}$. We define the (projective) *dimension* of the series to be $(\dim_{\mathbb{K}} V) - 1$. The linear series is *nondegenerate* if α is injective; in this case we think of V as a subspace of $\mathrm{H}^0(\mathscr{L})$, and write $(\mathscr{L}, V)$ for the linear series. Frequently we consider a linear series where the space V is the space $\mathrm{H}^0(\mathscr{L})$ and α is the identity. We call this the *complete linear series* defined by $\mathscr{L}$, and denote it by $|\mathscr{L}|$.

One can think of a linear series as a family of divisors on X parametrized by the nonzero elements of V: corresponding to $v \in V$ is the divisor which is the zero locus of the section $\alpha(v) \in \mathrm{H}^0(\mathscr{L})$. Since the divisor corresponding to v is the same as that corresponding to a multiple rv with $0 \neq r \in \mathbb{K}$, the family of divisors is really parametrized by the projective space of one-dimensional subspaces of V, which we think of as the projective space $\mathbb{P}(V^*)$. The simplest kind of linear series is the "hyperplane series" arising from a projective embedding $X \subset \mathbb{P}(V)$. It consists of the family of divisors that are hyperplane sections of X; more formally this series is $(\mathscr{O}_X(1), V, \alpha)$ where $\mathscr{O}_X(1)$ is the line bundle $\mathscr{O}_{\mathbb{P}(V)}(1)$ restricted to X and

$$\alpha : V = \mathrm{H}^0(\mathscr{O}_{\mathbb{P}(V)}(1)) \to \mathrm{H}^0(\mathscr{O}_X(1))$$

is the restriction mapping. This series is nondegenerate in the sense above if and only if X is nondegenerate in $\mathbb{P}(V)$ (that is, X is not contained in any hyperplane).

For example, if $X \cong \mathbb{P}^1$ is embedded in $\mathbb{P}^r$ as the rational normal curve of degree r, the hyperplane series is the complete linear series

$$|\mathscr{O}_{\mathbb{P}^1}(r)| = (\mathscr{O}_{\mathbb{P}^1}(r), \mathrm{H}^0(\mathscr{O}_{\mathbb{P}^1}(r)), \mathrm{id}),$$

where id is the identity map.

Not all linear series arise as the linear series of hyperplane sections of an embedded variety. For example, given $p \in \mathbb{P}^2$, we may describe the *linear series on $\mathbb{P}^2$ of conics through p* as follows: Let $\mathscr{L} = \mathscr{O}_{\mathbb{P}^2}(2)$. The global sections of $\mathscr{L}$ correspond to quadratic forms in three variables. Taking coordinates x, y, z, we choose p to be the point $(0, 0, 1)$, and we take V to be the vector space of quadratic forms vanishing at p:

$$V = \langle x^2, xy, xz, y^2, yz \rangle.$$

We call p a basepoint of the series $\mathscr{L}, V$). In general we define a *basepoint* of a linear series to be a point in the zero loci of all the sections in $\alpha(V) \subset \mathrm{H}^0(\mathscr{L})$. Equivalently, this is a point at which the sections of $\alpha(V)$ fail to generate $\mathscr{L}$; or, again, it is a point contained in all the divisors in the series. In the example above, p is the only basepoint. The linear series is called *basepoint-free* if it has no basepoints. The hyperplane series of any variety in $\mathbb{P}^r$ is basepoint-free because there is a hyperplane missing any given point.

Recall that a rational map from a variety X to a variety Y is a morphism defined on an open dense subset $U \subset X$. A nontrivial linear series $L = (\mathscr{L}, V, \alpha)$ always gives rise to a rational map from X to $\mathbb{P}(V)$. Let U be the set of points of X that are not basepoints of the series, and let $\Phi_L : U \to \mathbb{P}(V)$ be the map associating a point p to the hyperplane in V of sections $v \in V$ such that $\alpha(v)(p) = 0$. If L is basepoint-free, we get a morphism defined on all of X.

To justify these statements we introduce coordinates. Choose a basis $x_0, \ldots, x_r$ of V and regard the x_i as homogeneous coordinates on $\mathbb{P}(V) \cong \mathbb{P}^r$. Given $q \in X$, suppose that the global section $\alpha(x_j)$ generates $\mathscr{L}$ locally near q. There is a morphism from the open set $U_j \subset X$ where $\alpha(x_j) \neq 0$ to the open set $x_j \neq 0$ in $\mathbb{P}(V)$ corresponding to the ring homomorphism $\mathbb{K}[x_0/x_j, \ldots, x_r/x_j] \to \mathscr{O}_X(U)$ sending $x_i/x_j \mapsto \varphi(x_i)/\varphi(x_j)$. These morphisms glue together to form a morphism, from X minus the basepoint locus of L, to $\mathbb{P}(V)$. See [Hartshorne 1977, Section II.7] or [Eisenbud and Harris 2000, Section 3.2.5] for more details.

For example, we could have defined a rational normal curve in $\mathbb{P}^r$ to be the image of $\mathbb{P}^1$ by the complete linear series $|\mathscr{O}_{\mathbb{P}^1}(r)| = (\mathscr{O}_{\mathbb{P}^1}(r), \mathrm{H}^0(\mathscr{O}_{\mathbb{P}^1}(r)), \mathrm{id})$ together with an identification of $\mathbb{P}^r$ and $\mathbb{P}(V)$ — that is, a choice of basis of V.

On the other hand, the series of plane conics with a basepoint at $p = (0, 0, 1)$ above corresponds to the rational map from $\mathbb{P}^2$ to $\mathbb{P}^4$ sending a point (a, b, c) other than p to (a^2, ab, ac, b^2, bc). This map cannot be extended to a morphism on all of $\mathbb{P}^2$.

If $\Lambda \subset \mathbb{P}^s$ is a linear space of codimension $r + 1$, the *linear projection π_Λ from $\mathbb{P}^s$ to $\mathbb{P}^r$ with center Λ* is the rational map from $\mathbb{P}^s$ to $\mathbb{P}^r$ corresponding to the linear series of hyperplanes in $\mathbb{P}^s$ containing Λ. The next result shows that complete series are those not obtained in a nontrivial way by linear projection.

Proposition 6.5. *Let $L = (\mathscr{L}, V, \alpha)$ be a basepoint-free linear series on a variety X. The linear series L is nondegenerate (that is, the map α is injective) if and only if $\phi_L(X) \subset \mathbb{P}(V)$ is nondegenerate. The map α is surjective if and only if*

ϕ_L does not factor as the composition of a morphism from X to a nondegenerate variety in a projective space $\mathbb{P}^s$ and a linear projection π_Λ, where Λ is a linear space not meeting the image of X in $\mathbb{P}^s$.

Proof. A linear form on $\mathbb{P}(V)$ that vanishes on $\phi_L(X)$ is precisely an element of $\ker \alpha$, which proves the first statement. For the second, note that if ϕ_L factors through a morphism $\psi : X \to \mathbb{P}^s$ and a linear projection π_Λ to $\mathbb{P}^r$, where Λ does not meet $\psi(X)$, then the pullback of $\mathscr{O}_{\mathbb{P}^r}(1)$ to $\psi(X)$ is $\mathscr{O}_{\mathbb{P}^s}(1)|_{\psi(X)}$, so $\psi^*(\mathscr{O}_{\mathbb{P}^s}(1)) = \phi_L^*(\mathscr{O}_{\mathbb{P}^r}(1)) = \mathscr{L}$. If $\psi(X)$ is nondegenerate, then $H^0(\mathscr{L})$ is at least $(s+1)$-dimensional, so α cannot be onto. Conversely, if α is not onto, we can obtain a factorization as above where ψ is defined by the complete linear series $|\mathscr{L}|$. The plane Λ is defined by the vanishing of all the forms in $\alpha(V)$, and does not meet X because L is basepoint-free. $\qquad\square$

A variety embedded by a complete linear series is said to be *linearly normal*. In Corollary A1.13 it is shown that if $X \subset \mathbb{P}^r$ is a variety, the homogeneous coordinate ring S_X has depth 2 if and only if $S_X \to \bigoplus_{d\in\mathbb{Z}} H^0(\mathscr{O}_X(d))$ is an isormorphism. We can restate this condition by saying that, for every d, the linear series $(\mathscr{O}_X(d), H^0(\mathscr{O}_{\mathbb{P}^r}(d)), \alpha_d)$ is complete, where

$$\alpha_d : H^0(\mathscr{O}_{\mathbb{P}^r}(d)) \to H^0(\mathscr{O}_X(d))$$

is the restriction map. Using Theorem A2.28 we see that if X is normal and of dimension ≥ 1 (so that S_X is locally normal at any homogeneous ideal except the irrelevant ideal, which has codimension ≥ 2), then this condition is equivalent to saying that S_X is a normal ring. Thus the condition that $X \subset \mathbb{P}^r$ is linearly normal is the "degree-1 part" of the condition for the normality of S_X.

Ampleness

The linear series that interest us the most are those that provide embeddings. In general, a line bundle $\mathscr{L}$ is called *very ample* if $|\mathscr{L}|$ is basepoint-free and the morphism corresponding to $|\mathscr{L}|$ is an embedding of X in the projective space $\mathbb{P}(H^0(L))$. (The term *ample* is used for a line bundle for which some power is very ample.) In case X is a smooth variety over an algebraically closed field there is a simple criterion, which we recall here in the case of curves from [Hartshorne 1977, IV.3.1(b)].

Theorem 6.6. *Let X be a smooth curve over an algebraically closed field. A line bundle $\mathscr{L}$ on X is very ample if and only if*

$$h^0(\mathscr{L}(-p-q)) = h^0(\mathscr{L}) - 2$$

for every pair of points $p, q \in X$. $\qquad\square$

That is, $\mathscr{L}$ is very ample if and only if any two points of X (possibly equal to one another) impose independent conditions on the complete series $|\mathscr{L}|$.

Combining this theorem with the Riemann–Roch formula, we easily prove that any line bundle of high degree is very ample. In what follows we write $\mathscr{L}(D)$, where D is a divisor, for the line bundle $\mathscr{L} \otimes \mathscr{O}_X(D)$.

Corollary 6.7. *If X is a smooth curve of genus g, any line bundle of degree $\geq 2g+1$ on X is very ample. If $g = 0$ or $g = 1$, the converse is also true.*

Proof. For any points $p, q \in X$, $\deg \mathscr{L}(-p-q) > 2g-2 = \deg \omega_X$, so $\mathscr{L}$ and $\mathscr{L}(-p-q)$ are both nonspecial. Applying the Riemann Roch formula to each of these bundles we get

$$h^0(\mathscr{L}(-p-q)) = \deg \mathscr{L} - 2 - g + 1 = h^0(\mathscr{L}) - 2.$$

as reqired by Theorem 6.6.

Any very ample line bundle must have positive degree, so the converse is immediate for $g = 0$. For $g = 1$, we note that, by Riemann-Roch, $h^0(\mathscr{L}) = \deg \mathscr{L}$ as long as $\mathscr{L}$ has positive degree. Thus a linear series of degree 1 must map X to a point, and a linear series of degree 2 can at best map X to $\mathbb{P}^1$. Since $X \neq \mathbb{P}^1$, such a map is not very ample. □

The language of linear series is convenient for the following characterization:

Theorem 6.8. *Any nondegenerate curve $X \subset \mathbb{P}^r$ of degree r is a rational normal curve.*

Proof. Suppose that the embedding is given by the linear series $L = (\mathscr{L}, V, \alpha)$ on the curve X, so that $\mathscr{L}$ is the restriction to X of $\mathscr{O}_{\mathbb{P}^r}(1)$ and $\deg \mathscr{L} = r$. As X is nondegenerate, Lemma 6.5 shows that $h^0(\mathscr{L}) \geq r+1$.

We first prove that the existence of a line bundle $\mathscr{L}$ on X with $\deg \mathscr{L} \geq 1$ and $h^0(\mathscr{L}) \geq 1 + \deg \mathscr{L}$ implies that $X \cong \mathbb{P}^1$. To see this we do induction on $\deg \mathscr{L}$. If $\deg \mathscr{L} = 1$ we have $\deg \mathscr{L}(-p-q) = -1$ for any points $p, q \in X$, whence

$$h^0(\mathscr{L}(-p-q)) = 0 \leq h^0(\mathscr{L}) - 2.$$

In fact, we must have equality, since vanishing at two points can impose at most two independent linear conditions. Thus $\mathscr{L}$ is very ample and provides an isomorphism from X to $\mathbb{P}^1$.

If, on the other hand, $\deg \mathscr{L} > 1$, we choose a smooth point p of X, and consider the line bundle $\mathscr{L}(-p)$, which has degree $\deg \mathscr{L}(-p) = \deg \mathscr{L} - 1$. Since the condition of vanishing at p is (at most) one linear condition on the sections of $\mathscr{L}$, we see that $h^0(\mathscr{L}(-p)) \geq h^0(\mathscr{L}) - 1$, so $\mathscr{L}(-p)$ satisfies the same hypotheses as $\mathscr{L}$.

Returning to the hypotheses of the theorem, we conclude that $X \cong \mathbb{P}^1$. There is only one line bundle on $\mathbb{P}^1$ of each degree, so $\mathscr{L} \cong \mathscr{O}_{\mathbb{P}^1}(r)$. It follows that $h^0(\mathscr{L}) = 1 + r$. Thus the embedding is given by the complete linear series $|\mathscr{O}_{\mathbb{P}^1}(r)|$ and X is a rational normal curve. □

Corollary 6.9. (a) *If X is a nondegenerate curve of degree r in $\mathbb{P}^r$, then the ideal of X is generated by the 2×2 minors of a 1-generic, $2 \times r$ matrix of* linear forms and the minimal free resolution of S_X is the Eagon–Northcott complex of this matrix. In particular, S_X is Cohen–Macaulay.

(b) Conversely, if M is a 1-generic $2 \times r$ matrix of linear forms in $r+1$ variables, then the 2×2 minors of M generate the ideal of a rational normal curve.

Proof. (a) By Theorem 6.8, a nondegenerate curve of degree r in $\mathbb{P}^r$ is, up to change of coordinates, the standard rational normal curve. The desired matrix and resolution can be obtained by applying the same change of coordinates to the matrix $M_{r,1}$.

(b) By Theorem 6.4 the ideal P of minors is prime of codimension $r-1$, and thus defines a nondegenerate irreducible curve C in $\mathbb{P}^r$. Its resolution is the Eagon–Northcott complex, as would be the case for the ideal defining the standard rational normal curve X. Since the Hilbert polynomials of C and X can be computed from their graded Betti numbers, these Hilbert polynomials are equal; in particular C has the same degree, r, as X, and Theorem 6.8 completes the proof. $\square$

In fact any 1-generic $2 \times r$ matrix of linear forms in $r+1$ variables can be reduced to $M_{r,1}$ by row, column, and coordinate changes; see Exercise 6.6.

Matrices from Pairs of Linear Series

We have seen that the matrices produced from the multiplication table of the ring $\mathbb{K}[s,t]$ play a major role in the theory of the rational normal curve. Using linear series we can extend this idea to more general varieties.

Suppose that $X \subset \mathbb{P}^r$ is a variety embedded by the complete linear series $|\mathscr{L}|$ corresponding to some line bundle $\mathscr{L}$. Set $V = \mathrm{H}^0(\mathscr{L})$, the space of linear forms on $\mathbb{P}^r$. Suppose that we can factorize $\mathscr{L}$ as $\mathscr{L} = \mathscr{L}_1 \otimes \mathscr{L}_2$ for some line bundles $\mathscr{L}_1$ and $\mathscr{L}_2$. Choose ordered bases $y_1 \ldots y_m \in \mathrm{H}^0(\mathscr{L}_1)$ and $z_1 \ldots z_n \in \mathrm{H}^0(\mathscr{L}_2)$, and let

$$M(\mathscr{L}_1, \mathscr{L}_2)$$

be the matrix of linear forms on $\mathbb{P}(V)$ whose (i,j) element is the section $y_i \otimes z_j \in V = \mathrm{H}^0(\mathscr{L})$. (Of course this matrix is only interesting when it has at least two rows and two columns, that is, $\mathrm{h}^0 \mathscr{L}_1 \geq 2$ and $\mathrm{h}^0 \mathscr{L}_2 \geq 2$.) Each generalized row of $M(\mathscr{L}_1, \mathscr{L}_2)$ has entries $y \otimes z_1, \ldots, y \otimes z_n$ for some section $0 \neq y \in \mathrm{H}^0(\mathscr{L}_1)$, and a generalized entry of this row will have the form $y \otimes z$ for some section $0 \neq z \in \mathrm{H}^0(\mathscr{L}_2)$.

Proposition 6.10. *If X is a variety and $\mathscr{L}_1, \mathscr{L}_2$ are line bundles on X, then the matrix $M(\mathscr{L}_1, \mathscr{L}_2)$ is 1-generic, and its 2×2 minors vanish on X.*

Proof. With notation as above, a generalized element of M may be written $x = y \otimes z$ where y, z are sections of $\mathscr{L}_1, \mathscr{L}_2$ respectively. If $p \in X$ we may identify $\mathscr{L}_1$ and $\mathscr{L}_2$ with $\mathscr{O}_X$ in a neighborhood of p and write $x = yz$. Since $\mathscr{O}_{X,p}$ is an integral domain, x vanishes at p if and only if at least one of y and z vanish at p. Since X is irreducible, X is not the union of the zero loci of a nonzero y and a nonzero z, so no section $y \otimes z$ can vanish identically. This shows that M is 1-generic. On the other hand, any 2×2 minor of M may be written as

$$(y \otimes z)(y' \otimes z') - (y \otimes z')(y' \otimes z) \in H^0(\mathscr{L})$$

for sections $y, y' \in H^0(\mathscr{L}_1)$ and $z, z' \in H^0(\mathscr{L}_2)$. Locally near a point p of X we may identify $\mathscr{L}_1, \mathscr{L}_2$ and $\mathscr{L}$ with $\mathscr{O}_{X,p}$ and this expression becomes $(yz)(y'z') - (yz')(y'z)$ which is 0 because $\mathscr{O}_{X,p}$ is commutative and associative. $\qquad \square$

It seems that if both the line bundles $\mathscr{L}_1$ and $\mathscr{L}_2$ are "sufficiently positive" then the homogeneous ideal of X is generated by the 2×2 minors of $M(\mathscr{L}_1, \mathscr{L}_2)$. For example, we have seen that in the case where X is $\mathbb{P}^1$ it suffices that the bundles each have degree at least 1. For an easy example generalizing the case of rational normal curves see Exercise 6.10; for more results in this direction see [Eisenbud et al. 1988]. For less positive bundles, the 2×2 minors of $M(\mathscr{L}_1, \mathscr{L}_2)$ may still define an interesting variety containing X, as in Section 6D.

Using the idea introduced in the proof of Theorem 6.4 we can describe the geometry of the locus defined by the maximal minors of $M(\mathscr{L}_1, \mathscr{L}_2)$ in more detail. Interchanging $\mathscr{L}_1$ and $\mathscr{L}_2$ if necessary we may suppose that $n = h^0 \mathscr{L}_2 > h^0 \mathscr{L}_1 = m$ so $M(\mathscr{L}_1, \mathscr{L}_2)$ has more columns than rows. If $y = \sum r_i y_i \in H^0(\mathscr{L}_1)$ is a section, we write ℓ_y for the generalized row indexed by y. The maximal minors of $M(\mathscr{L}_1, \mathscr{L}_2)$ vanish at a point $p \in \mathbb{P}^r$ if and only if some row ℓ_y consists of linear forms vanishing at p; that is,

$$V\big(I_m(M(\mathscr{L}_1, \mathscr{L}_2))\big) = \bigcup_y V(\ell_y). \qquad (*)$$

The important point is that we can identify the linear spaces $V(\ell_y)$ geometrically.

Proposition 6.11. *Suppose $X \subset \mathbb{P}^r$ is embedded by a complete linear series, and assume that the hyperplane bundle $\mathscr{L} = \mathscr{O}_X(1)$ decomposes as the tensor product of two line bundles, $\mathscr{L} = \mathscr{L}_1 \otimes \mathscr{L}_2$. For each $y \in H^0 \mathscr{L}_1$ we have $V(\ell_y) = \bar{D}_y$, the linear span of D_y.*

Recall that the *linear span* of a divisor D on $X \subset \mathbb{P}^r$ is the smallest linear subspace of $\mathbb{P}^r$ containing D.

Proof. The linear span of D_y is the interesection of all the hyperplanes containing D_y, so we must show that the linear forms appearing in the row ℓ_y span the space of all linear forms vanishing on D_y. It is clear that every entry $y \otimes z_i$ of this row does in fact vanish where y vanishes.

Moreover, if $x \in H^0 \mathscr{L}$ is any section vanishing on D_y, and E is the divisor of x, then $\mathscr{O}_X(E - D) \otimes \mathscr{O}_X(D) = \mathscr{L}$, and multiplying by $\mathscr{O}_X(-D) = \mathscr{L}_1^{-1}$ we see that $\mathscr{O}_X(E - D) = \mathscr{L}_2$. Thus the divisor $E - D$ is represented by a section z of $\mathscr{L}_2$, and $x = y \otimes z$ up to a scalar, since both vanish on the same divisor. $\qquad \square$

Note that $V(\ell_y)$ and D_y do not change if we change y by a nonzero scalar multiple. Thus when we write D_y we may think of y as an element of $\mathbb{P}^{m-1}$. We can summarize the results of this section, in their most important special case, as follows.

Corollary 6.12. *Suppose that $X \subset \mathbb{P}^r$ is embedded by the complete linear series $|\mathscr{L}|$ and that $\mathscr{L}_1, \mathscr{L}_2$ are line bundles on X such that $\mathscr{L} = \mathscr{L}_1 \otimes \mathscr{L}_2$. Suppose that $h^0\mathscr{L}_1 = m \leq h^0\mathscr{L}_2$. If $y \in H^0\mathscr{L}_1$, write D_y for the corresponding divisor. If $\bar{D}_y$ denotes the linear span of D_y, the variety defined by the maximal minors of $M(\mathscr{L}_1, \mathscr{L}_2)$ is*

$$Y = V\big(I_m(M(\mathscr{L}_1, \mathscr{L}_2))\big) = \bigcup_{y \in \mathbb{P}^{m-1}} \bar{D}_y. \qquad \square$$

We may illustrate Corollary 6.12 with the example of the rational normal curve. Let $X = \mathbb{P}^1$ and let $\mathscr{L}_1 = \mathscr{O}_{\mathbb{P}^1}(1), \mathscr{L}_2 = \mathscr{O}_{\mathbb{P}^1}(r-1)$ so that

$$M(\mathscr{L}_1, \mathscr{L}_2) = M_{r,1} = \begin{pmatrix} x_0 & x_1 & \cdots & x_{r-1} \\ x_1 & x_2 & \cdots & x_r. \end{pmatrix}$$

The generalized row corresponding to an element $y = (y_1, y_2) \in \mathbb{P}^1$ has the form

$$\ell_y = (y_0 x_0 + y_1 x_1, y_0 x_1 + y_1 x_2, \ldots, y_0 x_{r-1} + y_1 x_r).$$

The linear space $V(\ell_y)$ is thus the set of solutions of the linear equations

$$y_0 x_0 + y_1 x_1 = 0,$$
$$y_0 x_1 + y_1 x_2 = 0,$$
$$\cdot \quad \cdot \quad \cdot \quad \cdot$$
$$y_0 x_{r-1} + y_1 x_r = 0.$$

Since these r equations are linearly independent, $V(\ell_y)$ is a single point. Solving the equations, we see that this point has coordinates $x_i = (-y_0/y_1)^i x_0$. Taking $y_0 = 1$, $x_0 = s^r$, $y_1 = -s/t$ we obtain the usual parametrization $x_i = s^{r-i}t^i$ of the rational normal curve.

Linear Subcomplexes and Mapping Cones

We have seen that if X is embedded by the complete linear series $|\mathscr{L}|$ and if $\mathscr{L} = \mathscr{L}_1 \otimes \mathscr{L}_2$ is a factorization, then by Theorem 6.4 and Proposition 6.10 the ideal $I = I_X$ of X contains the ideal of 2×2 minors of the 1-generic matrix $M = M(\mathscr{L}_1, \mathscr{L}_2)$. This has an important consequence for the free resolution of M.

Proposition 6.13. *Suppose that $X \subset \mathbb{P}^r$ is a variety embedded by a complete linear series $|\mathscr{L}|$, and that $\mathscr{L} = \mathscr{L}_1 \otimes \mathscr{L}_2$ for some line bundles $\mathscr{L}_1, \mathscr{L}_2$ on X. Let M' be a $2 \times h^0(\mathscr{L}_2)$ submatrix of $M(\mathscr{L}_1, \mathscr{L}_2)$, and let J be the ideal generated by the 2×2 minors of M'. If $\mathbf{F} : \cdots \to F_0 \to I_X$ is a minimal free resolution and $\mathbf{E} : \cdots \to E_0 \to J$ denotes the Eagon–Northcott complex of M', then $\mathbf{E}$ is a subcomplex of $\mathbf{F}$ in such a way that $F_i = E_i \oplus G_i$ for every i.*

Proof. Choose any map $\alpha : \mathbf{E} \to \mathbf{F}$ lifting the inclusion $J \subset I = I_X$. We will show by induction that $\alpha_i : E_i \to F_i$ is a split inclusion for every $i \geq 0$. Write

δ for the differentials — both of $\mathbf{E}$ and of $\mathbf{F}$. Write $P = (x_0, \ldots, x_r)$ for the homogeneous maximal ideal of S. It suffices to show that if $e \in E_i$ but $e \notin PE_i$ (so that e is a minimal generator) then $\alpha_i(e) \notin PF_i$.

Suppose on the contrary that $\alpha_i e \in PF_i$. If $i = 0$, we see that δe must be in $PI \cap J$. But the Eagon–Northcott complex $\mathbf{EN}(M')$ is a minimal free resolution, so δe is a nonzero quadratic form. As X is nondegenerate the ideal $I = I_X$ does not contain any linear form, so we cannot have $e \in PI$.

Now suppose $i > 0$, and assume by induction that α_{i-1} maps E_{i-1} isomorphically to a summand of F_{i-1}. Since $\mathbf{F}$ is a minimal free resolution the relation $\alpha_i \in PF_i$ implies that

$$\alpha_{i-1}\delta e = \delta\alpha_i e \in P^2 F_{i-1},$$

where δ is the differential of $\mathbf{EN}(M')$. However, the coefficients in the differential of the Eagon–Northcott complex are all linear forms. As $\mathbf{EN}(M')$ is a minimal free resolution we have $\delta e \neq 0$, so $\delta e \notin P^2 E_{i-1}$, a contradiction since E_{i-1} is mapped by α_{i-1} isomorphically to a summand of F_{i-1}. $\qquad\square$

You can verify that the idea just used applies more generally when one has a linear complex that is minimal in an appropriate sense and maps to the "least degree part" of a free resolution. We will pursue linear complexes further in the next chapter.

Proposition 6.13 is typically applied when $\mathscr{L}_1$ has just two sections — otherwise, to choose the $2 \times n$ submatrix M' one effectively throws away some sections, losing some information. It would be very interesting to have a systematic way of exploiting the existence of further sections, or more generally of exploiting the presence of many difference choices of factorization $\mathscr{L} = \mathscr{L}_1 \otimes \mathscr{L}_2$ with a choice of two sections of $\mathscr{L}_1$. In the next section we will see a case where we have in fact many such factorizations, but our analysis ignores the fact. See, however, [Kempf 1989] for an interesting case where the presence of multiple factorizations is successfully exploited.

The situation produced by Proposition 6.13 allows us to split the analysis of the resolution into two parts. Here is the general setup, which we will apply to a certain family of curves in the next section.

Proposition 6.14. Let $\mathbf{F} : \cdots \to F_0$ be a free complex with differential δ, and let $\mathbf{E} : \cdots \to E_0$ be a free subcomplex, with quotient complex $\mathbf{G} = \mathbf{F}/\mathbf{E} : \cdots \to G_0$. If E_i is a direct summand of F_i for each i, then $\mathbf{F}$ is the mapping cone of the map $\alpha : \mathbf{G}[-1] \to \mathbf{E}$ whose i-th component is the composite

$$G_{i+1} \subset G_{i+1} \oplus E_{i+1} = F_{i+1} \xrightarrow{\ \delta\ } F_i = G_i \oplus E_i \longrightarrow E_i.$$

Proof. Immediate from the definitions. $\qquad\square$

To reverse the process and construct $\mathbf{F}$ as a mapping cone, we need a different way of specifying the map from $\mathbf{G}[-1]$ to $\mathbf{E}$. In our situation the following observation is convenient. We leave to the reader the easy formalization for the most general case.

Proposition 6.15. *Suppose that $J \subset I$ are ideals of S. Let $\mathbf{G} : \cdots \rightarrow G_0$ be a free resolution of I/J as an S-module. Let $\mathbf{E} : \cdots \rightarrow E_1 \rightarrow S$ be a free resolution of S/J. If $\alpha : \mathbf{G} \rightarrow \mathbf{E}$ is a map of complexes lifting the inclusion $I/J \rightarrow S/J$, then the mapping cone, $\mathbf{F}$, of α is a free resolution of S/I. If matrices representing the maps $\alpha_i : G_i \rightarrow E_i$ have all nonzero entries of positive degree, and if both $\mathbf{E}$ and $\mathbf{G}$ are minimal resolutions, then $\mathbf{F}$ is also a minimal resolution.*

Proof. Denoting the mapping cylinder of α by $\mathbf{F}$, we have an exact sequence $0 \rightarrow \mathbf{E} \rightarrow \mathbf{F} \rightarrow \mathbf{G}[-1] \rightarrow 0$. Since $\mathbf{G}$ and $\mathbf{E}$ have no homology except at the right-hand end, we see from the long exact sequence in homology that $H_i\mathbf{F} = 0$ for $i \geq 2$. The end of this sequence has the form

$$\cdots \longrightarrow H_1\mathbf{E} \longrightarrow H_1\mathbf{F} \longrightarrow I/J \longrightarrow S/J \longrightarrow H_0\mathbf{F} \longrightarrow 0,$$

where the map $I/J \rightarrow S/J$ is the inclusion. It follows that $H_1\mathbf{F} = 0$ and $\mathbf{F} : \cdots \rightarrow F_1 \rightarrow S = F_0$ is a resolution of S/I. $\qquad\square$

6D Elliptic Normal Curves

Let X be a smooth, irreducible curve of genus 1, let $\mathscr{L}$ be a very ample line bundle on X, and let d be the degree of $\mathscr{L}$. By Corollary 6.7, $d \geq 3$, and by the Riemann–Roch formula, $h^0(\mathscr{L}) = d$. Thus the complete linear series $|\mathscr{L}|$ embeds X as a curve of degree d in $\mathbb{P}^r = \mathbb{P}^{d-1}$. We will call such an embedded curve an *elliptic normal curve* of degree d. (Strictly speaking, an *elliptic* curve is a smooth projective curve of genus 1 with a chosen point, made into an algebraic group in such a way that the chosen point is the origin. We will not need the chosen point for what we are doing, and we will accordingly not distinguish between an elliptic curve and a curve of genus 1.)

In this section we will use the ideas introduced above to study the minimal free resolution $\mathbf{F}$ of S_X, where $X \subset \mathbb{P}^r$ is an elliptic normal curve of degree d. Specifically, we will show that $\mathbf{F}$ is built up as a mapping cone from an Eagon–Northcott complex $\mathbf{E}$ and its dual, appropriately shifted and twisted. Further, we shall see that S_X is always Cohen–Macaulay, and of regularity 3.

The cases with $d \leq 4$ are easy and somewhat degenerate, so we will deal with them separately. If $d = 3$, then X is embedded as a cubic in $\mathbb{P}^2$, so the resolution has Betti diagram

	0	1
0	1	–
1	–	–
2	–	1

In this case the Eagon–Northcott complex in question would be that of the 2×2 minors of a 2×1 matrix — and thus isn't visible at all.

Next suppose $d = 4$. By the Riemann–Roch formula we have

$$h^0(\mathscr{L}^2) = 8 - g + 1 = 8,$$

while, since $r = 3$, the space of quadratic forms on $\mathbb{P}^r$ has dimension $\dim S_2 = 10$. It follows that the ideal I_X of X contains at least 2 linearly independent quadratic forms, Q_1, Q_2. If Q_1 were reducible then the quadric it defines would be the union of two planes. Since X is irreducible, X would have to lie entirely on one of them. But by hypothesis X is embedded by the complete series $|\mathscr{L}|$, so X is nondegenerate in $\mathbb{P}^3$. Thus Q_1 is irreducible, and $S/(Q_1)$ is a domain.

It follows that Q_1, Q_2 form a regular sequence. The complete intersection of the two quadrics corresponding to Q_1 and Q_2 has degree 4 by Bézout's Theorem, and it contains the degree 4 curve X, so it is equal to X. Since any complete intersection is unmixed (see Theorem A2.36), the ideal I_X is equal to (Q_1, Q_2). Since these forms are relatively prime, the free resolution of S_X has the form

$$0 \longrightarrow S(-4) \xrightarrow{\begin{pmatrix} Q_2 \\ -Q_1 \end{pmatrix}} S^2(-2) \xrightarrow{(Q_1, Q_2)} S,$$

with Betti diagram

	0	1	2
0	1	–	
1	–	2	–
2	–	–	1

In this case the Eagon–Northcott complex in question is that of the 2×2 minors of a 2×2 matrix. It has the form

$$0 \longrightarrow S(-2) \xrightarrow{Q_1} S.$$

In both these cases, the reader can see from the Betti diagrams that S_X is Cohen–Macaulay of regularity 3 as promised. Henceforward, we will freely assume that $d \geq 5$ whenever it makes a difference.

To continue our analysis, it is helpful to identify the surface Y. Let D be a divisor consisting of 2 points on X. We have $h^0(\mathcal{O}_X(D)) = 2$ and $h^0(\mathscr{L}(-D)) = d-2$, so from the theory of the previous section we see that $M = M(\mathcal{O}_X(D), \mathscr{L}(-D))$ is a $2 \times (d-2)$ matrix of linear forms on $\mathbb{P}^r$ that is 1-generic, and the ideal J of 2×2 minors of M is contained in the ideal of X. Moreover, we know from Theorem 6.4 that J is a prime ideal of codimension equal to $(d-2)-2+1 = r-2$; that is, $J = I_Y$ is the homogeneous ideal of an irreducible surface Y containing X. The surface Y is the union of the lines spanned by the divisors linearly equivalent to D in X. Since Y is a surface, X is a divisor on Y.

We can now apply Proposition 6.13 and Proposition 6.15 to construct the free resolution of I from the Eagon–Northcott resolution of J and a resolution of I/J. To this end we must identify I/J. We will show that it is a line bundle on Y.

Although it is not hard to continue this analysis in general, the situation is slightly simpler when $D = 2p$ and $\mathscr{L} = \mathcal{O}_X(dp)$ for some point $p \in X$. This case will suffice for the analysis of any elliptic normal curve because of the following:

Theorem 6.16. *If $\mathscr{L}$ is a line bundle of degree k on a smooth projective curve of genus 1 over an algebraically closed field, then $\mathscr{L} = \mathscr{O}_X(kp)$ for some point $p \in X$.*

Proof. The result follows from the construction of a group law on X: Choose a point $q \in X$ to act as the identity. By the Riemann–Roch Theorem, any line bundle of degree 1 on X has a unique section, so there is a one-to-one correspondence between the group of divisor classes of degree 0 on X and X itself, taking a divisor class D to the unique point where the global section of $\mathscr{O}_X(q+D)$ vanishes, and taking a point $p \in X$ to the class of $p-q$. This makes X into an algebraic group.

Multiplication by k is a nonconstant map of projective curves $X \to X$, and is thus surjective. It follows that there is a divisor $p-q$ such that $D-kq \sim k(p-q)$, and thus $D \sim kp$ as claimed. $\qquad\square$

Returning to our elliptic normal curve X embedded by $|\mathscr{L}|$, we see from Theorem 6.16 that we may write $\mathscr{L} = \mathscr{O}_X(dp)$ for some $p \in X$, and we choose $D = 2p$. To make the matrix $M(\mathscr{O}_X(2p), \mathscr{O}_X((d-2)p))$ explicit, we choose bases of the global sections of $\mathscr{O}_X(dp)$ and $\mathscr{O}_X(2p)$.

In general the global sections of $\mathscr{O}_X(kp)$ may be thought of as rational functions on X having no poles except at p, and a pole of order at most k at p. Thus there is a sequence of inclusions $\mathbb{K} = H^0\mathscr{O}_X \subseteq H^0\mathscr{O}_X(p) \subseteq H^0\mathscr{O}_X(2p) \subseteq \cdots \subseteq H^0\mathscr{O}_X(kp) \subseteq \cdots$. Moreover, we have seen that $h^0\mathscr{O}_X(kP) = k$ for $k \geq 1$. It follows that $1 \in H^0(\mathscr{O}_X) = H^0(\mathscr{O}_X(p))$ may be considered as a basis of either of these spaces. But there is a new section $\sigma \in H^0(\mathscr{O}_X(2p))$, with a pole at p of order exactly 2, and in addition to 1 and σ a section $\tau \in H^0(\mathscr{O}_X(3p))$ with order exactly 3. The function σ^2 has a pole of order 4, and continuing in this way we get:

Proposition 6.17. *If p is a point of the smooth projective curve X of genus 1 and $d \geq 1$ is an integer, the rational functions σ^a for $0 \leq a \leq d/2$ and $\sigma^a\tau$, for $0 \leq a \leq (d-3)/2$, form a basis of $H^0(\mathscr{O}_X(d))$.*

Proof. The function $\sigma^a\tau^b$ has pole of order $2a+3b$ at p, so the given functions are all sections, and are linearly independent. Since the dimension of $H^0(\mathscr{O}_X(dp))$ is $d = 1 + \lfloor d/2 \rfloor + \lfloor (d-1)/2 \rfloor = (1 + \lfloor d/2 \rfloor) + (1 + \lfloor (d-3)/2 \rfloor)$, the number of sections given, this suffices. $\qquad\square$

Corollary 6.18. *Let X be an elliptic curve, and let $p \in X$ be a point. If $d \geq 2$ and $e \geq 3$ are integers, the multiplication map*

$$H^0(\mathscr{O}_X(dp)) \otimes H^0(\mathscr{O}_X(ep)) \to H^0(\mathscr{O}_X((d+e)p))$$

is surjective. In particular, if $\mathscr{L}$ is a line bundle on X of degree ≥ 3, and $X \subset \mathbb{P}^r$ is embedded by the complete linear series $|\mathscr{L}|$, then S_X is Cohen–Macaulay and normal.

We will see Corollaries 8.2 and 8.4 that the regularity of S_X (which is 2) can be deduced easily from Corollary 6.18. In this chapter we will deduce it from an explicit free resolution.

Proof. The sections of $H^0(\mathcal{O}_X(dp))$ exhibited in Proposition 6.17 include sections with every vanishing order at p from 0 to d except for 1, and similarly for $H^0(\mathcal{O}_X(dp))$. When we multiply sections we add their vanishing orders at p, so the image of the multiplication map contains sections with every vanishing order from 0 to $d+e$ except 1, a total of $d+e$ distinct orders. These elements must be linearly independent, so they span the $d+e$-dimensional space $H^0(\mathcal{O}_X((d+e)p)$.

For the second statement we may first extend the ground field if necessary until it is algebraically closed, and then use Theorem 6.16 to rewrite $\mathcal{L}$ as $\mathcal{O}_X(dp)$ for some $d \geq 3$. From the first part of the Corollary we see that the multiplication map

$$H^0 \mathcal{O}_X(d) \otimes H^0 \mathcal{O}_X(md) \to H^0 \mathcal{O}_X((m+1)d)$$

is surjective for every $m \geq 0$. From Corollary A1.13 we see that S_X has depth 2. Since S_X is a two-dimensional ring, this implies in particular that it is Cohen–Macaulay. $\square$

For example, consider an elliptic normal cubic $X \subset \mathbb{P}^2$. By Theorem 6.16 the embedding is by a complete linear series $|\mathcal{O}_X(3p)|$ for some point $p \in X$. Let $S = \mathbb{K}[x_0, x_1, x_2] \to S_X = \bigoplus_n H^0(\mathcal{O}_X(3np))$ be the map sending x_0 to 1, x_1 to σ and x_2 to τ. By Corollary 6.18 this map is a surjection. To find its kernel, the equation of the curve, consider $H^0(\mathcal{O}_X(6p))$, the first space for which we can write down an "extra" section τ^2. We see that there must be a linear relation among 1, $\sigma, \sigma^2, \sigma^3, \tau, \sigma\tau$ and τ^2, and since σ^3 and τ^2 are the only two sections on this list with a triple pole at p, each must appear with a nonzero coefficient. From this we get an equation of the form $\tau^2 = f(\sigma) + \tau g(\sigma)$, where f is a polynomial of degree 3 and g a polynomial of degree ≤ 1. This is the affine equation of the embedding of the open subset $X \setminus \{p\}$ of X in $\mathbb{A}^2$ with coordinates σ, τ corresponding to the linear series $|\mathcal{O}_X(3p)|$. Homogenizing, we get an equation of the form

$$x_0 x_2^2 = F(x_0, x_1) + x_0 x_2 G(x_0, x_1),$$

where F and G are the homogenizations of f and g respectively. Since $3p$ is a hyperplane section, the point p goes to a flex point of X, and the line at infinity is the flex tangent. When the characteristic of $\mathbb{K}$ is not 2 or 3, further simplification yields the Weierstrass normal form $y^2 = x^3 + ax + b$ for the equation in affine coordinates.

In general, the table giving the multiplication between the sections of $\mathcal{O}_X(2p)$, and the sections of $\mathcal{O}_X((d-2)p)$, with the choice of bases above, can be written as

	1	σ	$\ldots$	σ^{n-1}	τ	$\sigma\tau$	$\ldots$	$\sigma^{m-1}\tau$
1	1	σ	$\ldots$	σ^{n-1}	τ	$\sigma\tau$	$\ldots$	$\sigma^{m-1}\tau$
σ	σ	σ^2	$\ldots$	σ^n	$\sigma\tau$	$\sigma^2\tau$	$\ldots$	$\sigma^m\tau,$

where $n = \lfloor d/2 \rfloor$ and $m = \lfloor (d-3)/2 \rfloor$, so that $(m+1)+(n+1) = r+1 = d$. Taking x_i to be the linear form on $\mathbb{P}^r$ corresponding to σ^i and y_j to be the linear form corresponding to $\sigma^j \tau$, the matrix $M = M(\mathscr{O}_X(2p), \mathscr{O}_X((d-2)p)$ takes the form

$$M = \begin{pmatrix} x_0 & x_1 & \cdots & x_{n-1} & y_0 & y_1 & \cdots & y_{m-1} \\ x_1 & x_2 & \cdots & x_n & y_1 & y_2 & \cdots & y_m \end{pmatrix},$$

where the vertical line indicates the division of M into two parts, which we will call M' and M''. The reader should recognize the matrices M' and M'' from Section 6A: their ideals of 2×2 minors define rational normal curves X' and X'' of degrees n and m in the disjoint subspaces L' defined by $y_0 = \cdots = y_m$ and L'' defined by $x_0 = \cdots = x_n$ respectively.

Let Y be the vanishing locus of the 2×2 minors of M, the union of the linear spaces defined by the vanishing of the generalized rows of M. Since M is 1-generic each generalized row consists of linearly independent linear forms — that is, its vanishing locus is a line. Moreover, the intersection of the line with the subspace L_x is the the point on the rational normal curve in that space given by the vanishing of the corresponding generalized row of M', and similarly for L_y. Thus M defines an isomorphism $\alpha : X' \to X''$, and in terms of this isomorphism the surface Y is the union of the lines joining $p \in X'$ to $\alpha(p) \in X''$. Such a surface is called a rational normal scroll; the name is justified

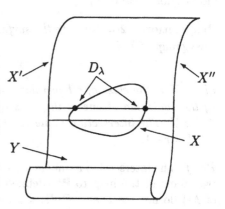

by the picture on the right:

In the simplest interesting case, $r = 3$, we get $m = 2$ and $n = 0$ so

$$M = \begin{pmatrix} x_0 & x_1 \\ x_1 & x_2 \end{pmatrix}.$$

In this case Y is the cone in $\mathbb{P}^3$ over the irreducible conic $x_0 x_2 = x_1^2$ in $\mathbb{P}^2$, and the lines F are the lines through the vertex on this cone. When $r \geq 4$, however, we will show that Y is smooth.

Proposition 6.19. *Suppose that $d \geq 5$, or equivalently that $r \geq 4$. The surface Y defined by the 2×2 minors of the matrix $M(\mathscr{O}_X(2p), \mathscr{O}_X((d-2)p))$ is smooth.*

Proof. As we have already seen, Y is the union of the lines defined by the generalized rows of the matrix $M(\mathscr{O}_X(2p), \mathscr{O}_X((d-2)p))$. To see that no two of these lines can intersect, note that any two distinct generalized rows span the space of all generalized rows, and thus any two generalized rows contain linear forms that span the space of all linear forms on $\mathbb{P}^r$. It follows that the set on which the linear forms in both generalized rows vanish is the empty set.

We can parametrize Y on the open set where $x_0 \neq 0$ as the image of $\mathbf{A}^2$ by the map sending $f : (t, u) \mapsto (1, t, \ldots, t^m, u, ut, \ldots, ut^n)$. The differential of f is nowhere vanishing, so f is an immersion. It is one-to-one because, from our previous argument, the lines $t = c_1$ and $t = c_2$ are distinct for any distinct constants c_1, c_2. A similar argument applies to the open set $y_m \neq 0$, and these two sets cover Y.   □

One can classify the 1-generic matrices of size $2 \times m$ completely using the classification of matrix pencils due to Kronecker and Weierstrass. The result shows that the varieties defined by the 2×2 minors of such a matrix are all rational normal scrolls of some dimension; for example, if such a variety is of dimension 1 then it is a rational normal curve. See [Eisenbud and Harris 1987] for details and many more properties of these interesting and ubiquitous varieties.

To identify X as a divisor, we use a description of the Picard group and intersection form of Y.

Proposition 6.20. *Let Y be the surface defined in Proposition 6.19. The divisor class group of Y is*

$$\operatorname{Pic} Y = \mathbf{Z}H \oplus \mathbf{Z}F,$$

where H is the class of a hyperplane section and F is the class of a line defined by the vanishing of one of the rows of the matrix $M(\mathcal{O}_X(D), \mathcal{L}(-D))$ used to define Y. The intersection numbers of these classes are $F \cdot F = 0$, $F \cdot H = 1$, and $H \cdot H = r - 1$.

Proof. The intersection numbers are easy to compute: We have $F \cdot F = 0$ because two fibers of the map to $\mathbf{P}^1$ (defined by the vanishing of the generalized rows of M) do not meet, and $F \cdot H = 1$ because F is a line, which meets a general hyperplane transversely in a single point. Since Y is a surface the number $H \cdot H$ is just the degree of the surface.

Modulo the polynomial $x_{m+1} - y_0$, the matrix M becomes the matrix whose 2×2 minors define the rational normal curve of degree $m + n + 2 = r - 1$. Thus the hyperplane section of Y is this rational normal curve, and the degree of Y is also $r - 1$. The fact that the intersection matrix

$$\begin{pmatrix} 0 & 1 \\ 1 & r-1 \end{pmatrix}$$

that we have just computed has rank 2 shows that the divisor classes of F and H are linearly independent. The proof that they generate the group is outlined in Exercise 6.8.   □

We can now identify a divisor by computing its intersection numbers with the classes H and F:

Proposition 6.21. *In the basis introduced above, the divisor class of X on the surface Y is $2H - (r-3)F$.*

Proof. By Proposition 6.20 we can write the class of X as $[X] = aH + bF$ for some integers a, b. From the form of the intersection matrix we see that $a = X.F$ and $b = X.H - (r-1)a$. Since the lines F on the surface are the linear spans of divisors on X that are linearly equivalent to D, and thus of degree 2, we have $a = 2$. On the other hand $X.H$ is the degree of X as a curve in $\mathbb{P}^r$, that is, $r+1$. Thus $b = r + 1 - (r-1)2 = -(r-3)$. $\qquad\square$

By this proposition, the sheaf of ideals $\widetilde{I/J} = \mathcal{I}_{X/Y}$ defining X in Y is the sheaf

$$\widetilde{I/J} = \mathcal{O}_Y((r-3)F - 2H) = \mathcal{O}_Y((r-3)F)(-2)$$

and thus the homogeneous ideal I/J of X in Y is, up to a shift of grading,

$$\bigoplus_{n \geq 0} H^0 \mathcal{O}_Y((r-3)F)(n).$$

Here is a first step toward identifying this module and its free resolution.

Proposition 6.22. *The cokernel K of the matrix*

$$M = M(\mathcal{O}_X(2p), \, \mathcal{O}_X((r-1)p))$$

has associated sheaf on $\mathbb{P}^r$ equal to $\tilde{K} = \mathcal{O}_Y(F)$.

Proof. Let $\tilde{K}$ be the sheaf on $\mathbb{P}^r$ that is associated to the module K. We will first show that $\tilde{K}$ is an invertible sheaf on Y. The entries of the matrix M span all the linear forms on $\mathbb{P}^r$ so locally at any point $p \in \mathbb{P}^r$ one of them is invertible, and we may apply the following result.

Lemma 6.23. *If N is a $2 \times n$ matrix over a ring R and M has one invertible entry, the cokernel of N is isomorphic to R modulo the 2×2 minors of N.*

Proof. Using row and column operations we may put N into the form

$$N' = \begin{pmatrix} 1 & 0 & \cdots & 0 \\ 0 & r_2 & \cdots & r_n \end{pmatrix}$$

for some $r_i \in R$. The result is obvious for this N', which has the same cokernel and same ideal of minors as N. $\qquad\square$

Continuing the proof of Proposition 6.22, we note that the module K is generated by degree 0 elements e_1, e_2 with relations $x_i e_1 + x_{i+1} e_2 = 0$ and $y_i e_1 + y_{i+1} e_2 = 0$. The elements e_i determine sections σ_i of $\tilde{K}$. Thus if $p \in Y$ is a point where some linear form in the second row of M is nonzero, then σ_1 generates $\tilde{K}$ locally at p. As the second row vanishes precisely on the fiber F, this shows that the zero locus of σ_1 is contained in F.

Conversely, suppose $p \in F$ so that the second row of M vanishes at p. Since the linear forms in M span the space of all linear forms on $\mathbb{P}^r$, one of the linear forms in the first row of M is nonzero at p. Locally at p this means $m_1 \sigma_1 + m_2 \sigma_2 = 0$ in

$\tilde{K}_p$ where m_1 is a unit in $\mathcal{O}_{Y,p}$, the local ring of Y at p, and m_2 is in the maximal ideal $\mathfrak{m}_{Y,p} \subset \mathcal{O}_{Y,p}$. Dividing by m_1 we see that $\sigma_1 \in \mathfrak{m}_{Y,p}\tilde{K}_p$. Since $\mathfrak{m}_{Y,p}$ is the set of functions vanishing at p, we see that σ_1 vanishes at p when considered as a section of a line bundle. Since this holds at all $p \in F$ we obtain $\tilde{K} = \mathcal{O}_Y(F)$.   □

To apply Proposition 6.21, we wish to find a free resolution (as S-module) of the ideal $I_{X/Y} \subset S/I_Y$, that is, of the module of twisted global sections of the sheaf $\mathcal{O}_Y((r-3)F)(-2)$. This sheaf is the sheafification of the module $K^{\otimes(r-3)}(-2)$, but one can show that for $r \geq 5$ this module has depth 0, so it differs from the module of twisted global sections. A better module — in this case the right one — is given by the symmetric power.

Proposition 6.24. *Let L be an S-module. If the sheaf $\mathcal{L} = \tilde{L}$ on $\mathbb{P}^r$ is locally generated by at most one element, then the sheafification $\mathcal{L}^{\otimes k}$ of $L^{\otimes k}$ is also the sheafification of $\mathrm{Sym}_k L$. In particular, this is the case when $\mathcal{L}$ is a line bundle on some subvariety $Y \subset \mathbb{P}^r$.*

Proof. Since the formation of tensor powers and symmetric powers commutes with localization, and with taking degree 0 parts, it suffices to do the case where L is a module over a ring R such that L is generated by at most one element. In this case, $L \cong R/I$ for some ideal I. If r_i are elements of R/I then

$$r_1 \otimes r_2 = r_1 r_2 (\bar{1} \otimes \bar{1}) = r_2 \otimes r_1 \in R/I \otimes R/I.$$

Since $\mathrm{Sym}_2 L$ is obtained from $L \otimes L$ by factoring out the submodule generated by elements of the form $r_1 \otimes r_2 - r_2 \otimes r_1$, we see that $L \otimes L = \mathrm{Sym}_2 L$. The same argument works for products with k factors.   □

We return to the module $K = \mathrm{coker}\, M$, and study $\mathrm{Sym}_{r-3} K$.

Proposition 6.25. *With notation as above, $\bigoplus_d \mathrm{H}^0(\mathcal{L}^{\otimes(r-3)}(d)) = \mathrm{Sym}_{r-3} K$ as S-modules. The free resolution of these modules is, up to a shift of degree, given by the dual of the Eagon–Northcott complex of M.*

Proof. We use the exact sequence of Corollary A1.12,

$$0 \to \mathrm{H}^0_\mathfrak{m}(\mathrm{Sym}_{r-3} K) \to \mathrm{Sym}_{r-3} K \to \bigoplus_d \mathrm{H}^0(\mathcal{L}(d)) \to \mathrm{H}^1_\mathfrak{m}(\mathrm{Sym}_{r-3} K) \to 0.$$

Thus we want to show that $\mathrm{H}^0_\mathfrak{m}(\mathrm{Sym}_{r-3} K) = \mathrm{H}^1_\mathfrak{m}(\mathrm{Sym}_{r-3} K) = 0$. By Proposition A1.16 it suffices to prove that the depth of K is at least 2. Equivalently, by the Auslander–Buchsbaum Formula A2.15 it suffices to show that the projective dimension of $\mathrm{Sym}_{r-3} K$ is at most $r-1$.

From the presentation $S^{r-1}(-1) \xrightarrow{\varphi} S^2 \to K \to 0$, we can derive a presentation

$$S^{r-1} \otimes \mathrm{Sym}_{r-4} S^2(-1) \xrightarrow{\varphi \otimes 1} \mathrm{Sym}_{r-3} S^2 \longrightarrow \mathrm{Sym}_{r-3} K \to 0;$$

see [Eisenbud 1995, Proposition A2.2d]. This map is, up to some identifications and a twist, the dual of the last map in the Eagon–Northcott complex associated to M_μ, namely

$$0 \to (\mathrm{Sym}_{r-3} S^2)^* \otimes \textstyle\bigwedge^{r-1} S^{r-1}(-r+1) \to (\mathrm{Sym}_{r-4} S^2)^* \otimes \textstyle\bigwedge^{r-2} S^{r-1}(-r+2).$$

To see this we use the isomorphisms $\bigwedge^i S^{r-1} \simeq (\bigwedge^{r-1-i} S^{r-1})^*$ (which depend on an "orientation", that is, a choice of basis element for $\bigwedge^{r-1} S^{r-1}$). Since the Eagon–Northcott complex is a free resolution of the Cohen–Macaulay S-module S/I, its dual is again a free resolution, so we see that the module $\mathrm{Sym}_{r-3} K$ is also of projective dimension $r-1$. $\qquad\square$

By Proposition 6.15, there is an S-free resolution of the homogeneous coordinate ring S/I of the elliptic normal curve X obtained as a mapping cone of the Eagon–Northcott complex of the matrix M, which is a resolution of J, and the resolution of the module I/J. The proof of Proposition 6.25 shows that the dual of the Eagon–Northcott complex, appropriately shifted, is a resolution of $\mathrm{Sym}_{r-3} K$, while $I/J \cong \mathrm{Sym}_{r-3} K(-2)$. Thus the free resolution of I/J is isomorphic to the dual of the Eagon–Northcott complex with a different shift in degrees. Using an orientation as above, it may be written as

$$0 \to (\textstyle\bigwedge^2 S^2)^*(-r-1) \longrightarrow (\textstyle\bigwedge^2 S^{r-1})^*(-r+1) \longrightarrow \cdots$$
$$\cdots \longrightarrow S^{r-1} \otimes \mathrm{Sym}_{r-4} S^2(-3) \xrightarrow{\varphi \otimes 1} \mathrm{Sym}_{r-3} S^2(-2).$$

The resolution constructed this way is minimal:

Theorem 6.26. *The minimal free resolution of an elliptic normal curve in $\mathbb{P}^r$ has the form*

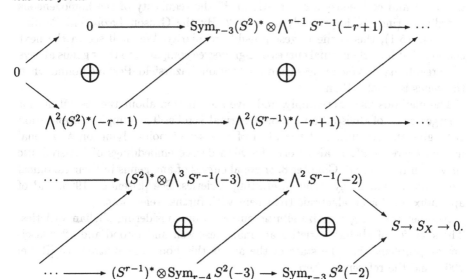

It has Betti diagram of the form

	0	1	2	$\cdots$	$r-2$	$r-1$
0	1	0	0	$\cdots$	0	0
1	0	b_1	b_2	$\cdots$	b_{r-2}	0
2	0	0	0	$\cdots$	0	1

with

$$b_i = i\binom{r-1}{i+1} + (r-i-1)\binom{r-1}{i-1}.$$

In particular, $\operatorname{reg} X = 3$.

The terms of the resolution are symmetric about the middle. A closer analysis shows that the i-th map in the resolution can be taken to be the dual of the $(r-1-i)$-th map, and if $r \cong 0 \pmod 4$ then the middle map can be chosen to be skew symmetric, while if $r \cong 2 \pmod 4$ then the middle map can be chosen to be symmetric. See [Buchsbaum and Eisenbud 1977] for the beginning of this theory.

Proof. We have already shown that the given complex is a resolution. Each map in the complex goes from a free module generated in one degree to a free module generated in a lower degree. Thus the differentials are represented by matrices of elements of strictly positive degree, and the complex is minimal. Given this, the value for the regularity follows by inspection. $\square$

The regularity statement says that for an elliptic normal curve X (degree $d = r+1$ and codimension $c = r-1$) in $\mathbb{P}^r$ the regularity of the homogeneous coordinate ring S_X is precisely $d - c = 2$. By the Gruson–Lazarsfeld–Peskine Theorem (5.1), this is the largest possible regularity. We shall see in the next chapter that linearly normal curves of high degree compared to their genus always have regularity 3, which is less than the Gruson–Lazarsfeld–Peskine bound when the genus is greater than 1.

The methods used here apply, and give information about the resolution, for a larger class of divisors on rational normal scrolls. The simplest application is to give the resolution of the ideal of any set of points lying on a rational normal curve in $\mathbb{P}^r$. It also works for high degree embeddings of hyperelliptic curves (in the sense of Chapter 8, trigonal curves of any genus in their canonical embeddings, and many other interesting varieties. See [Eisenbud 1995, end of appendix A2] for an algebraic treatment with further references.

Another way to generalize elliptic curves is by considering abelian varieties. The syzygies of abelian varieties are much less well understood and offer fascinating problems. For the state of the art as this book was written see [Rubei 2001] and the references there.

6E   Exercises

The catalecticant matrix. (The results of Exercises 1 and 2 were proved by a different method, requiring characteristic 0, in [Gruson and Peskine 1982], following the observation by T. G. Room [1938] that these relations held set-theoretically. The simple proof in full generality sketched here was discovered by Conca [1998].)

1. Prove that $I_e(M_{r,d}) = I_e(M_{r,e-1})$ for all d with $e \leq d+1$ and $e \leq r-d+1$ and thus the ideal $I_e(M_{r,d})$ is prime of codimension $r-2e+1$, with free resolution given by the Eagon–Northcott complex associated to $M_{r,e-1}$. In particular, the ideal of the rational normal curve may be written as $I_2(M_{r,e})$ for any $e \leq r-d$. You might follow these steps:

   (a) Using the fact that the transpose of $M_{r,d}$ is $M_{r,r-d}$, reduce the problem to proving $I_e(M_{r,d}) \subset I_e(M_{r,d+1})$ for $e-1 \leq d < d+1 \leq r-e+1$.

   (b) If $a = (a_1, \ldots, a_s)$ with $0 \leq a_1, \ldots, a_s$ and $b = (b_1, \ldots, b_s)$ with $0 \leq b_1, \ldots, b_s$ with $a_i + b_j \leq r$ for every i, j, then we write $[a, b]$ for the determinant of the submatrix involving rows $a_1, \ldots, a_s$ and columns $b_1, \ldots, b_s$ of the triangular array

$$
\begin{array}{ccccc}
x_0 & x_1 & \cdots & x_{r-1} & x_r \\
x_1 & x_2 & \cdots & x_r & \\
\vdots & \vdots & & & \\
x_{r-1} & x_r & & & \\
x_r & & & &
\end{array}
$$

   Let e be the vector of length s equal to $(1, \ldots, 1)$. Prove that $[a+e, b] = [a, b+e]$ whenever this makes sense.

   (c) Generalize the previous identity as follows: for $I \subset \{1, \ldots, s\}$ write $\#I$ for the cardinality of I, and write $e(I)$ for the *characteristic vector* of I, which has a 1 in the i-th place if and only if $i \in I$. Show that for each k between 1 and s we have

$$
\sum_{\#I=k} [a+e(I), b] = \sum_{\#J=k} [a, b+e(J)].
$$

   (Hint: Expand each minor $[a + e(I), b]$ on the left-hand side along the collection of rows indexed by I, as

$$
[a+e(I), b] = \sum_{\#J=k} \pm 1 [a_I + e(I)_I, b_J] [a_{I^c} + e(I^c)_I, b_{J^c}],
$$

   where $|I| = \sum_{i \in I} i$, the superscript ^{c} denotes complements and a_I denotes the subvector involving only the indices in I. Expand the right-hand side similarly using along the set of columns from J, and check that the two expressions are the same.)

2. Let M be any matrix of linear forms in S. We can think of M as defining a linear space of matrices parametrized by $\mathbb{K}^{r+1}$ by associating to each point p in $\mathbb{K}^{r+1}$ the scalar matrix $M(p)$ whose entries are obtained by evaluating the entries of M at p. A property of a matrix that does not change when the matrix is multiplied by a scalar then corresponds to a subset of $\mathbb{P}^r$, namely the set of points p such that $M(p)$ has the given property, and these are often algebraic sets. For example the locus of points p where $M(p)$ has rank at most k is the algebraic set defined by the $(k+1) \times (k+1)$ minors of M.

(a) From the fact that the sum of k rank 1 matrices has rank at most k, show that the locus where $M(p)$ has rank $\leq k$ contains the k-secant locus of the locus where $M(p)$ has rank at most 1. (The k-*secant locus* of a set $X \subset \mathbb{P}^r$ is the closure of the union of all linear spans of k-point sets in X.)

(b) If $M = M_{r,d}$ is the catalecticant matrix, show that the rank k locus of M is actually equal to the k-secant locus of the rational normal curve $X \subset \mathbb{P}^r$ of degree r as follows: First show that two generic k-secant planes with $k < r/2$ cannot meet (if they did they would span a $2k$-secant $(2k-2)$-plane, whereas any set of d points on X spans a $d-1$-plane as long as $d \leq r$.) Use this to compute the dimension of the k-secant locus. Use Exercise 6.1 above, together with Theorem 6.4, to show that the ideal of $(e+1) \times (e+1)$ minors of $M_{r,d}$ is the defining ideal of the e-secant locus of X.

3. We can identify $\mathbb{P}^r$ with the set of polynomials of degree r in 2 variables, up to scalar. Show (in characteristic 0) that the points of the rational normal curve may be identified with the set of r-th powers of linear forms, and a sufficiently general point of the k-secant locus may thus be identified with the set of polynomials that can be written as a sum of just k pure r-th powers. The general problem of writing a form as a sum of powers is called Waring's problem. See, for example, [Geramita 1996], and [Ranestad and Schreyer 2000] for more information.

4. Use Theorem 6.4 to reprove Proposition 6.1 by comparing the codimensions of the (necessarily prime) ideal generated by the minors and the prime ideal defining the curve.

5. Let $X = \{p_1, \ldots, p_{r+3}\} \subset \mathbb{P}^r$ be a set of $r+3$ points in linearly general position. Show that there is a unique rational normal curve in $\mathbb{P}^r$ containing X, perhaps as follows:

(a) *Existence.* We will use Corollary 6.9. We look for a 1-generic matrix of linear forms

$$M = \begin{pmatrix} a_0 & \cdots & a_{r-1} \\ b_0 & \cdots & b_{r-1} \end{pmatrix}$$

whose minors all vanish on X. We choose a linear form a_i that vanishes on $p_1, \ldots, \hat{p}_i, \ldots, p_n, p_{n+1}$, and also a linear form b_i that vanishes on

$p_1, \ldots, \hat{p}_i, \ldots, p_n, p_{n+3}$. These forms are unique up to scalars, so we may normalize them to make all the rational functions a_i/b_i take the value 1 at p_{n+2}. Show that with these choices the matrix M is 1-generic and that its minors vanish at all the points of X.

For example, let X be the set of $r+3$ points p_i with homogeneous coordinates given by the rows of the matrix

$$\begin{pmatrix} 1 & 0 & \cdots & 0 \\ 0 & 1 & \cdots & 0 \\ & & \ddots & \\ 0 & 0 & \cdots & 1 \\ 1 & 1 & \cdots & 1 \\ t_0 & t_1 & \cdots & t_r \end{pmatrix}$$

Show that these points are in linearly general position if and only if the $t_i \in \mathbb{K}$ are all nonzero and are pairwise distinct, and that any set of $r+3$ points in linearly general position can we written this way in suitable coordinates. Show that the 2×2 minors of the matrix

$$M = \begin{pmatrix} x_0 & \cdots & x_{r-1} \\ \dfrac{t_n x_0 - t_0 x_n}{t_n - t_0} & \cdots & \dfrac{t_n x_{n-1} - t_{n-1} x_n}{t_n - t_{n-1}} \end{pmatrix}$$

generate the ideal of a rational normal curve containing these points. See [Griffiths and Harris 1978, p. 530] for a more classical argument, and [Harris 1995] for further information.

(b) *Uniqueness.* Suppose that C_1, C_2 are distinct rational normal curves containing X. Show by induction on r that the projections of these curves from p_{r+3} into $\mathbb{P}^{r-1}$ are equal. In general, suppose that C_1, C_2 are two rational normal curves through p_{r+3} that project to the same curve in $\mathbb{P}^{r-1}$, so that C_1, C_2 both lie on the cone F over a rational normal curve in $\mathbb{P}^{r-1}$.

Let F' be the surface obtained by blowing up this cone at p_{r+3}, let $E \subset F'$ be the exceptional divisor, a curve of self-intersection $-r+1$, and let $R' \subset F'$ be the preimage of a ruling of the cone F. (See figure on next page.) See for example [Hartshorne 1977, Section V.2] for information about such surfaces, and [Eisenbud and Harris 2000, Section VI.2] for information about blowups in general.

Show that F' is a minimal rational surface, ruled by lines linearly equivalent to R', and $E.E = -r+1$. Let $C_1', C_2' \subset F'$ be the strict transforms of C_1, C_2. Compute the intersection numbers $C_i'.E$ and $C_i'.R$, and conclude that $C_i' \sim E + rR$ so $C_1'.C_2' = r+1$. Deduce that the number of distinct points in $C_1 \cap C_2$ is at most $r+2$, so that $C_1 \cap C_2$ cannot contain X, a contradiction.

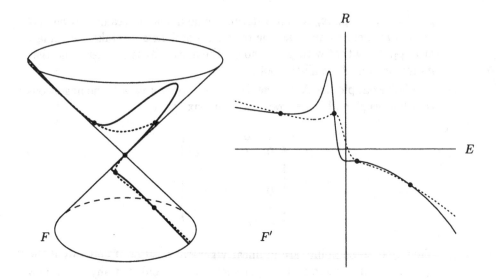

FIGURE 6.1. (See Exercise 5b.) Two rational normal curves in $\mathbb{P}^r$ meet in at most $r+2$ points. The picture on the left shows two twisted cubics in $\mathbb{P}^3$, lying on a quadric cone F, while the picture on the right shows the strict transforms of these curves, lying on the surface F' obtained by blowing up the cone at the vertex. The horizontal line E represents the exceptional divisor, while the vertical line R is the strict transform of a line on the cone, which is a ruling of the ruled surface F'.

6. Let M be a 1-generic $2 \times r$ matrix of linear forms on $\mathbb{P}^r$, and let $X \cong \mathbb{P}^1$ be the rational normal curve defined by the 2×2 minors of M. Suppose that M' is any other 1-generic $2 \times r$ matrix of linear forms on $\mathbb{P}^r$ whose minors are contained in the ideal of X. Show that the sheaf associated to the S-module coker M is isomorphic to the line bundle $\mathscr{O}_X(p)$ for any point $p \in X$, and that M is a minimal free presentation of this module. Deduce from the uniqueness of minimal free resolutions that M and M' differ by invertible row and column transformations and a change of variable.

7. (For those who know about Gröbner bases.) Let $<$ be the reverse lexicographic order on the monomials of S with $x_0 < \cdots < x_r$. For $1 \leq e \leq d+1 \leq r$ show that the initial ideal, with respect to the order $<$, of the ideal $I_e(M_{r,d})$, is the ideal $(x_{e-1}, \ldots, x_{r-e})^e$. This gives another proof of the formula for the codimension of $I_e(M_{r,d})$ above, and also for the vector space dimension of the degree e component of $I_e(M_{r,d})$. Use this and Theorem 5.1 to give another proof of the fact that $I_2(M_{r,1})$ is the ideal of the rational normal curve.

8. With notation as in Proposition 6.22, show that the two sections $\mathscr{O}_Y(F)$ corresponding to generators of coker M define a morphism π of Y to $\mathbb{P}^1$. The fibers are the linear spaces defined by rows of M, thus projective spaces, and Y is a projective space bundle; in fact, $Y = \mathrm{Proj}(\pi_*(\mathscr{O}_Y(1)))$ (we could show this is $\mathscr{O}_{\mathbb{P}^1}(m) \oplus \mathscr{O}_{\mathbb{P}^1}(n)$.) From [Hartshorne 1977, V.2.3] it follows that

$\mathrm{Pic}(Y) = \mathbb{Z} \oplus \pi^* \mathrm{Pic}(\mathbb{P}^1) = \mathbb{Z} \oplus \mathbb{Z}$. Since the determinant of the intersection form on the sublattice spanned by F and H is 1, these two elements must be a basis.

9. (For those who know about schemes.) Generalize Theorem 6.6 as follows: Let X be a smooth projective variety over an algebraically closed field and let $L = (\mathscr{L}, V, \alpha)$ be a linear series on X. Show that L is very ample if, for each finite subscheme Y of length 2 in X, the space of sections in $\alpha(V)$ vanishing on Y has codimension 2 in $\alpha(V)$.

10. Here is the easiest case of the (vague) principle that embeddings of varieties by sufficiently positive bundles are often defined by ideals of 2×2 minors: Suppose that the homogeneous ideal I of X in $\mathbb{P}^r$ is generated by equations of degrees $\leq d$, and let Y_e be the image of X in $\mathbb{P}(\mathrm{H}^0(\mathscr{O}_X(e)))$ under the complete series $|\mathscr{O}_X(e)|$. Choose $e \geq d$, and let $e_1 \geq 1$ and $e_2 \geq 1$ be integers with $e_1 + e_2 = e$. Show that the ideal of Y_e is generated by the 2×2 minors of $M(\mathscr{O}_X(e_1), \mathscr{O}_X(e_2))$. (Hint: Start with the case $X = \mathbb{P}^r$.)

11. Theorem 6.8 shows that any nondegenerate, reduced irreducible curve of degree r in $\mathbb{P}^r$ is equivalent by a linear automorphism to the rational normal curve (we usually say: *is* a rational normal curve.) One can be almost as explicit about curves of degree $r + 1$. Use the Riemann–Roch theorem and Clifford's theorem [Hartshorne 1977, Theorem IV.5.4] to prove:

Proposition 6.27. *If X is a nondegenerate reduced irreducible curve of degree $r + 1$ in $\mathbb{P}^r$ over an algebraically closed field, then X is either*

- *a smooth elliptic normal curve,*
- *a rational curve with one double point (also of arithmetic genus 1), or*
- *a smooth rational curve.*

Moreover, up to linear transformations of $\mathbb{P}^r$ each singular curve (type 2) is equivalent to the image of one of the two maps

$$\mathbb{P}^1 \ni (s, t) \mapsto (s^{r+1}, s^{r-1}t, s^{r-2}t^2, \ldots, t^{r+1}) \in \mathbb{P}^r, \quad or$$

$$\mathbb{P}^1 \ni (s, t) \mapsto (s^{r+1} + t^{r+1}, st \cdot s^{r-2}t, st \cdot s^{r-3}t^2, \ldots, st \cdot t^{r-1}) \in \mathbb{P}^r.$$

Unlike for the singular case there are moduli for the embeddings of a smooth rational curve of degree $r + 1$ (third case in the proposition above), and several different Betti diagrams can appear. However, in all of these cases, the curve lies on a rational normal scroll and its free resolution can be analyzed in the manner of the elliptic normal curves (see [Eisenbud and Harris 1987] for further information.)

7
Linear Complexes and the Linear Syzygy Theorem

Minimal free resolutions are built out of linear complexes, and in this chapter we study a canonical linear subcomplex (the *linear strand*) of a free resolution. We start with an elementary version of the Bernstein–Gelfand–Gelfand correspondence (BGG) and use it to prove Green's Linear Syzygy Theorem. In brief, BGG allows us to translate statements about linear complexes over a polynomial ring S into statements about modules over an exterior algebra E. The Linear Syzygy Theorem bounds the length of the linear part of the minimal free resolution of a graded S-module M. Its translation is that a certain E-module is annihilated by a particular power of the maximal ideal. This is proved with a variant of Fitting's Lemma, which gives a general way of producing elements that annihilate a module.

The proof presented here is a simplification of that in Green's original paper [Green 1999]. Our presentation is influenced by the ideas of [Eisenbud et al. 2003a] and [Eisenbud and Weyman 2003]. In Chapter 8 we will apply the Linear Syzygy Theorem to the ideals of curves in $\mathbb{P}^r$.

The last section of the chapter surveys some other aspects of BGG, including the connection between Tate resolutions and the cohomology of sheaves.

Throughout this chapter, we denote the polynomial ring in $r+1$ variables by $S = \mathbb{K}[x_0, \ldots, x_r]$. We write $W = S_1$ for the space of linear forms, and V or $\widehat{W}$ for its dual $\mathrm{Hom}_{\mathbb{K}}(W, \mathbb{K})$. (In this chapter we will use $\widehat{}$ to denote the vector space dual $\mathrm{Hom}_{\mathbb{K}}(-, \mathbb{K})$, reserving $*$ for the dual of a module over a larger ring.)

We let $E = \bigwedge V$ be the exterior algebra of V.

7A   Linear Syzygies

The Linear Strand of a Complex

One natural way to study the minimal resolution of a graded S-module is as an iterated extension of a sequence of linear complexes. In general, suppose that

$$\mathbf{G}: \cdots \longrightarrow G_i \xrightarrow{d_i} G_{i-1} \longrightarrow \cdots$$

is a complex of graded free S-modules, whose i-th term G_i is generated in degrees $\geq i$, and suppose, moreover that $\mathbf{G}$ is *minimal* in the sense that $d_i(G_i) \subset WG_{i-1}$ (for example $\mathbf{G}$ might be a minimal free resolution, or a free sub- or quotient-complex of a minimal free resolution of a module generated in degrees ≥ 0.) Let $F_i \subset G_i$ be the submodule generated by all elements of degree precisely i. Since i is the minimal degree of generators of G_i, the submodule F_i is free. Since $d_i(F_i)$ is generated in degree i and is contained in WG_{i-1}, it must in fact be contained in WF_{i-1}. In particular the F_i form a free subcomplex $\mathbf{F} \subset \mathbf{G}$, called the *linear strand* of $\mathbf{G}$. The Betti diagram of $\mathbf{F}$ is simply the 0-th row of the Betti diagram of $\mathbf{G}$. The linear strand sometimes isolates interesting information about $\mathbf{G}$.

For an arbitrary free complex $\mathbf{G}$, we define the linear strand to the be the linear strand of the complex $\mathbf{G}(i)$ where $i = \sup\{\operatorname{reg} G_j - j\}$, the least twist so that $\mathbf{G}(i)$ satisfies the condition that the j-th free module is generated in degrees $\geq j$. (The case where $\mathbf{G}$ is infinite and the supremum is infinity will not concern us.)

Since $\mathbf{F}$ is a subcomplex of $\mathbf{G}$ we can factor it out and start again with the quotient complex $\mathbf{G}/\mathbf{F}$. The linear strand of $\mathbf{G}/\mathbf{F}(1)$, shifted by -1, is called the *second linear strand* of $\mathbf{G}$. Continuing in this way we produce a series of linear strands, and we see that $\mathbf{G}$ is built up from them as an iterated extension. The Betti diagram of the i-th linear strand is the i-th row of the Betti diagram of $\mathbf{G}$.

For example Theorem 3.16 shows that there is a set X of 9 points in $\mathbb{P}^2$ whose ideal $I = I_X$ has minimal free resolution $\mathbf{G}$ with Betti diagram

	0	1
3	2	1
4	1	–
5	–	1

From this Betti diagram we see that the ideal of X is generated by two cubics and a quartic and that its syzygy matrix has the form

$$d = \begin{pmatrix} q & 0 \\ f_1 & \ell_1 \\ f_2 & \ell_2 \end{pmatrix},$$

where q has degree 2, the ℓ_i are linear forms and the f_i have degree 3.

Let p be the intersection of the lines L_1 and L_2 defined by ℓ_1 and ℓ_2. We claim that p is a point and that the nine points consist of p together with the 8 points

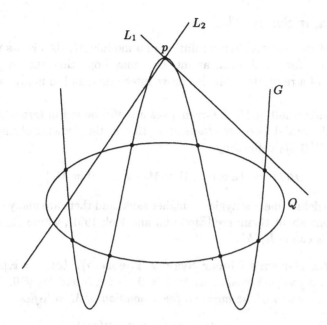

Nine points whose ideal is generated by two cubics and a quartic,
the equations of the curves $Q \cup L1$, $Q \cup L2$, and G.

of intersection of the conic Q and the quartic G defined by q and by

$$g = \det \begin{pmatrix} f_1 & \ell_1 \\ f_2 & \ell_2 \end{pmatrix}$$

respectively (counted with appropriate multiplicities).

Indeed, the Hilbert–Burch Theorem 3.2 shows that I is minimally generated by the 2×2 minors $\ell_1 q$, $\ell_2 q$, and g of the matrix d, so ℓ_1 and ℓ_2 must be linearly independent. At the point p both ℓ_1 and ℓ_2 vanish, so all the forms in the ideal of X vanish, whence $p \in X$. Away from p, the equations $\ell_1 q = 0, \ell_2 q = 0$ imply $q = 0$, so the other points of X are in $Q \cap G$ as required.

On the other hand, the Betti diagram of the linear strand of the resolution $\mathbf{G}$ of I is

	0	1
3	2	1

and the matrix representing its differential is

$$d|_{\mathbf{F}} = \begin{pmatrix} \ell_1 \\ \ell_2 \end{pmatrix}.$$

Thus the linear strand of the resolution captures a subtle fact: a set of 9 distinct points in $\mathbb{P}^2$ with resolution as above contains a distinguished point. In this case the second and third linear strands of $\mathbf{G}$ have trivial differential; the remaining information about the maps of $\mathbf{G}$ is in the extension data.

Green's Linear Syzygy Theorem

The length of the minimal free resolution of a module M, that is, its projective dimension, is a fundamental invariant. One may hope that the length of the linear strand of a resolution will also prove interesting, and in many examples it does.

The following result of Mark Green gives a useful bound in terms of a simple property of the *rank*-1 *linear relations of* M, that is, the elements of the algebraic set $R(M) \subset W \otimes M_0$ defined by

$$R(M) := \{w \otimes m \in W \otimes M_0 \mid wm = 0 \text{ in } M_1\}.$$

One can also define linear syzygies of higher rank, and there are many interesting open questions about them; see [Eisenbud and Koh 1991], where the set $R(M)$ just defined is called $R_1(M)$.

Theorem 7.1 (Green's Linear Syzygy Theorem). *Let $S = \mathbb{K}[x_0, \ldots, x_r]$ and let M be a graded S-module with $M_i = 0$ for $i < 0$ and $M_0 \neq 0$. The length n of the linear strand of the minimal free resolution of M satisfies*

$$n \leq \max\left(\dim M_0 - 1, \ \dim R(M)\right).$$

See Exercise 7.3 for a way to see the maximum as the dimension of a single natural object.

We postpone the proof, which will occupy most of this chapter, to study some special cases. First, we give examples illustrating that either term in the max of the theorem can be achieved.

Example 7.2. Consider first the Koszul complex

$$\mathbf{K}(x_1, \ldots, x_n): \ 0 \to S(-n) \to \cdots \to S(-1)^n \to S \to 0,$$

which is the resolution of $S/(x_1, \ldots, x_n)$. It is linear, and has length n. We have $\dim M_0 = \dim \mathbb{K} = 1$, but the variety R is all of $W \otimes M_0 = W \otimes \mathbb{K}$, which has dimension precisely n.

Example 7.3. For the other possibility, let $r = n+2$ and consider the $2 \times (n+2)$ matrix

$$N = \begin{pmatrix} x_0 & \cdots & x_{r-1} \\ x_1 & \cdots & x_r \end{pmatrix}$$

whose minors define the rational normal curve in $\mathbb{P}^r$, or more generally any $2 \times (n+2)$ 1-generic matrix of linear forms

$$N = \begin{pmatrix} \ell_{1,1} & \cdots & \ell_{1,n+2} \\ \ell_{2,1} & \cdots & \ell_{2,n+2} \end{pmatrix}.$$

It follows from Theorem 6.4 that the ideal $I = I_2(N)$ has codimension $n+1$, the largest possible value. In this case we know from Theorem A2.60 that the minimal free resolution of S/I is the Eagon–Northcott complex of N,

$$\mathbf{EN}(N):\ 0 \longrightarrow \widehat{\mathrm{Sym}_n}\, S^2 \otimes \bigwedge^{n+2} S^{n+2}(-n-2) \longrightarrow \cdots$$
$$\longrightarrow \widehat{\mathrm{Sym}_0}\, S^2 \otimes \bigwedge^2 S^{n+2}(-2) \xrightarrow{\ \bigwedge^2 N\ } \bigwedge^2 S^2 \longrightarrow 0,$$

with Betti diagram

	0	1	$\cdots$	$n+1$
0	1	$-$	$\cdots$	$-$
1	$-$	$\binom{n+2}{2}$	$\cdots$	$n+1$

The dual of $\mathbf{EN}(N)$ is a free resolution of a module ω; see Theorem A2.60. (This module is, up to a shift of degrees, the canonical module of S/I, though we shall not need this here; see [Bruns and Herzog 1998, Chapter 3].) Let $\mathbf{G}$ be the dual of $\mathbf{EN}(N)$, so that $\mathbf{G}$ has Betti diagram

	0	$\cdots$	n	$n+1$
$-n-2$	$n+1$	$\cdots$	$\binom{n+2}{2}$	$-$
$-n-1$	$-$	$\cdots$	$-$	1

We see that the linear part of $\mathbf{G}$ has length n. The module ω requires $n+1$ generators, so equality holds with the first term of the max in Theorem 7.1. In this case we claim that $R(\omega) = 0$ (see also Exercise 7.4).

To see this, note first that $\omega = \mathrm{Ext}_S^{n+1}(S/I, S)$ is annihilated by I. If a nonzero element $m \in \omega$ were annihilated by a nonzero linear form x then it would be annihilated by $I+(x)$. By Theorem 6.4 I is a prime ideal of codimension $n+1$, so $I+(x)$ has codimension greater than $n+1$. It follows that some associated prime (= maximal annihilator of an element) of ω would have codimension greater than $n+1$, and thus ω would have projective dimension greater than $n+1$ by Theorem A2.16. Since we have exhibited a resolution length $n+1$, this is a contradiction.

The phenomenon we saw in the second example is the one we will apply in the next chapter. Here is a way of codifying it.

Corollary 7.4. *Let $X \subset \mathbb{P}^r$ be a reduced, irreducible variety that is not contained in a hyperplane, let $\mathscr{E}$ be a vector bundle on X, and let $M \subset \bigoplus_{i \geq 0} \mathrm{H}^0\mathscr{E}(i)$ be a submodule of the S-module of nonnegatively twisted global sections. If $M_0 \neq 0$, the linear strand of the minimal free resolution of M, as an S-module, has length at most $\dim M - 1$.*

Proof. Let $R(M) \subset M_0 \otimes W$ be the variety defined in 7.1. If $w \in W$ and $m \in M_0 = \mathrm{H}^0\mathscr{E}$ with $wm = 0$ then X would be the union of the subvariety of X defined by the vanishing of w and the subvariety of X defined by the vanishing of m. Since X is irreducible and not contained in any hyperplane, this can only happen if $w = 0$ or $m = 0$. Thus $R(M) = 0$, and Theorem 7.1 gives the result. $\square$

The history is this: Corollary 7.4 was proved in [Green 1984a]. In trying to understand and extend it algebraically, Eisenbud, Koh and Stillman were lead to conjecture the truth of the Theorem 7.1, as well as some stronger results in this direction [Eisenbud and Koh 1991]. Green [1999] proved the given form; as of this writing the stronger statements are still open.

7B   The Bernstein–Gelfand–Gelfand Correspondence

Graded Modules and Linear Free Complexes

Recall that $V = \widehat{W}$ denotes the vector space dual to W, and $E = \bigwedge V$ denotes the exterior algebra. If $e_0, \ldots, e_r$ is a dual basis to $x_0, \ldots, x_r$ then $e_i^2 = 0$, $e_i e_j = -e_j e_i$, and the algebra E has a vector space basis consisting of the square-free monomials in the e_i. Since we think of elements of W as having degree 1, we will think of elements of V as having degree -1.

Although E is not commutative, it is *skew-commutative* (or *strictly commutative*): that is, homogeneous elements $e, f \in E$ satisfy $ef = (-1)^{\deg e \deg f} fe$, and E behaves like a commutative local ring in many respects. For example, any one-sided ideal is automatically a two-sided ideal. The algebra E has a unique maximal ideal, generated by the basis $e_0, \ldots, e_r$ of V; we will denote this ideal by (V). The analogue of Nakayama's Lemma is almost trivially satisfied (and even works for modules that are not finitely generated, since (V) is nilpotent). It follows for example that any graded E-module P has unique (up to isomorphism) minimal free graded resolution $\mathbf{F}$, and that $\mathrm{Tor}^E(P, \mathbb{K}) = \mathbf{F} \otimes_E \mathbb{K}$ as graded vector spaces. The same proofs work as in the commutative case.

Also, just as in the commutative case, any graded left E-module P can be naturally regarded as a graded right E-module, but we must be careful with the signs: if $p \in P$ and $e \in E$ are homogeneous elements then $pe = (-1)^{\deg p \deg e} ep$. We will work throughout with left E-modules.

An example where this change-of-sides is important comes from duality. If $P = \bigoplus P_i$ is a finitely generated left-E-module, then the vector space dual $\widehat{P} := \bigoplus \widehat{P_i}$, where $\widehat{P_i} := \mathrm{Hom}_{\mathbb{K}}(P_i, \mathbb{K})$, is naturally a right E-module, where the product $\phi \cdot e$ is the functional defined by $(\phi \cdot e)(p) = \phi(ep)$ for $\phi \in \widehat{P_i}$, $e \in E_{-j}$, and $p \in P_{i+j}$. As a graded left module, with $(\widehat{P})_{-i} = \widehat{P_i}$ in degree $-i$, we have

$$(e\phi)(p) = (-1)^{\deg e \deg \phi}(\phi e)(p) = (-1)^{\deg e \deg \phi}\phi(ep).$$

Let P be any graded E-module. We will make $S \otimes_{\mathbb{K}} P$ into a complex of graded free S-modules,

$$\mathbf{L}(P): \quad \cdots \longrightarrow S \otimes_{\mathbb{K}} P_i \xrightarrow{d_i} S \otimes_{\mathbb{K}} P_{i-1} \longrightarrow \cdots$$

$$1 \otimes p \longmapsto \sum x_i \otimes e_i p$$

where the term $S \otimes P_i \cong S(-i)^{\dim P_i}$ is in homological degree i, and is generated in degree i as well. The identity

$$d_{i-1} d_i p = \sum_j \sum_i x_j x_i \otimes e_j e_i p = \sum_{i \leq j} x_j x_i \otimes (e_j e_i + e_i e_j) p = 0$$

follows from the associative and commutative laws for the E-module structure of P. Thus $\mathbf{L}(P)$ is a *linear free complex*.

If we choose bases $\{p_s\}$ and $\{p'_t\}$ for P_i and P_{i-1} respectively we can represent the differential d_i as a matrix, and it will be a matrix of linear forms: writing $e_m p_s = \sum_t c_{m,s,t} p'_t$ the matrix of d_i has (t,s)-entry equal to the linear form $\sum_m c_{m,s,t} x_m$.

It is easy to see that $\mathbf{L}$ is actually a functor from the category of graded E-modules to the category of linear free complexes of S-modules. Even more is true.

Proposition 7.5. *The functor $\mathbf{L}$ is an equivalence from the category of graded E-modules to the category of linear free complexes of S-modules.*

Proof. We show how to define the inverse, leaving to the reader the routine verification that it is the inverse. For each $e \in V = \operatorname{Hom}(W, \mathbb{K})$, and any vector space P there is a unique linear map $e : W \otimes P \to P$ satisfying $e(x \otimes p) = e(x)p$. If now

$$\cdots \longrightarrow S \otimes_{\mathbb{K}} P_i \xrightarrow{d_i} S \otimes_{\mathbb{K}} P_{i-1} \longrightarrow \cdots$$

is a linear free complex of S-modules, then $d(P_i) \subset W \otimes P_{i-1}$ so we can define a multiplication $V \otimes_{\mathbb{K}} P_i \to P_{i-1}$ by $e \otimes p \mapsto e(d(p))$. Direct computation shows that the associative and anti-commutative laws for this multiplication follow from the identity $d_{i-1} d_i = 0$. (See Exercise 7.9 for a basis-free approach to this computation.) □

Example 7.6. Take $P = E$, the free module of rank 1. The complex $\mathbf{L}(E)$ has the form

$$\mathbf{L}(E): \ 0 \to S \otimes \mathbb{K} \to S \otimes V \to \cdots \to S \otimes {\textstyle\bigwedge}^r V \to S \otimes {\textstyle\bigwedge}^{r+1} V \to 0,$$

since $\bigwedge^{r+2} V = 0$. The differential takes $s \otimes f$ to $\sum x_i s \otimes e_i f$. This is one way to write the Koszul complex of $x_0, \ldots, x_r$, though we must shift the degrees to regard $\bigwedge^{r+1} V \cong S$ as being in homological degree 0 and as being generated in degree 0 if we wish to have a graded resolution of $\mathbb{K}$ (see [Eisenbud 1995, Section 17.4]). Usually the Koszul complex is written as the dual of this complex:

$$\mathbf{K}(x_0, \ldots, x_r) = \operatorname{Hom}_S(\mathbf{L}(E), S):$$

$$0 \to {\textstyle\bigwedge}^{r+1} W \otimes \mathbb{K} \to S \otimes {\textstyle\bigwedge}^r W \to \cdots \to S \otimes {\textstyle\bigwedge}^1 W \to S \otimes \mathbb{K} \to 0,$$

where we have exploited the identifications $\bigwedge^k W = \operatorname{Hom}_{\mathbb{K}}(\bigwedge^k V, \mathbb{K})$ coming from the identification $W = \operatorname{Hom}_{\mathbb{K}}(V, \mathbb{K})$. It is useful to note that

$$\operatorname{Hom}_S(\mathbf{L}(E), S) = \mathbf{L}(\operatorname{Hom}_{\mathbb{K}}(E, \mathbb{K})) = \mathbf{L}(\widehat{E})$$

(and more generally $\mathbf{L}(\widehat{P}) = \mathrm{Hom}_S(\mathbf{L}(P), S)$ for any graded E-module P, as the reader is asked to verify in Exercise 7.7. From Theorem 7.5 and the fact that the Koszul complex is isomorphic to its own dual, it now follows that $\widehat{E} \cong E$ as E-modules. For a more direct proof, see Exercise 7.6.

There are other ways of treating linear complexes and the linear strand besides BGG. One approach is given by [Eisenbud et al. 1981]. Another is the Koszul homology approach of Green — see, for example, [Green 1989]. The method we follow here is implicit in the original paper of Bernstein, Gelfand, and Gelfand [Bernstein et al. 1978] and was made explicit by Eisenbud, Fløystad, and Schreyer [Eisenbud et al. 2003a].

What It Means to Be the Linear Strand of a Resolution

We see from Proposition 7.5 that there must be a dictionary between properties of linear free complexes over S and properties of graded E-modules. When is $\mathbf{L}(P)$ a minimal free resolution? When is it a subcomplex of a minimal resolution? When is it the whole *linear strand* of a resolution? It turns out that these properties are most conveniently characterized in terms of the dual E-module $\widehat{P}$ introduced above. For simplicity we normalize and assume that $\mathbf{L}(P)$ has no terms of negative homological degree, or equivalently that $P_i = 0$ for $i < 0$. For the proof of Green's Theorem 7.1 we will use part 3 of the following dictionary.

Theorem 7.7 (Dictionary Theorem). *Let P be a finitely generated, graded E-module with no component of negative degree, and let*

$$\mathbf{F} = \mathbf{L}(P) : \cdots \xrightarrow{d_2} S \otimes_{\mathbb{K}} P_1 \xrightarrow{d_1} S \otimes_{\mathbb{K}} P_0 \longrightarrow 0$$

be the corresponding finite linear free complex of S-modules.

1. $\mathbf{F}$ *is a free resolution (of* $\mathrm{coker}\, d_1$*) if and only if* $\widehat{P}$ *is generated in degree 0 and has a linear free resolution.*
2. $\mathbf{F}$ *is a subcomplex of the minimal free resolution of* $\mathrm{coker}\, d_1$ *if and only if* $\widehat{P}$ *is generated in degree 0.*
3. $\mathbf{F}$ *is the linear strand of the free resolution of* $\mathrm{coker}\, d_1$ *if and only if* $\widehat{P}$ *is linearly presented (that is,* $\widehat{P}$ *is generated in degree 0 and has relations generated in degree -1.)*

In Example 7.6 above we saw that $\mathbf{L}(E)$ and $\mathbf{L}(\widehat{E})$ are both linear free resolutions. By part 1 of Theorem 7.7, this statement is equivalent to saying that both E and $\widehat{E}$ have linear free resolutions as E-modules. Since E is itself free, and $\widehat{E} \cong E$, this is indeed satisfied.

We will deduce Theorem 7.7 from a more technical result expressing the graded components of the homology of $\mathbf{L}(P)$ in terms of homological invariants of $\widehat{P}$.

Theorem 7.8. *Let P be a finitely generated graded module over the exterior algebra E. For any integers $i \geq 0$ and k the vector space $\mathrm{H}_k(\mathbf{L}(P))_{i+k}$ is dual to* $\mathrm{Tor}_i^E(\widehat{P}, \mathbb{K})_{-i-k}$.

We postpone the proof of Theorem 7.8 until the end of this section.

Proof of Theorem 7.7 from Theorem 7.8. Let P be a finitely generated graded E-module such that $P_i = 0$ for $i < 0$ as in Theorem 7.7, and set $M = \operatorname{coker} d_1 = H_0(\mathbf{L}(P))$.

The module $\widehat{P}$ is generated in degree 0 and has a linear free resolution if and only if $\operatorname{Tor}_i^E(\widehat{P}, \mathbb{K})_{-i-k} = 0$ for $k \neq 0$. By Theorem 7.8 this occurs if and only if $\mathbf{L}(P)$ has vanishing homology except at the 0-th step; that is, $\mathbf{L}(P)$ is a free resolution of M. This proves part 1.

For part 2, note that $\widehat{P}$ is generated as an E-module in degree 0 if and only if

$$\operatorname{Tor}_0^E(\widehat{P}, \mathbb{K})_{-k} = 0$$

for $k \neq 0$. By Theorem 7.8 this means that $H_k(\mathbf{L}(P))_k = 0$ for $k \neq 0$. Since $\mathbf{L}(P)_{k+1}$ is generated in degree $-k-1$, this vanishing is equivalent to the statement that, for every k, the map of P_k to the kernel of $W \otimes P_{k-1} \to S_2(W) \otimes P_{k-2}$ is a monomorphism.

Suppose that

$$\mathbf{L}(P)_{\leq k-1} : \ S \otimes P_{k-1} \to S \otimes P_{k-2} \to \cdots$$

is a subcomplex of the minimal free resolution $\mathbf{G}$ of M (this is certainly true for $k = 1$). In order for $\mathbf{L}(P)_{\leq k}$ to be a subcomplex of $\mathbf{G}$, it is necessary and sufficient that $1 \otimes P_k \subset S \otimes P_k$ maps monomorphically to the linear relations in $\ker S \otimes P_{k-1} \to S \otimes P_{k-2}$, and this is the same condition as above. This proves 2.

For part 3, notice that $\widehat{P}$ is linearly presented if, in addition to being generated in degree 0, it satisfies $\operatorname{Tor}_1^E(\widehat{P}, \mathbb{K})_{-1-k} = 0$ for $k \neq 0$. By Theorem 7.8 this additional condition is equivalent to the statement that $H_k(\mathbf{L}(P))_{1+k} = 0$ for all k, or in other words that the image of P_k generates the linear relations in $\ker S \otimes P_{k-1} \to S \otimes P_{k-2}$, making $\mathbf{L}(P)$ the linear part of the minimal resolution of M. □

To prove Theorem 7.8 we will compute $\operatorname{Tor}^E(\widehat{P}, \mathbb{K})$ using the *Cartan complex*, which we will show to be the minimal free resolution of $\mathbb{K}$ as an E-module. Define $\widehat{S}$ to be the S-module

$$\widehat{S} := \bigoplus \operatorname{Hom}_{\mathbb{K}}(S_i, \mathbb{K}) = \bigoplus_i \widehat{S}_i.$$

We regard $\widehat{S}_i$ as a graded vector space concentrated in degree $-i$. The Cartan resolution is an infinite complex of the form

$$\mathbf{C}: \ \cdots \ \xrightarrow{d_2} \ E \otimes_{\mathbb{K}} \widehat{S}_1 \ \xrightarrow{d_1} \ E \otimes_{\mathbb{K}} \widehat{S}_0,$$

where the free E-module $E \otimes_{\mathbb{K}} \widehat{S}_i$, which is generated in degree $-i$, has homological degree i.

To define the differential $d_i : E \otimes \widehat{S}_i \to E \otimes \widehat{S}_{i-1}$ we regard $\widehat{S}$ as a graded S-module, taking multiplication by $s \in S$ to be the dual of the multiplication on

S, and we choose dual bases $\{e_j\}$ and $\{w_j\}$ of V and W. If $p \in E$ and $f \in \widehat{S}_i$, we set

$$d_i(p \otimes f) = \sum_j p e_j \otimes w_j f \in E \otimes \widehat{S}_{i-1}. \qquad (*)$$

It is easy to check directly that $d_{i-1} d_i = 0$, so that $\mathbf{C}$ is a complex of free E-modules, and that d_i is independent of the choice of dual bases; as with the differential of the Koszul complex, this occurs because the differential is really right multiplication by the element $\sum_j e_j \otimes w_j \in E \otimes S$, and this well-defined element squares to zero.

Proposition 7.9. *If P is a finitely generated graded E-module then, for any integers i, k the vector space $H_i(P \otimes_E \mathbf{C})_{-i-k}$ is dual to $H_k(\mathbf{L}(\widehat{P}))_{i+k}$.*

Proof. The i-th term of $P \otimes_E \mathbf{C}$ is

$$P \otimes_E E \otimes_{\mathbb{K}} \widehat{S}_i = P \otimes_{\mathbb{K}} \widehat{S}_i,$$

and the differential $P \otimes_E d_i$ is expressed by the formula $(*)$ above (but now we take $p_i \in P$). We will continue to denote it d_i. Taking graded components we see that $H_i(P \otimes_E \mathbf{C})_{-i-k}$ is the homology of the sequence of vector spaces

$$P_{-k+1} \otimes \widehat{S}_{i+1} \xrightarrow{d_{i+1}} P_{-k} \otimes \widehat{S}_i \xrightarrow{d_i} P_{-k-1} \otimes \widehat{S}_{i-1}.$$

Its dual is the homology of the dual sequence

$$\widehat{P}_{k-1} \otimes S_{i+1} \xleftarrow{\hat{d}_{i+1}} \widehat{P}_k \otimes S_i \xleftarrow{\hat{d}_i} \widehat{P}_{k+1} \otimes S_{i-1}$$

which is the component of degree $i + k$ of the complex $\mathbf{L}(\widehat{P})$ at homological degree k. $\qquad \square$

Corollary 7.10. *The Cartan complex $\mathbf{C}$ is the minimal E-free resolution of the residue field $\mathbb{K} = E/(V)$.*

Proof. By the Proposition, it suffices to show that $H_0(\mathbf{L}(\widehat{E})) = \mathbb{K}$ in degree 0, while $H_k(\mathbf{L}(\widehat{E})) = 0$ for $k > 0$; that is, $\mathbf{L}(\widehat{E})$ is a free resolution of $\mathbb{K}$ as an S-module. But we have already seen that $\mathbf{L}(\widehat{E})$ is the Koszul complex, the minimal free resolution of $\mathbb{K}$, as required. $\qquad \square$

Proof of Theorem 7.8. By Corollary 7.10, $\mathrm{Tor}_i^E(\widehat{P}, \mathbb{K})_{-i-k} = H_i(\widehat{P} \otimes_E \mathbf{C})_{-i-k}$. By Proposition 7.9, $H_i(\widehat{P} \otimes_E \mathbf{C})_{-i-k}$ is dual to $H_k(\mathbf{L}(P))_{i+k}$. $\qquad \square$

Identifying the Linear Strand

Given a graded S-module M we can use part 3 of the Dictionary Theorem to identify the E-module Q such that $\mathscr{L}(\widehat{Q})$ is the linear strand of the minimal free resolution of M. If we shift grading so that M "begins" in degree 0, the result is the following:

Corollary 7.11. *Let $M = \sum_{i \geq 0} M_i$ be a graded S-module with $M_0 \neq 0$. The linear strand of the minimal free resolution of M as an S-module is $\mathbf{L}(\widehat{Q})$, where Q is the E-module with free presentation*

$$E \otimes \widehat{M}_1 \xrightarrow{\alpha} E \otimes \widehat{M}_0 \longrightarrow Q \longrightarrow 0$$

where the map α is defined on the generators $1 \otimes \widehat{M}_1 = \widehat{M}_1$ by the condition that

$$\alpha|_{\widehat{M}_1} : \widehat{M}_1 \to V \otimes \widehat{M}_0$$

is the dual of the multiplication map $\mu : W \otimes M_0 \to M_1$.

Proof. By Proposition 7.5 we may write the linear part of the resolution of M as $\mathbf{L}(P)$ for some E-module P, so

$$\mathbf{L}(P) : \quad \cdots \longrightarrow S \otimes P_1 \longrightarrow S \otimes P_0 \longrightarrow M.$$

It follows that $P_0 = M_0$, and $P_1 = \ker \mu : W \otimes M_0 \to M_1$, that is, $P_1 = R$. Dualizing, we get a right-exact sequence $\widehat{M}_1 \to V \otimes \widehat{M}_0 \to \widehat{R} \to 0$; that is, the image of $\widehat{M}_1$ generates the linear relations on $Q = \widehat{P} = \cdots \oplus \widehat{R} \oplus \widehat{M}$. By part 3 of Theorem 7.7, Q is linearly presented, so α is the presentation map as claimed. $\square$

Using Corollary 7.11 we can explain the relationship between the linear strand of the free resolution of a module M over the polynomial ring $S = \operatorname{Sym} W$ and the linear strand of the resolution of M when viewed, by "restriction of scalars", as a module M' over a smaller polynomial ring $S' = \operatorname{Sym} W'$ for a subspace $W' \subset W$. Write $V' = W'^{\perp} \subset V = \widehat{W}$ for the annihilator of W', and let $E' = E/(V') = \bigwedge(V/V')$, so that $E' = \bigwedge \widehat{W}'$.

Corollary 7.12. *With notation as above, the linear part of the S'-free resolution of M' is $\mathbf{L}(P')$, where P' is the E'-module $\{p \in P \mid V'p = 0\}$.*

Proof. The dual of the multiplication map $\mu' : W' \otimes M_0 \to M_1$ is the induced map $\widehat{M}_1 \to (V/V') \otimes \widehat{M}_0$, and the associated map of free modules $E' \otimes \widehat{M}_1 \to E' \otimes \widehat{M}_0$ is obtained by tensoring the one for M with E'. Its cokernel is $Q' = Q/V'Q$. By Corollary 7.11 the linear part of the S'-free resolution of M' is $\mathbf{L}(P')$, where $P' = \widehat{Q}'$ is the set of elements of $\widehat{Q}$ annihilating $V'Q$. This is the same as the set of elements of $\widehat{Q}$ annihilated by V'. $\square$

One concrete application is to give a bound on the length of the linear part that will be useful in the proof of Green's Theorem.

Corollary 7.13. *With notation as in Corollary 7.12, suppose that the codimension of W' in W is c. If the length of the linear strand of the minimal free resolution of M as an S' module is n, then the length of the linear strand of the minimal free resolution of M is at most $n+c$.*

Proof. By an obvious induction, it suffices to do the case $c = 1$. Write the linear strand of the minimal S-free resolution of M as $\mathbf{L}(P)$ for some E-module P. Suppose that W' is spanned by $e \in V$, so that $P' = \{p \in P \mid ep = 0\} \supset eP$. Because the degree of e is -1, there is a left exact sequence

$$0 \longrightarrow P' \longrightarrow P \overset{e}{\longrightarrow} P(-1).$$

The image of the right-hand map is inside $P'(-1)$. Thus if $P'_i = 0$ for $i > n$ then $P_i = 0$ for $i > n+1$ as required. $\square$

7C   Exterior Minors and Annihilators

From Theorem 7.7 we see that the problem of bounding the length of the linear part of a free resolution over S is the same as the problem of bounding the number of nonzero components of a finitely generated E-module P that is linearly presented. Since P is generated in a single degree, the number of nonzero components is $\leq n$ if and only if $(V)^n P = 0$. Because of this, the proof of Theorem 7.1 depends on being able to estimate the annihilator of an E-module.

Over a commutative ring such as S we could do this with Fitting's Lemma, which says that if a module M has free presentation

$$\phi : S^m \overset{\phi}{\longrightarrow} S^d \longrightarrow M \longrightarrow 0$$

then the $d \times d$ minors of ϕ annihilate M (see Section A2G on page 220.) The good properties of minors depend very much on the commutativity of S, so this technique cannot simply be transplanted to the case of an E-module. But Green discovered a remarkable analogue, the *exterior minors*, that works in the case of a matrix of linear forms over an exterior algebra. (The case of a matrix of forms of arbitrary degrees is treated in [Eisenbud and Weyman 2003].) We will first give an elementary description, then a more technical one that will allow us to connect the theory with that of ordinary minors.

It is instructive to look first at the case $m = 1$. Consider an E-module P with linear presentation

$$E(1) \overset{\begin{pmatrix} e_1 \\ \vdots \\ e_d \end{pmatrix}}{\longrightarrow} E^d \longrightarrow P \longrightarrow 0.$$

where the $e_i \in V$ are arbitrary. We claim that $(e_1 \wedge \cdots \wedge e_d)P = 0$. Indeed, if the basis of E^d maps to generators $p_1, \ldots, p_d \in P$, so that $\sum_i e_i p_i = 0$, then

$$(e_1 \wedge \cdots \wedge e_d)p_i = \pm(e_1 \wedge \cdots \wedge e_{i-1} \wedge e_{i+1} \wedge \cdots \wedge e_d) \wedge e_i p_i$$

$$= \mp(e_1 \wedge \cdots \wedge e_{i-1} \wedge e_{i+1} \wedge \cdots \wedge e_d) \sum_{j \neq i} e_j p_i = 0,$$

since $e_j^2 = 0$ for all j.

When the presentation matrix ϕ has many columns, it follows that the product of the elements in any one of the columns of ϕ is in the annihilator of P, and the same goes for the elements of any generalized column of ϕ—that is, of any column that is a scalar linear combination of the columns of ϕ. These products are particular examples of exterior minors. We shall see in Corollary 7.16 that all the exterior minors are linear combinations of exterior minors of this type, at least over an infinite field

In general, suppose that ϕ is a $p \times q$ matrix with entries $e_{i,j} \in V \subset E$. Given a collection of columns numbered $c_1, \ldots, c_k$, with multiplicities $n_1, \ldots, n_k$ adding up to d, and any collection of d rows $r_1, \ldots r_d$, we will define an $d \times d$ *exterior minor*

$$\phi[r_1, \ldots, r_d \mid c_1^{(n_1)}, \ldots, c_k^{(n_k)}] \in \bigwedge^d V$$

to be the sum of all products of the form $e_{r_1, j_1} \wedge \cdots \wedge e_{r_d, j_d}$ where precisely n_i of the numbers j_s are equal to c_i.

For example, if the multiplicities n_i are all equal to 1, the exterior minor is the *permanent* of the $d \times d$ submatrix of ϕ with the given rows and columns. On the other hand, if we take a single column with multiplicity d, then $\phi[r_1, \ldots, r_d \mid c_1^{(d)}]$ is the product of d entries of column number c_1, as above.

With general multiplicities, but in characteristic zero,

$$\phi[r_1, \ldots, r_d \mid c_1^{(n_1)} \cdots c_k^{(n_k)}]$$

is the permanent of the $d \times d$ matrix whose columns include n_i copies of c_i, divided by the product $n_1! \cdots n_k!$. (The *permanent* of a $d \times d$ matrix with entries x_{ij} is the sum, over all permutations σ on d indices, of the products $\prod_{i=1}^{d} x_{i,\sigma_i}$—the same products that appear in the determinant, but not multiplied by alternating signs.) If we think of the rows and columns as being vectors in V, the exterior minor is alternating in the rows and symmetric in the columns. We have chosen the notation $i^{(n_i)}$ to suggest a divided power; see for example [Eisenbud 1995, Appendix 2].

Description by Multilinear Algebra

We next give an invariant treatment, which also relates the exterior minors of ϕ to the ordinary minors of a closely related map ϕ'.

We first write the transpose $\phi^* : E^p(1) \to E^q$ of ϕ without using bases as a map $\phi^* : E \otimes_K A \to E \otimes_K B$ where A and B are vector spaces of dimensions p and q generated in degrees -1 and 0, respectively. Thus the rows of ϕ (columns of ϕ^*) correspond to elements of A while the columns of ϕ (rows of ϕ^*) correspond to elements of $\widehat{B}$.

The map ϕ^* (and with it ϕ) is determined by its restriction to the generating set $A = 1 \otimes A \subset E \otimes A$, and the image of A is contained in $V \otimes B$. Let

$$\psi : A \to V \otimes B,$$

be the restriction of ϕ^*. We can recover ϕ from ψ; explicitly,

$$\phi' : \wedge V \otimes \widehat{B} \longrightarrow \wedge V \otimes \widehat{A}$$
$$1 \otimes \widehat{b} \longmapsto \sum_i v_i \otimes (\widehat{v}_i \otimes \widehat{b}) \circ \psi,$$

where $\{v_i\}$ and $\{\widehat{v}_i\}$ are dual bases of V and $\widehat{V}$.

Taking the d-th exterior power of ψ, we get a map

$$\textstyle\bigwedge^d \psi : \bigwedge^d A \to \bigwedge^d (V \otimes B).$$

Because any element $x \in V \otimes B \subset (\wedge V) \otimes \operatorname{Sym} B$ satisfies $x^2 = 0$, the identity map on $V \otimes B$ extends uniquely to an algebra map $\wedge(V \otimes B) \to (\wedge V) \otimes \operatorname{Sym} B$. The degree-$d$ component m of this map is given by

$$\textstyle\bigwedge^d (V \otimes B) \xrightarrow{\; m \;} \bigwedge^d V \otimes \operatorname{Sym}_d B$$
$$(v_1 \otimes b_1) \wedge \cdots \wedge (v_1 \otimes b_d) \longmapsto (v_1 \wedge \cdots \wedge v_d) \otimes (b_1 \cdots\cdots b_d).$$

We will see that $m \circ \bigwedge^d \psi$ may be regarded as the matrix of exterior minors of ϕ, so to speak.

On the other hand, we could equally consider ψ as specifying a map of free modules in which "variables" are elements of B, and columns correspond to elements of $\widehat{V}$, with rows corresponding to elements of A as before. This could in fact be done over any algebra containing the vector space B. We take the algebra to be the new polynomial ring $\operatorname{Sym} B$ and define

$$\phi' : \operatorname{Sym} B \otimes \widehat{V} \longrightarrow \operatorname{Sym} B \otimes \widehat{A}$$
$$1 \otimes \widehat{v} \longmapsto \sum_i b_i \otimes (\widehat{v} \otimes \widehat{b}_i) \circ \psi,$$

where $\{b_i\}$ and $\{\widehat{b}_i\}$ are dual bases of B and $\widehat{B}$.

If $a_1, \ldots, a_d \in A$ and $\widehat{v}_1, \ldots, \widehat{v}_d \in \widehat{V}$, we write

$$\phi'(a_1, \ldots, a_d \,|\, \widehat{v}_1 \ldots \widehat{v}_d) \in \operatorname{Sym}_d B$$

for the $d \times d$ minor of ϕ' involving the rows corresponding to $a_1, \ldots, a_d$ and the columns corresponding to $v_1, \ldots, v_d$.

We will similarly extend our previous notation $\phi[r_1, \ldots, r_d \,|\, c_1^{(n_1)}, \ldots, c_k^{(n_k)}]$ to allow the r_i to be elements of A and to allow the c_i to be elements of $\widehat{B}$ instead of row and column numbers.

We can now show that the map $m \circ \bigwedge^d \psi$ expresses both the exterior minors of ϕ and the ordinary minors of ϕ'.

Proposition 7.14. *With notation as above, let $\{v_0, \ldots, v_r\}$ and $\{\widehat{v}_0, \ldots, \widehat{v}_r\}$ be dual bases for V and $\widehat{V}$, and let $\{b_1, \ldots, b_q\}$ and $\{\widehat{b}_1, \ldots, \widehat{b}_q\}$ be dual bases for B*

and $\widehat{B}$. The map $m \circ \bigwedge^d \psi$ is given by the formulas

$$m \circ \bigwedge^d \psi(a_1 \wedge \cdots \wedge a_d) = \sum_{0 \leq i_1 < \cdots < i_d \leq r} v_{i_1} \wedge \cdots \wedge v_{i_d} \otimes \phi'(a_1, \ldots, a_d \mid \hat{v}_{i_1}, \ldots, \hat{v}_{i_d})$$

$$= \sum_{\substack{1 \leq i_1 \leq \ldots \leq i_k \leq q \\ \sum n_j = d,\, 0 < n_j}} \phi[a_1, \ldots, a_d \mid \hat{b}_{i_1}^{(n_1)} \cdots \hat{b}_{i_k}^{(n_k)}] \otimes b_{i_1}^{n_1} \cdots b_{i_k}^{n_k}$$

Proof. Let $\psi(a_t) = \sum_{i,j} c_{i,j,t} v_i \otimes b_j$ with coefficients $c_{i,j,t} \in \mathbb{K}$. Let G be the symmetric group on $\{1, \ldots, d\}$.

For the first equality, set $\ell_{i,t} = \sum_j c_{i,j,t} b_j \in B = \mathrm{Sym}_1 B$, so that $(\phi')^*$ has (i,t)-entry equal to $\ell_{i,t}$ and $\psi(a_t) = \sum_i v_i \otimes \ell_{i,t}$. We have

$$m \circ \bigwedge^d \psi(a_1 \wedge \cdots \wedge a_d) = m\left(\sum_i (v_i \otimes \ell_{i,1}) \wedge \cdots \wedge \sum_i (v_i \otimes \ell_{i,d}) \right)$$

$$= m\left(\sum_{0 \leq i_1, \ldots, i_d \leq r} (v_{i_1} \otimes \ell_{i_1,1}) \wedge \cdots \wedge (v_{i_d} \otimes \ell_{i_d,d}) \right)$$

$$= \sum_{0 \leq i_1, \ldots, i_d \leq r} v_{i_1} \wedge \cdots \wedge v_{i_d} \otimes \ell_{i_1,1} \cdots \ell_{i_d,d}.$$

Gathering the terms corresponding to each (unordered) set of indices $\{i_1, \ldots, i_d\}$, we see that this sum is equal to the first required expression:

$$\sum_{0 \leq i_1 < \cdots < i_d \leq r,\, \sigma \in G} v_{i_1} \wedge \cdots \wedge v_{i_d} \otimes (\mathrm{sign}\,\sigma) \ell_{i_{\sigma(1)},1} \cdots \ell_{i_{\sigma(d)},d}$$

$$= \sum_{0 \leq i_1 < \cdots < i_d \leq r} v_{i_1} \wedge \cdots \wedge v_{i_d} \otimes \phi'(a_1, \ldots, a_d \mid \hat{v}_{i_1}, \ldots, \hat{v}_{i_d}).$$

The proof that $m \circ \bigwedge^d \psi(a_1 \wedge \cdots \wedge a_d)$ is given by the second expression is completely parallel once we write $m_{j,t} = \sum_i c_{i,j,t} v_i \in V = \bigwedge^1(V)$, so that $(\phi)^*$ has (j,t)-entry equal to $m_{j,t}$ and $\psi(a_t) = \sum_j m_{j,t} \otimes b_j$. $\quad\square$

How to Handle Exterior Minors

Here are some results that illustrate the usefulness of Proposition 7.14.

Corollary 7.15. *With the notation above, the span of the $d \times d$ exterior minors of ϕ is the image of a map*

$$m_d : \bigwedge^d A \otimes \overbrace{\mathrm{Sym}_d}^{d} B \to \bigwedge^d V$$

that depends only on ϕ as a map of free modules, and not on the matrix chosen. In particular, if $v_1, \ldots, v_d$ are the elements of any generalized column of ϕ, then $v_1 \wedge \cdots \wedge v_d$ is in this span.

Proof. The map m_d is defined by saying that it sends $a \otimes g \in \bigwedge^d A \otimes \widehat{\operatorname{Sym}_d} B$ to

$$(1 \otimes g)(m \circ \bigwedge^d \psi(a)).$$

Since we can replace one of the columns of ϕ by a generalized column without changing the map of free modules, the second statement follows from our original description of the exterior minors.   □

Corollary 7.15 suggests a different approach to the the exterior minors. In particular, if we take $V = A \otimes \widehat{B}$ and if ϕ is the generic matrix of linear forms over the ring E, then the span of the $d \times d$ exterior minors of ϕ is invariant under the product of linear groups $\operatorname{GL}(A) \times \operatorname{GL}(B)$, and is the (unique) invariant submodule of $\bigwedge(A \otimes B)$ isomorphic to $\bigwedge^d A \otimes \widehat{\operatorname{Sym}_d} B$. For further information see [Eisenbud and Weyman 2003].

Corollary 7.16. *If $\mathbb{K}$ is an infinite field and ϕ is a $d \times m$ matrix of linear forms over E, then the vector space generated by all the $d \times d$ exterior minors of ϕ is in fact generated by all elements of the form $e_1 \wedge \cdots \wedge e_d$, where $e_1, \ldots, e_d$ are the elements of a generalized column of ϕ.*

Proof. A (generalized) column of ϕ corresponds to an element $\hat{b} : B \to \mathbb{K}$. Such an element induces a map $\operatorname{Sym} B \to \operatorname{Sym} \mathbb{K} = \mathbb{K}[x]$, and thus for every d it induces a map $\operatorname{Sym}_d B \to \mathbb{K} \cdot x^d = \mathbb{K}$ that we will call $\hat{b}^{(d)}$. This notation is compatible with our previous notation because

$$\phi[a_1, \ldots, a_d | \hat{b}^{(d)}] = e_1 \wedge \cdots \wedge e_d = m(a_1 \wedge \cdots \wedge a_d \otimes \hat{b}^{(d)}).$$

By Corollary 7.15 the span of the exterior minors of ϕ is the image of

$$m_d : \bigwedge^d A \otimes \widehat{\operatorname{Sym}_d} B \to \bigwedge^d V.$$

Thus to show that the special exterior minors that are products of the elements in a generalized column span all the exterior minors, it suffices to show that the elements $\hat{b}^{(d)}$ span $\widehat{\operatorname{Sym}_d} B$. Equivalently, it suffices to show that there is no element in the intersection of the kernels of the projections $\hat{b}^{(d)} : \operatorname{Sym}_d B \to \mathbb{K}$. But this kernel is the degree d part of the ideal generated by the kernel of $\hat{b}$. If we think of this ideal as the ideal of the point in projective space $\mathbb{P}(B)$ corresponding to $\hat{b}$, the desired result follows because the only polynomial that vanishes on all the points of a projective space over an infinite field is the zero polynomial.   □

The next two corollaries are the keys to the proof of the Linear Syzygy Theorem to be given in the next section.

Corollary 7.17. (Exterior Fitting Lemma) *If ϕ is a $d \times m$ matrix of linear forms over the exterior algebra E then the cokernel of ϕ is annihilated by the exterior minors of ϕ.*

Proof. We may harmlessly extend the field $\mathbb{K}$, and thus we may suppose that $\mathbb{K}$ is infinite. By Corollary 7.16 it suffices to prove the result for the special exterior minors that are products of the elements in generalized columns. The proof in this case is given at the beginning of Section 7C. □

Corollary 7.18. *Let* $\phi : E \otimes \widehat{B} \to E \otimes \widehat{A}$ *and* $\phi' : \operatorname{Sym} B \otimes \widehat{V} \to \operatorname{Sym} B \otimes \widehat{A}$ *be maps of free modules coming from a single map of vector spaces* $\psi : A \to V \otimes B$ *as above. If* $\dim_\mathbb{K} A = d$, *then the dimension of the span of the* $d \times d$ *exterior minors of* ϕ *is the same as the dimension of the span of the (ordinary)* $d \times d$ *minors of* ϕ'.

Proof. Let $a_1, \dots, a_d$ be a basis of A. The element

$$f = m \circ \bigwedge\nolimits^d \psi(a_1 \wedge \cdots \wedge a_d) \in \bigwedge\nolimits^d V \otimes \operatorname{Sym}_d B$$

may be regarded as a map $\bigwedge^d V \to \operatorname{Sym}_d B$ or as a map $\widehat{\operatorname{Sym}_d B} \to \bigwedge^d V$. These maps are dual to one another, and thus have the same rank. By Proposition 7.14 the image of the first is the span of the ordinary minors of ϕ', while the image of the second is the span of the exterior minors of ϕ. □

7D Proof of the Linear Syzygy Theorem

We now turn to the proof of the Linear Syzygy Theorem 7.1 itself. Let $M = M_0 \oplus M_1 \oplus \cdots$ be an S-module with $M_0 \neq 0$, and let $m_0 = \dim M_0$. We must show that the length of the linear strand of the minimal free resolution of M is at most $\max(m_0 - 1, \dim R)$, where $R = \{w \otimes a \in W \otimes M_0 \mid wa = 0\}$. We may harmlessly extend the ground field if necessary and assume that $\mathbb{K}$ is algebraically closed.

Suppose first that $\dim R \leq m_0 - 1$. In this case we must show that the length of the linear strand is $\leq m_0 - 1$. From Theorem 7.5 and Corollary 7.11 we know that the linear strand has the form $\mathbf{L}(P)$, where $P = \widehat{Q}$ and

$$Q = \operatorname{coker}\left(E \otimes \widehat{M}_1 \xrightarrow{\ \alpha\ } E \otimes \widehat{M}_0\right).$$

Here α is the dual of the multiplication map $\mu : W \otimes M_0 \to M_1$. Since Q is generated in degree 0, it will suffice to show that Q is annihilated by $(V)^{m_0}$, and by Corollary 7.17 it suffices in turn to show that the $m_0 \times m_0$ exterior minors of α span all of E_{m_0}, a space of dimension $\binom{r+1}{m_0}$.

By Corollary 7.18, the dimension of the span of the exterior minors of α is the same as the dimension of the span of the ordinary $m_0 \times m_0$ minors of the map of $\operatorname{Sym} M_1$-modules

$$\phi' : \operatorname{Sym} M_1 \otimes W \to \operatorname{Sym} M_1 \otimes \widehat{M}_0$$

corresponding to the map $W \to M_1 \otimes \widehat{M}_0$ adjoint to the multiplication map $W \otimes M_0 \to M_1$.

Perhaps the reader is by now lost in the snow of dualizations, so it may help to remark that ϕ' is represented by an $m_0 \times (r+1)$ matrix whose rows are indexed by a basis of M_0 and whose columns indexed by a basis of W. The entry of this matrix corresponding to $m \in M_0$ and $w \in W$ is simply the element $wm \in M_1$. It suffices to prove that the $m_0 \times m_0$ minors of ϕ' span a linear space of dimension $\binom{r+1}{m_0}$ — that is, these minors are linearly independent.

Using the Eagon–Northcott complex as in Corollary A2.61 it is enough to show that the $m_0 \times m_0$ minors of ϕ' vanish only in codimension $r + 1 - m_0 + 1 = r + 2 - m_0$. The vanishing locus of these minors is the union of the loci where the generalized rows of ϕ' vanish, so we consider these rows. Let $B_e \subset M_0$ be the set of elements m such that the corresponding generalized row vanishes in codimension e. This means that m is annihilated by an $r + 1 - e$ dimensional space $W_m \subset W$. The tensors $w \otimes m$ with $w \in W_m$ and $m \in B_e$ form a $\dim B_e + (r + 1 - e) - 1 = \dim B_e + r - e$-dimensional family in R. By hypothesis, $\dim R \leq m_0 - 1$, so $\dim B_e \leq m_0 - 1 - (r - e) = m_0 - r + e - 1$.

Two elements of B_e that differ by a scalar correspond to rows with the same vanishing locus. Thus the union of the vanishing loci of the generalized rows corresponding to elements of B_e has codimension at least $e - (\dim B_e - 1) \geq r + 2 - m_0$. Since this is true for each e, the union of the B_e, which is the set defined by the $m_0 \times m_0$ minors of ϕ', has codimension at least $r + 2 - m_0$, as required. This completes the proof in the case $\dim R \leq m_0 - 1$.

Finally, suppose that $\dim R \geq m_0$. By induction and the proof above, we may assume that the Theorem has been proved for all modules with the same value of m_0 but smaller $\dim R$.

The affine variety R is a union of lines through the origin in the vector space $W \otimes M_0$. Let $\bar{R}$ be the corresponding projective variety in $\mathbb{P}(\widehat{W \otimes M_0})$. The set of pure tensors $w \otimes a$ corresponds to the Segre embedding of $\mathbb{P}(\widehat{W}) \times \mathbb{P}(\widehat{M}_0)$, so $\bar{R}$ is contained in this product. Each hyperplane $W' \subset W$ corresponds to a divisor

$$\mathbb{P}(\widehat{W'}) \times \mathbb{P}(\widehat{M}_0) \subset \mathbb{P}(\widehat{W}) \times \mathbb{P}(\widehat{M}_0),$$

and the intersection of all such divisors is empty. Thus we can find a hyperplane W' such that $\dim R \cap (\mathbb{P}(\widehat{W'}) \times \mathbb{P}(M_0)) \leq \dim R - 1$.

Let M' be the $S' = \operatorname{Sym} W'$-module obtained from M by restriction of scalars. By Corollary 7.13, the length of the linear strand of the minimal free resolution of M' is shorter than that of M by at most 1. By induction Theorem 7.1 is true for M', whence it is also true for M. $\qquad\square$

7E   More about the Exterior Algebra and BGG

In this section we will go a little further into the the module theory over the exterior algebra $E = \bigwedge V$ and then explain some more about the Bernstein–Gelfand–Gelfand correspondence. Our approach to the latter is based on [Eisenbud et al. 2003a].

The Gorenstein Property and Tate Resolutions

We have already introduced the duality functor $P \mapsto \widehat{P} = \mathrm{Hom}_{\mathbb{K}}(P, \mathbb{K})$ for finitely generated E-modules. Since $\mathbb{K}$ is a field the duality functor $P \mapsto \widehat{P}$ is exact, so it takes projective modules to injective modules. Just as in the commutative local case, Nakayama's Lemma implies that every projective E-module is free (even the nonfinitely generated modules are easy here because the maximal ideal (V) of E is nilpotent). It follows that every finitely generated injective E-module is a direct sum of copies of the module $\widehat{E}$. We gave an ad hoc proof, based on the self-duality of the Koszul complex, that $\widehat{E} \cong E$ as E-modules, but the isomorphism is noncanonical and does not preserve the grading. Here is a more precise statement, with an independent proof; note that by Theorem 7.5 it implies the self-duality of the Koszul complex.

Proposition 7.19 (Gorenstein property). *The rank 1 free E-module E has a unique minimal nonzero ideal, and is injective as an E-module. Thus it is an injective envelope of the simple E-module and is isomorphic to $\widehat{E}$ as an E-module (with a shift in grading.) Moreover, $\widehat{E} \cong E \otimes_{\mathbb{K}} \bigwedge^{r+1} W$ canonically.*

Proof. In fact the minimal nonzero ideal is the one-dimensional vector space $\bigwedge^{r+1} V = E_r$, generated by the product of the elements of any basis of V. To see this, we show that any nonzero element of E generates an ideal containing $\bigwedge^{r+1} V$. If e is a nonzero element of E we can write $e = a \cdot e_{i_1} e_{i_2} \cdots e_{i_t} + e'$ with respect to a basis e_i of V, where $0 \neq a \in \mathbb{K}$, $i_1 < \cdots < i_t$, and e' consists of other monomials of degree t as well (perhaps) as monomials of degree exceeding t. Let J be the complement of $i_1, \dots, i_t$ in $0, \dots, r$. It follows that every monomial of e' is divisible by one of the elements e_j with $j \in J$, so

$$e \cdot \prod_{j \in J} e_j = \pm a \cdot e_0 \cdots e_r$$

is a generator of $\bigwedge^{r+1} V$, as required.

From this we see that $\widehat{E}$ is generated by the one-dimensional vector space $\bigwedge^{r+1} V = \bigwedge^{r+1} W$, so there is a canonical surjection $E \otimes \bigwedge^{r+1} W \to \widehat{E}$. Since E and $\widehat{E}$ have the same dimension, this surjection is an isomorphism. It follows that E is injective, so E is the injective envelope of its submodule $(\bigwedge^{r+1} V)$. $\square$

As a consequence we can give another view of the duality functor $P \mapsto \widehat{P}$ for finitely generated E-modules:

Corollary 7.20. *There is a natural isomorphism $\widehat{P} \cong \mathrm{Hom}_E(P, E) \otimes \bigwedge^{r+1} W$. In particular, $\mathrm{Hom}_E(-, E)$ is an exact functor.*

Proof. Since $E \otimes -$ is left adjoint to the forgetful functor from E-modules to $\mathbb{K}$-modules we have

$$\mathrm{Hom}_{\mathbb{K}}(P, \mathbb{K}) = \mathrm{Hom}_E(E \otimes_{\mathbb{K}} P, \mathbb{K}) = \mathrm{Hom}_E(P, \mathrm{Hom}_{\mathbb{K}}(E, \mathbb{K})) = \mathrm{Hom}_E(P, \widehat{E}),$$

and by the Gorenstein property (Proposition 7.19) $\mathrm{Hom}_E(P,\widehat{E}) = \mathrm{Hom}_E(P,E) \otimes \bigwedge^{r+1} W$.

The last statement follows from this (or directly from the fact that E is injective as an E-module.) $\qquad\square$

Over any ring we can combine a projective resolution $\mathbf{F}$ and an injective resolution $\mathbf{I}$ of a module P into a *Tate resolution*:

$$\cdots \longrightarrow F_1 \longrightarrow F_0 \overset{\searrow\quad\nearrow}{\underset{P}{}} I_0 \longrightarrow I_1 \longrightarrow \cdots$$

Over a ring like E the I_j are also free. In fact, we may take $\mathbf{F}$ to be a minimal free resolution and $\mathbf{I}$ to be the dual of a minimal free resolution of $\widehat{P}$, and we get a unique minimal Tate resolution, a doubly infinite exact free complex as above where the image of the 0-th differential is isomorphic to P.

For example, if we take $P = E/(V) = \mathbb{K}$ to be the residue field of E, then we already know that the minimal free resolution of P is the Cartan resolution. Since P is self-dual, the minimal injective resolution is the dual of the Cartan resolution, and the Tate resolution has the form

$$\to E \otimes \widehat{\mathrm{Sym}_2 W} \to E \otimes \widehat{W} \to E \longrightarrow \widehat{E} \to \widehat{E} \otimes W \to \widehat{E} \otimes \mathrm{Sym}_2 W \to$$

The sum of the terms on the right is $\widehat{E} \otimes S$; we shall see in the next section that this is not an accident. Tate resolutions over E appear rather naturally in algebraic geometry.

It is not hard to show that Gröbner basis methods apply to the exterior algebra just as to the commutative polynomial ring (in fact, there are some advantages to computation that come from the finite dimensionality of E.) Thus it is possible to compute Tate resolutions — or at least bounded portions of them — explicitly in a program such as Macaulay 2 [Grayson and Stillman 1993–].

Where BGG Leads

The Bernstein–Gelfand–Gelfand correspondence was stated in [Bernstein et al. 1978] as an equivalence between the derived categories of bounded complexes of finitely generated graded S-modules and graded E-modules, or between the bounded derived categories of coherent sheaves on $\mathbb{P}^r$ and the graded E-modules modulo free modules. A different way of describing this equivalence was discovered at the same time in [Beĭlinson 1978]. Both these papers were inspired by a lecture of Manin. BGG was the first appearance of the "derived equivalences" between various module and sheaf categories that now play an important role in representation theory (for example [Ringel 1984]), algebraic geometry (for example [Bridgeland 2002]) and mathematical physics (for example [Polishchuk and Zaslow 1998]). Here we will explain a little more about the BGG equivalence, and describe one of its recent applications.

The functor $\mathbf{L}$ from graded E-modules to linear free complexes of S-modules has a version $\mathbf{R}$ that goes "the other way" from graded S-modules to linear free E complexes: it takes a graded S-module $M = \oplus M_i$ to the complex

$$\mathbf{R}(M): \quad \cdots \longrightarrow \widehat{E} \otimes_{\mathbb{K}} M_i \longrightarrow \widehat{E} \otimes_{\mathbb{K}} M_{i-1} \longrightarrow \cdots$$
$$f \otimes m \longmapsto \sum_i f e_i \otimes x_i m,$$

where $\{x_i\}$ and $\{e_i\}$ are dual bases of W and V. We think of M_i as being a vector space concentrated in degree i, and the term $\widehat{E} \otimes M_i$ as being in cohomological degree i ($\equiv$ homological degree $-i$). For any vector space N we have $\widehat{E} \otimes_{\mathbb{K}} N = \mathrm{Hom}_{\mathbb{K}}(E, N)$, so thinking of $\mathbf{R}(M)$ as a differential graded E-module, we could simply write $\mathbf{R}(M) = \mathrm{Hom}_{\mathbb{K}}(E, M)$, just as we can write $\mathbf{L}(P) = S \otimes_{\mathbb{K}} P$.

This suggests that the two functors might somehow be adjoint. However, they do not even go between the same pair of categories! To repair this, we extend the functor $\mathbf{L}$ from the category of modules to the category of complexes: If $\cdots \to A \to B \to \cdots$ is a complex of graded S-modules, then $\cdots \to \mathbf{L}(A) \to \mathbf{L}(B) \to \cdots$ is naturally a double complex, and we can take its total complex to get a complex of S-modules. Thus $\mathbf{L}$ goes from the category of complexes of E-modules to the category of complexes of S-modules. Similarly, $\mathbf{R}$ may be extended to a functor going the other way. These two functors are adjoint. Moreover, they pass to the derived categories and are inverse equivalences there. See for example [Gelfand and Manin 2003].

We will not pursue this line of development further. Instead we want to point out a source of interesting Tate resolutions connected with the functor $\mathbf{R}$. An argument similar to the proof of Theorem 7.8 (see also Exercise 7.10) yields:

Proposition 7.21. *If M is a graded S-module, the homology of the complex $\mathbf{R}(M)$ is*

$$\mathrm{H}^j(\mathbf{R}(M))_{i+j} = \mathrm{Tor}_i(\mathbb{K}, M)_{i+j}.$$

This shows in particular that $\mathbf{R}(M)$ is exact far out to the right. The key invariant is, once again, the Castelnuovo–Mumford regularity of M:

Corollary 7.22. $\mathrm{reg}\, M \leq d$ *if and only if* $\mathrm{H}^i(M) = 0$ *for all* $i > d$.

Proof. The condition $\mathrm{reg}\, M = d$ means that $\mathrm{Tor}_i(\mathbb{K}, M)_{i+j} = 0$ for $j > d$. $\qquad \square$

Now suppose that M is a finitely generated graded S-module of regularity n. By Corollary 7.22 the free complex

$$\cdots \longrightarrow 0 \longrightarrow \widehat{E} \otimes M_n \xrightarrow{d^n} \widehat{E} \otimes M_{n+1} \xrightarrow{d^{n+1}} \cdots$$

is exact except at $\widehat{E} \otimes M_n$.

We will truncate this complex at $\widehat{E} \otimes M_{n+1}$ and then adjoin a minimal free resolution of $\ker d^{n+1}$. The result is a Tate resolution

$$\mathbf{T}(M): \cdots T^{n-1} \longrightarrow T^n \longrightarrow \widehat{E} \otimes M_{n+1} \xrightarrow{d^{n+1}} \cdots.$$

The truncation at M_{n+1} is necessary in order to ensure minimality (as we will see in the proof of the Proposition 7.23.)

The resolution $\mathbf{T}(M)$ obviously depends only on the truncation $M_{\geq n+1}$, but even more is true:

Proposition 7.23. *Let $\mathscr{F}$ be a coherent sheaf on $\mathbb{P}^r$, and let M be a finitely generated graded S-module whose sheafification is $\mathscr{F}$. The Tate resolution $\mathbf{T}(M)$ depends, up to noncanonical isomorphism, only on the sheaf $\mathscr{F}$.* $\square$

Proof. The sheaf $\mathscr{F}$ determines M up to finite truncation, so it suffices to show that if $m \geq n = \operatorname{reg} M$ then the Tate resolution

$$\mathbf{T}(M_{\geq m}): \cdots T'^{m-1} \longrightarrow T'^m \longrightarrow \widehat{E} \otimes M_{m+1} \xrightarrow{d^{m+1}} \cdots.$$

is isomorphic to $\mathbf{T}(M)$. By the definition of $\mathbf{T}(M_{\geq m})$ and the uniqueness of minimal resolutions, it suffices to show that

$$\widehat{E} \otimes M_{n+1} \longrightarrow \cdots \xrightarrow{d^{m-1}} \widehat{E} \otimes M_m \qquad (*)$$

is the beginning of a minimal free resolution of $\operatorname{coker} d^{m-1} = \ker d^{m+1}$. By Corollary 7.22 it is at least a resolution, and this would be so even if we extended it one more step to $\widehat{E} \otimes M_n$. But the differentials in the complex

$$\widehat{E} \otimes M_n \xrightarrow{d_n} \cdots \xrightarrow{d^{m-1}} \widehat{E} \otimes M_m$$

are all minimal (their matrices have entries of degree 1), so for all $i > n$ the module $\widehat{E} \otimes M_i$ is the minimal free cover of $\ker d^{i+1}$. $\square$

Henceforward, when $\mathscr{F}$ is a coherent sheaf on $\mathbb{P}^r$, we will write $\mathbf{T}(\mathscr{F})$ for the Tate resolution $\mathbf{T}(M)$ associated with any finitely generated S-module having sheafification $\mathscr{F}$, and call it the *Tate resolution of $\mathscr{F}$*.

For example, let X be the standard twisted cubic curve in $\mathbb{P}^3$ with structure sheaf $\mathcal{O}_X$ and homogeneous coordinate ring S_X. To simplify notation write a, b, c, d for the homogeneous coordinates of $\mathbb{P}^3$, instead of $x_0, \ldots x_3$. We have reg $S_X \leq 1$ by Gruson–Lazarsfeld–Peskine (Theorem 5.1), and in fact the resolution is the Eagon–Northcott complex of

$$\begin{pmatrix} a & b & c \\ b & c & d \end{pmatrix}$$

with Betti diagram

	0	1	2
0	1	–	–
1	–	3	2

so reg $S_X = 1$. The values of the Hilbert function $H_{S_X}(n)$ are $1, 4, 7, \ldots$, and $\mathbf{R}(S_X)$ is the complex

$$\widehat{E} \xrightarrow{\begin{pmatrix} a \\ b \\ c \\ d \end{pmatrix}} \widehat{E}^4(-1) \xrightarrow{d^2 = \begin{pmatrix} a & 0 & 0 & 0 \\ b & a & 0 & 0 \\ c & b & a & 0 \\ d & c & b & a \\ 0 & d & c & b \\ 0 & 0 & d & c \\ 0 & 0 & 0 & d \end{pmatrix}} \widehat{E}^7(-2) \longrightarrow \cdots .$$

Corollary 7.22 shows that $\mathbf{R}(S_X)$ is *not* exact at

$$\widehat{E}^4(-1) = \widehat{E} \otimes (S_X)_1,$$

but we can see this in a more primitive way. It suffices to show that $\widehat{\mathbf{R}(S_X)}$ is not exact at $E^4(1)$. But the first map in $\widehat{\mathbf{R}(S_X)}$ is the same as the first map in the Cartan resolution $\widehat{\mathbf{R}(S)}$, while the second map has source $E^7(2)$ instead of the

$$E \otimes \operatorname{Sym}_2 W = E^{10}(2)$$

that occurs in the Cartan resolution. Since the Cartan resolution is minimal, this proves the inexactness.

It turns out that $\ker d^2$ has three minimal generators: the given linear one and two more, which have quadratic coefficients. The map d^1 of the Tate resolution may be represented by the matrix

$$d^1 = \begin{pmatrix} a & 0 & 0 \\ b & ad & ac \\ c & bd & bc+ad \\ d & cd & bd \end{pmatrix} .$$

(It is obvious that the columns of this matrix are in the kernel, and that no two of them could generate it; to prove that they actually generate it requires either an (easy) computation with Gröbner bases or an application of Theorem 7.24 below.) The rest of the Tate resolution of $\mathscr{O}_X$ has the form

$$\widehat{E} \xrightarrow{\begin{pmatrix} a \\ b \\ c \\ d \end{pmatrix}} \widehat{E}^4(-1) \xrightarrow{d^2 = \begin{pmatrix} a & 0 & 0 & 0 \\ b & a & 0 & 0 \\ c & b & a & 0 \\ d & c & b & a \\ 0 & d & c & b \\ 0 & 0 & d & c \\ 0 & 0 & 0 & d \end{pmatrix}} \widehat{E}^7(-2) \longrightarrow$$

$$\longrightarrow \widehat{E}^8(3) \xrightarrow{\begin{pmatrix} d & c & b & a & 0 & 0 & 0 & 0 \\ 0 & d & c & b & a & 0 & 0 & 0 \\ 0 & 0 & d & c & b & a & 0 & 0 \\ 0 & 0 & 0 & d & c & b & a & 0 \\ 0 & 0 & 0 & 0 & d & c & b & a \end{pmatrix}} \widehat{E}^5(2) \xrightarrow{\begin{pmatrix} d & c & b & a & 0 \\ 0 & d & c & b & a \\ 0 & 0 & da & ca & ba \end{pmatrix}} \widehat{E} \oplus \widehat{E}^2(1) \xrightarrow{d^1}$$

The reader with a background in algebraic geometry may have noticed that the ranks of the free modules with generators in various degrees in the Tate resolution of $\mathscr{O}_X$ are precisely the numbers $h^i(\mathscr{O}_X(n))$, as suggested in this table:

n	-3	-2	-1	0	1	2
$h^1\mathscr{O}_X(n)$	8	5	2	0	0	0
$h^0\mathscr{O}_X(n)$	0	0	0	1	4	7

The terms of the Tate resolution are reflected in the pairs of numbers on the diagonals of this table: for example, $\widehat{E} \oplus \widehat{E}^2(1)$ corresponds to the terms 2 (upper row) and 1 (lower row).

Here is the general result:

Theorem 7.24. *Let $\mathscr{F}$ be a coherent sheaf on $\mathbb{P}^r$. The free module T^i in cohomological degree i of the Tate resolution $\mathbf{T}(\mathscr{F})$ is*

$$T^i = \bigoplus_j \widehat{E} \otimes H^j(\mathscr{F}(i-j)),$$

where $H^j(\mathscr{F}(i-j))$ is regarded as a vector space concentrated in degree $i-j$.

For the proof we refer to [Eisenbud et al. 2003a]. For further applications see [Eisenbud et al. 2003b], and for an exposition emphasizing how to use these techniques in computation see [Decker and Eisenbud 2002]. We close this section by interpreting Theorem 7.24 in the case of the Tate resolution of the residue field, the Cartan resolution.

We claim that the Tate resolution of $\mathbb{K} = E/(V)$ derived above by putting the Cartan resolution together with its dual is precisely the Tate resolution of the sheaf $\mathscr{O}_{\mathbb{P}^r}$. In fact, S is a module whose sheafification is $\mathscr{O}_{\mathbb{P}^r}$, and the regularity of S (as an S-module) is 0, so from Corollary 7.22 we can deduce again what we already knew: the dual

$$\mathbf{R}(S): \; \widehat{E} \longrightarrow \widehat{E} \otimes W \longrightarrow \widehat{E} \otimes \mathrm{Sym}_2 W \longrightarrow \cdots$$

of the Cartan resolution is exact starting from $\widehat{E} \otimes W$. Since $\widehat{E}$ is a minimal cover of the next map, we may complete it to a Tate resolution $\mathbf{T}(\mathscr{O}_{\mathbb{P}^r})$ by adjoining a minimal free resolution of the kernel $\mathbb{K}$ of $\widehat{E} \to \widehat{E} \otimes W$. This gives us the Tate resolution of $\mathbb{K}$ as claimed.

Comparing the free modules T^i with Theorem 7.24 we deduce the well-known formula

$$\mathrm{H}^i \mathscr{O}_{\mathbb{P}^r}(n) = \begin{cases} \mathrm{Sym}_n W & \text{if } i = 0, \\ 0 & \text{if } 0 < i < r, \\ \widehat{\mathrm{Sym}_{n-r-1} W}, & \text{if } i = r. \end{cases}$$

See [Hartshorne 1977, III.3.1], and also Corollary A1.6.

7F Exercises

1. Let F be a finitely generated free graded module. Show that, for any i, the submodule of F generated by all elements of degree $\leq i$ is free.

2. Let $\mathbf{F}: \; \cdots \longrightarrow F_i \xrightarrow{\phi_i} F_{i-1} \longrightarrow \cdots$ be the linear strand of a minimal free resolution. Show that when F_i is nonzero, no generalized column of ϕ_i can have all entries equal to zero.

3. With hypotheses as in the Linear Syzygy Theorem 7.1, let

$$A = \{(w, \overline{m}) \in W \times \mathbb{P}(M_0^*) \mid wm = 0\},$$

where $\overline{m}$ denotes the one-dimensional subspace spanned by a nonzero element $m \in M_0$.

Show that the statement of Theorem 7.1 is equivalent to the statement that the length of the linear strand of the free resolution of M is $\leq \dim A$.

4. Consider Example 7.3. Show that if the linear forms $\ell_{i,j}$ span all of W, then the variety X defined by the minors of N is nondegenerate. Since N is 1-generic, the module ω is the module of twisted global sections of a line bundle, so the hypotheses of Corollary 7.4 apply.

5. Show that over any local Artinian ring, any free submodule of a free module is a summand. Deduce that the only modules of finite projective dimension are the free modules. Over the exterior algebra, show that any free submodule of any module is a summand.

6. Though E is a noncommutative ring, it is so close to commutative that commutative proofs can usually be used almost unchanged. Following the ideas at the beginning of [Eisenbud 1995, Chapter 21], give a direct proof that $\widehat{E} \cong E$ as E-modules.

7. Show that $\mathbf{L}(\widehat{P}) = \mathrm{Hom}_S(\mathbf{L}(P), S)$ as complexes.

8. Let $e \in E$ be an element of degree -1. Show that the periodic complex

$$\cdots \xrightarrow{e} E \xrightarrow{e} E \xrightarrow{e} E \xrightarrow{e} \cdots$$

is exact. In fact, it is the Tate resolution of a rather familiar sheaf. What is the sheaf?

9. Here is a basis free approach to the equivalence in Proposition 7.5.

(a) If V is finite-dimensional vector space and P_i, P_{i-1} are any vector spaces over $\mathbb{K}$, show that there is a natural isomorphism

$$\mathrm{Hom}_{\mathbb{K}}(V \otimes P_i, P_{i-1}) \cong \mathrm{Hom}_{\mathbb{K}}(P_i, W \otimes P_{i-1}),$$

where W is the dual of V, taking a map $\mu : V \otimes P_i \to P_{i-1}$ to the map

$$d : P_i \to W \otimes P_{i-1}; \qquad d(p) = \sum_i x_i \otimes \mu(e_i \otimes p).$$

Maps that correspond under this isomorphism are said to be *adjoint* to one another.

(b) Suppose that $\mu_i : V \otimes P_i \to P_{i-1}$ and $\mu_{i-1} : V \otimes P_{i-1} \to P_{i-2}$ are adjoint to d_i and d_{i-1}, and write $s : W \otimes W \to \mathrm{Sym}_2 W$ for the natural projection. Show that $P_i \oplus P_{i-1} \oplus P_{i-2}$ is an $E = \bigwedge V$-module (the associative and anti-commutative laws hold) if and only if the map $V \otimes V \otimes P_i \to P_{i-2}$ factors through the natural projection $V \otimes V \otimes P_i \to \bigwedge^2 V \otimes P_i$.

(c) Show that the maps

$$S \otimes P_i \longrightarrow S \otimes P_{i-1} \longrightarrow S \otimes P_{i-2}$$

induced by d_i and d_{i-1} compose to zero if and only if the composite map

$$P_i \xrightarrow{d_i} W \otimes P_{i-1} \xrightarrow{1 \otimes d_{i-1}} W \otimes W \otimes P_{i-2} \xrightarrow{s \otimes 1} \mathrm{Sym}_2 W \otimes P_{i-2}$$

is zero.

(d) Show that the composite map

$$P_i \xrightarrow{d_i} W \otimes P_{i-1} \xrightarrow{1 \otimes d_{i-1}} W \otimes W \otimes P_{i-2} \xrightarrow{s \otimes 1} \mathrm{Sym}_2 W \otimes P_{i-2}$$

is adjoint to the composite map

$$(\mathrm{Sym}_2 W)^* \otimes P_i \xrightarrow{s^*} W^* \otimes W^* \otimes P_i \xrightarrow{1 \otimes \mu_i} W^* \otimes P_{i-1} \xrightarrow{\mu_{i-1}} P_{i-2}.$$

Deduce that the first of these maps is 0 if and only if the second is zero if and only if the map

$$W^* \otimes W^* \otimes P_i \xrightarrow{1 \otimes \mu_i} W^* \otimes P_{i-1} \xrightarrow{\mu_{i-1}} P_{i-2}$$

factors through $\bigwedge^2(V) \otimes P_i$.

(e) Deduce Proposition 7.5.

10. Prove Proposition 7.21 by examining the sequence of vector spaces whose homology is $\mathrm{H}^i(\mathbf{R}(M))_{i+j}$, as in Theorem 7.8.

8

Curves of High Degree

Let X be a curve of genus g. We know from Corollary 6.7 that any line bundle of degree at least $2g+1$ on X is very ample. By an embedding *of high degree* we will mean any embedding of X by a complete linear series of degree $d \geq 2g+1$, and by a *curve of high degree* we mean the image of such an embedding.

In Chapter 6 we gave an account of the free resolutions of curves of genus 0 and 1, embedded by complete linear series, constructing them rather explicitly. For curves of genus $g = 0$, we had embeddings of any degree ≥ 1. For curves of genus $g = 1$, only linear series of degree ≥ 3 could be very ample, so these were all curves of high degree. In this chapter we will see that many features of the free resolutions we computed for curves of genus 0 and 1 are shared by all curves of high degree.

To study these matters we will introduce some techniques that play a central role in current research: the restricted tautological subbundle, Koszul cohomology, the property N_p and the strands of the resolution. We will see that the form of the free resolution is related to special varieties containing X, and also to special sets of points on the curve in its embedding.

For simplicity we will use the word *curve* to mean a smooth irreducible one-dimensional variety over an algebraically closed field $\mathbb{K}$, though the sophisticated reader will see that many of the results can be extended to Gorenstein one-dimensional subschemes over any field. Recall that the *canonical sheaf* ω_X of a curve X is the sheaf of differential forms associated to the cotangent bundle of X. If X is embedded in some projective space $\mathbb{P}^r$, it is convenient to use a different characterization: ω_X is the sheaf associated to the module

$$\operatorname{Ext}_S^{r-1}(S_X, S(-r-1)).$$

For this and much more about canonical sheaves, see [Altman and Kleiman 1970]. We will denote by K_X the (class of) a canonical divisor — the divisor of a section of ω_X.

8A   The Cohen–Macaulay Property

Theorem 8.1. *Let $X \subset \mathbb{P}^r$ be a smooth irreducible curve of arithmetic genus g over an algebraically closed field $\mathbb{K}$, embedded by a complete linear series as a curve of degree d. If $d \geq 2g+1$, then the homogeneous coordinate ring S_X is Cohen–Macaulay.*

This result first appeared in [Castelnuovo 1893]; subsequent proofs were published in [Mattuck 1961], [Mumford 1970] ,and [Green and Lazarsfeld 1985]. Here we follow a method of Green and Lazarsfeld because it works in all characteristics and generalizes easily to singular curves. In Exercises 8.16–8.20 we give an attractive geometric argument that works most smoothly in characteristic 0.

Before giving the proof, we deduce the Castelnuovo–Mumford regularity and Hilbert function of S_X:

Corollary 8.2. *Let $X \subset \mathbb{P}^r$ be an irreducible smooth curve of genus g over an algebraically closed field $\mathbb{K}$, embedded by a complete linear series as a curve of degree $d \geq 2g+1$. If $g = 0$ then reg $S_X = 1$; otherwise reg $S_X = 2$.*

Proof of Corollary 8.2. Since S_X is Cohen–Macaulay of dimension 2 we have $H_m^0(S_X) = H_m^1(S_X) = 0$, so S_X is m-regular if and only if $H_m^2(S_X)_{m-1} = 0$. By Corollary A1.12 this is equivalent to the condition that $H^1 \mathcal{O}_X(m-1) = 0$. Serre duality says that $H^1(\mathcal{O}_X(m-1))$ is dual to $H^0(K_X(-m+1))$, where K_X is the canonical divisor of X. Since the degree of $\mathcal{O}_X(1) = \mathscr{L}$ is at least $2g+1$, we have $\deg K_X(-1) \leq 2g-2-(2g+1) < 0$. Thus $H^0(K_X(-1)) = 0$, and S_X is 2-regular. On the other hand S_X is 1-regular if and only if $h^1(\mathcal{O}_X) = 0$. Since $h^1(\mathcal{O}_X) = g$, this concludes the proof.   □

Classically, the Cohen–Macaulay property of S_X was described as a condition on linear series. The degree n part of the homogeneous coordinate ring $(S_X)_n$ of X is the image of $H^0 \mathcal{O}_{\mathbb{P}^r}(n)$ in $H^0 \mathcal{O}_X(n)$. Thus the linear series $(\mathcal{O}_X(n), (S_X)_n)$ may be described as the *linear series cut out by hypersurfaces of degree n on X*. We may compare it to the *complete series* $(\mathcal{O}_X(n), H^0 \mathcal{O}_X(n))$. To prove Theorem 8.1 we will use the following criterion.

Proposition 8.3. *Let X be a curve in $\mathbb{P}^r$. The homogeneous coordinate ring S_X of X is Cohen–Macaulay if and only if the series of hypersurfaces of degree n in $\mathbb{P}^r$ is complete for every n; that is, the natural monomorphism*

$$S_X \rightarrow \bigoplus_n H^0 \mathcal{O}_X(n)$$

is an isomorphism.

Proof. The ring S_X has dimension 2, so it is Cohen–Macaulay if and only if it has has depth 2. By Proposition A1.16 this is the case if and only if $H^0_m S_X = 0 = H^1_m S_X$. Since S_X has no nilpotent elements we have $H^0_m S_X = 0$ in any case. The conclusion of the proposition follows from the exactness of the sequence

$$0 \longrightarrow H^0_m S_X \longrightarrow S_X \longrightarrow \bigoplus_n H^0 \mathcal{O}_X(n) \longrightarrow H^1_m S_X \longrightarrow 0$$

and from Corollary A1.12.   □

Corollary 8.4. *Let $X \subset \mathbb{P}^r$ be a smooth irreducible curve of arithmetic genus g over an algebraically closed field $\mathbb{K}$, embedded by a complete linear series as a curve of degree $d = 2g + 1 + p \geq 2g + 1$. If x, y are linear forms of S that do not vanish simultaneously anywhere on X, the Hilbert functions of S_X, $S_X / x S_X$ and $S_X / (x, y) S_X$ are as follows:*

	n	$M = $	$S_X/(x,y)S_X$	S_X/xS_X	S_X
	0		1	1	1
	1		$d-g+1$	$d-g$	$g+p$
$H_M(n)$:	2		$2d-g+1$	d	g
	3		$3d-g+1$	d	0
	$\vdots$		$\vdots$	$\vdots$	$\vdots$
	n		$nd-g+1$	d	0

In particular, if $\Gamma = H \cap X$ is a hyperplane section of X consisting of d distinct points, the points of Γ impose independent linear conditions on forms of degree ≥ 2, and the "last" graded Betti number of X is $\beta_{r-1,r+1}(S_X) = g$.

Proof. By Theorem 8.1 and Proposition 8.3 we have $(S_X)_n = H^0(\mathcal{O}_X(n))$. Furthermore, $H^1 \mathcal{O}_X(n) = 0$ for $n > 0$ because $d \geq 2g - 2$. For $M = S_X$, the value $H_M(n)$ is thus given by the Riemann–Roch formula,

$$H_M(n) = h^0 \mathcal{O}_X(n) = h^0 \mathcal{O}_X(n) - h^1 \mathcal{O}_X(n)$$
$$= \deg \mathcal{O}_X(n) - g + 1 = dn - g + 1.$$

These are the values in the left-hand column of the table. Since S_X is Cohen–Macaulay, the elements x, y for a regular sequence on S_X and we get short exact sequences

$$0 \longrightarrow S_X(-1) \longrightarrow S_X \longrightarrow S_X/xS_X \longrightarrow 0,$$
$$0 \longrightarrow (S_X/xS_X)(-1) \longrightarrow (S_X/xS_X) \longrightarrow S_X/(x,y)S_X \longrightarrow 0.$$

From these we see that the Hilbert functions of S_X/xS_X and $S_X/(x,y)S_X$ can be obtained from that of S_X by taking first and second differences, giving the rest of the values in the table.

If a hyperplane H has equation $x = 0$, then for any variety Y the homogeneous ideal of the hyperplane section $H \cap Y$ is the saturation of the homogeneous ideal

$I_Y + (x)$ defining S_Y/xS_Y. Since S_X is Cohen–Macaulay, S_X/xS_X has depth 1, and the ideal $I_X + (x)$ is already saturated. Thus the homogeneous coordinate ring $S_{H\cap X}$ is equal to S_X/xS_X. To say that the points of $\Gamma = H \cap X$ impose d linearly independent conditions on quadrics means that for $M = S_{H\cap X}$ we have $H_M(2) = d$, and the second column of the table shows that this is so.

Finally, to compute the "last" graded Betti number, we use the idea of Section 2B. If $x, y \in S_1$ form a regular sequence on S_X as above, then by Lemma 3.15 graded Betti numbers of S_X, as a module over S, variables are the same as those of $S_X/(x,y)S_X$, as a module over $S/(x,y)S$. The last column of the table gives us the Hilbert function of $S_X/(x,y)S_X$. By Proposition 2.7, $\beta_{r-1,r+1}(S_X)$ is the dimension of the homology of the complex $0 \to \mathbb{K}^g \to 0$, which is obviously g.
□

In fact, if X is a linearly normal curve and the points of a hyperplane section of X impose independent conditions on quadrics, then S_X is Cohen–Macaulay (Exercise 8.16). The alternate proof given in the Exercises relies on this.

Proof of Theorem 8.1. We will use the criterion in Proposition 8.3, and check that for each n the map $\alpha_n : (S_X)_n \to H^0\mathscr{O}_X(n)$ is surjective. Any effective divisor has nonnegative degree, so for $n < 0$ we have $H^0\mathscr{O}_X(n) = 0$ (see Exercise 8.6 for a generalization). Since the curve X in Theorem 8.1 is projective and connected, $H^0(\mathscr{O}_X)$ consists of the constant functions [Hartshorne 1977, Theorem I.3.4(a)]. Thus α_0 is an isomorphism, while α_1 is an isomorphism by our assumption that X is embedded by a complete linear series.

We now do induction and prove the surjectivity of α_{n+1} given the surjectivity of α_n with $n \geq 1$. There is a commutative diagram

$$
\begin{array}{ccc}
(S_X)_1 \otimes (S_X)_n & \xrightarrow{\alpha_1 \otimes \alpha_n} & H^0\mathscr{O}_X(1) \otimes H^0\mathscr{O}_X(n) \\
\downarrow & & \downarrow \mu_n \\
(S_X)_{n+1} & \xrightarrow{\alpha_{n+1}} & H^0\mathscr{O}_X(n+1).
\end{array}
$$

Since α_n is surjective, so is $\alpha_1 \otimes \alpha_n$. Thus it suffices to show that μ_n is surjective for each $n \geq 1$.

For $n \geq 2$ the surjectivity can be proved by the "basepoint-free pencil trick" of Castelnuovo; see Exercise 4.13. This is presumably the origin of the idea of Castelnuovo–Mumford regularity. For the case $n = 1$ we need a new tool, which in fact works in all cases.

8A.1   The Restricted Tautological Bundle

For simplicity we return to the notation $\mathscr{L} = \mathscr{O}_X(1)$. The map μ_n is the map on cohomology induced by the multiplication map of sheaves $H^0(\mathscr{L}) \otimes_\mathbb{K} \mathscr{L}^n \to \mathscr{L}^{n+1}$ where $\mathscr{L}^n$ means $\mathscr{L} \otimes \cdots \otimes \mathscr{L}$ with n factors). Thus μ_n is the map on cohomology induced by the tensor product of the identity map on $\mathscr{L}^n$ with the multiplication map $H^0(\mathscr{L}) \otimes_K \mathscr{O}_X \to \mathscr{L}$. We set

$$\mathscr{M}_X = \ker(H^0(\mathscr{L}) \otimes_K \mathscr{O}_X \to \mathscr{L}).$$

Thus $\mathscr{M}_X$ is the restriction to X of the tautological subbundle on $\mathbb{P}^r$ (see page 77).

Tensoring with $\mathscr{L}^n$ we obtain an exact sequence

$$0 \to \mathscr{M}_X \otimes \mathscr{L}^n \to H^0(\mathscr{L}) \otimes_K \mathscr{L}^n \to \mathscr{L}^{n+1} \to 0.$$

Taking cohomology, we see that the surjectivity of the map μ_n would follow from the vanishing of $H^1(\mathscr{M}_X \otimes \mathscr{L}^n)$. We will prove this vanishing by analyzing $\mathscr{M}_X$.

We first generalize. For any sheaf $\mathscr{F}$ on X we define

$$\mathscr{M}_{\mathscr{F}} = \ker(H^0(\mathscr{F}) \otimes_K \mathscr{O}_X \to \mathscr{F}),$$

so that $\mathscr{M}_X = \mathscr{M}_{\mathscr{L}}$. Thus we have a tautological left exact sequence

$$\epsilon_{\mathscr{F}} : 0 \to \mathscr{M}_{\mathscr{F}} \to H^0 \mathscr{F} \otimes \mathscr{O}_X \to \mathscr{F} \to 0.$$

which is right exact if and only if $\mathscr{F}$ is generated by global sections. This construction is functorial in $\mathscr{F}$. For any effective divisor $D \subset X$, the short exact sequence

$$0 \to \mathscr{L}(-D) \to \mathscr{L} \to \mathscr{L}|_D \to 0$$

gives rise to a diagram (whose rows and columns may not be exact!)

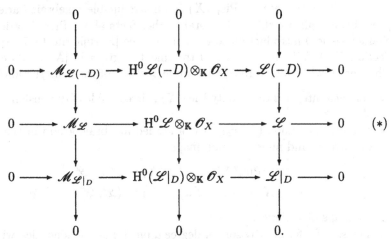

Whenever we can prove that the left-hand column is exact and analyze the sheaves $\mathscr{M}_{\mathscr{L}(-D)}$ and $\mathscr{M}_{\mathscr{L}|_D}$ we will get useful information about $\mathscr{M}_X = \mathscr{M}_{\mathscr{L}}$.

We will do exactly that for the case where D is the sum of $d - g - 1$ general points of X. For this we need some deeper property of linear series, expressed in part 6 of the following Lemma. Parts 1–3 will be used in the proof of part 6. We will leave parts 4 and 5, which we will not use, for the reader's practice (Exercise 8.11).

Theorem 8.5. *Suppose that X is a smooth curve of arithmetic genus g over an algebraically closed field, and let d be an integer.*

1. *If $d \geq g-1$ then the set of line bundles $\mathscr{L}' \in \mathrm{Pic}_d(X)$ with $\mathrm{h}^1\mathscr{L}' = 0$ is open and dense.*
2. *If $\mathscr{L}'$ is any line bundle of degree $\geq g$ then $\mathscr{L}' = \mathcal{O}_X(D)$ for some effective divisor D on X.*
3. *If $\mathscr{L}'$ is a general line bundle of degree $\geq g+1$ then $|\mathscr{L}'|$ is base point free. In particular, if $\deg \mathscr{L}' = g+1$, then $\mathrm{h}^2\mathscr{L}' = 0$, $\mathrm{h}^0\mathscr{L}' = 2$, and $|\mathscr{L}'|$ exhibits X as a $(g+1)$-fold cover of $\mathbb{P}^1$.*
4. *If $\mathscr{L}'$ is a general line bundle of degree $\geq g+2$ then $|\mathscr{L}'|$ maps X birationally. If $\deg \mathscr{L}' = g+2$, the image is a curve of degree $g+2$ with at worst ordinary nodes in $\mathbb{P}^2$.*
5. *If $\mathscr{L}'$ is a general line bundle of degree $\geq g+3$ then $\mathscr{L}'$ is very ample; that is, $|\mathscr{L}'|$ embeds X. In particular, if $\deg \mathscr{L}' = g+3$, then $|\mathscr{L}'|$ embeds X as a curve of degree $g+3$ in $\mathbb{P}^3$.*
6. *If $\mathscr{L}$ is a line bundle of degree $d \geq 2g+1$ and D is a general effective divisor of degree $d-g-1$ then $\mathscr{L}' = \mathscr{L}(-D)$ has $\mathrm{h}^1(\mathscr{L}') = 0$, $\mathrm{h}^0(\mathscr{L}') = 2$, and $|\mathscr{L}'|$ is basepoint-free.*

Here, when we say that something is true for "a general effective divisor of degree m," we mean that there is a dense open subset $U \subseteq X^m = X \times X \times \cdots \times X$ such that the property holds for all divisors $D = \sum_1^m p_i$ with $(p_1, \ldots, p_m) \in U$. To say that something holds for a general line bundle of degree m makes sense in the same way because $\mathrm{Pic}_m(X)$ is an irreducible algebraic variety. In the proof below will use this and several further facts about Picard varieties. For a characteristic 0 introduction to the subject, see [Hartshorne 1977, Appendix B, Section 5]. A full characteristic 0 treatment is given in [Arbarello et al. 1985, Chapter 1], while [Serre 1988] gives an exposition of the construction in general.

- For each integer d the variety $\mathrm{Pic}_d(X)$ is irreducible of dimension g, the genus of X.
- The disjoint union $\bigcup \mathrm{Pic}_d(X)$ is a graded algebraic group in the sense that the inverse and multiplication maps

$$\mathrm{Pic}_d(X) \to \mathrm{Pic}_{-d}(X): \quad \mathscr{L} \mapsto \mathscr{L}^{-1}$$
$$\mathrm{Pic}_d(X) \times \mathrm{Pic}_e \to \mathrm{Pic}_{d+e}(X): \quad (\mathscr{L}, \mathscr{L}') \mapsto \mathscr{L} \otimes \mathscr{L}'$$

are maps of varieties.
- The set of effective divisors of degree d on X may be identified with the d-th *symmetric power* $X^{(d)} := X^d/G$, where $X^d = X \times \cdots \times X$ is the direct product of d copies of X and G is the symmetric group on d elements, permuting the factors. The identification is given by

$$X^d \ni (x_1, \ldots, x_d) \mapsto x_1 + \cdots + x_d.$$

Since X^d is a projective variety of dimension d and G is a finite group, $X^{(d)}$ is also a projective variety of dimension d.

- The map of sets $X^{(d)} \to \mathrm{Pic}_d(X)$ sending $x_1 + \cdots + x_d$ to the line bundle $\mathcal{O}_X(x_1 + \cdots + x_d)$ is a map of algebraic varieties, called the Abel–Jacobi map. Its fiber over a line bundle $\mathcal{L}$ is thus isomorphic to the projective space of global sections of $\mathcal{L}$, modulo nonzero scalars.

Proof of Theorem 8.5. Part 1: By Serre duality, $h^1 \mathcal{L}' = h^0(\omega_X \otimes \mathcal{L}'^{-1})$. Further, if $\deg \mathcal{L}' = d \geq g-1$ then $\deg(\omega_X \otimes \mathcal{L}'^{-1}) = 2g-2-d \leq g-1$. The map $\mathrm{Pic}_d(X) \to \mathrm{Pic}_{2g-2-d}(X)$ taking $\mathcal{L}'$ to $\omega_X \otimes \mathcal{L}'^{-1}$ is a morphism. Its inverse is given by the same formula, so it is an isomorphism. Thus it suffices to show the set of line bundles $\mathcal{L}'' \in \mathrm{Pic}_{2g-2-d}(X)$ of with $h^0 \mathcal{L}'' = 0$ is open and dense. Let $e = 2g-2-d \leq g-1$. The complementary set, the set of bundles $\mathcal{L}'' \in \mathrm{Pic}_e(X)$ with nonzero sections, is the image of the Abel–Jacobi map $X^{(e)} \to \mathrm{Pic}_e$. Since $X^{(e)}$ is projective, the image is closed and of dimension at most $\leq \dim X^{(e)} = e < g = \dim \mathrm{Pic}_e(X)$. Thus the set of bundles of degree e without sections is nonempty and open; it is dense since $\mathrm{Pic}_e(X)$ is irreducible.

Part 2: Let x be a point of X. For any integer d the morphism

$$\mathrm{Pic}_d(X) \ni \mathcal{L}' \mapsto \mathcal{L}'(p) = \mathcal{L}' \otimes \mathcal{O}_X(p) \in \mathrm{Pic}_{d+1}(X)$$

is an isomorphism (its inverse is $\mathcal{L}'' \mapsto \mathcal{L}''(-p)$). Thus it suffices to show that every line bundle of degree exactly g can be written as $\mathcal{O}_X(D)$ for some $D \in X^{(g)}$. That is, it suffices to show that the Abel–Jacobi map $X^{(g)} \to \mathrm{Pic}_g(X)$ is surjective. These varieties both have dimension g. Since $X^{(g)}$ is a projective variety its image is closed, so it suffices to show that the image has dimension g, or equivalently, that the general fiber is finite. The fiber through a general divisor D consists of the set of divisors linearly equivalent to D, so it suffices to show that there are none except D — that is, $h^0(\mathcal{O}_X(D)) = 1$.

By the Riemann–Roch theorem and Serre duality,

$$h^0(\mathcal{O}_X(D)) = \deg D - g + 1 + h^1(\mathcal{O}_X(D)) = 1 + h^0(\omega_X(-D)).$$

If $\mathcal{F}$ is any sheaf on X with $H^0 \mathcal{F} \neq 0$ then the set of sections of $\mathcal{F}$ vanishing at a general point of X is a proper linear subspace of $H^0 \mathcal{F}$. We may write D as the sum of g general points, $D = p_1 + \cdots + p_g$. Since $h^0(\omega_X) = g$, we have $h^0(\omega_X(-p_1 - \cdots - p_g)) = 0$ as required.

Part 3: Suppose $d \geq g+1$ and let $U \subset \mathrm{Pic}_d(X)$ be set of line bundles $\mathcal{L}'$ with $h^1(\mathcal{L}') = 0$, which is open and dense by part 8.5. Let

$$U' = \{(\mathcal{L}', p) \subset U \times X \mid p \text{ is a basepoint of } \mathcal{L}'\},$$

and let $\pi_1 : U' \to U$ and $\pi_2 : U' \to X$ be the projections. The set of line bundles of degree d without basepoints contains the complement of $\pi_1(U')$. It thus suffices to show that $\dim U' < g$.

Consider the map

$$\phi : U' \to \mathrm{Pic}_{d-1}(X); \quad (\mathcal{L}', p) \mapsto \mathcal{L}'(-p).$$

The fiber $\phi^{-1}(\mathcal{L}'')$ over any line bundle $\mathcal{L}''$ is contained in the set $\{(\mathcal{L}''(p), p) \mid p \in X\}$ parametrized by X, so $\dim \phi^{-1}(\mathcal{L}'') \leq 1$. On the other hand, the image $\phi(U')$ consists of line bundles $\mathcal{L}'(-p)$ such that $h^0(\mathcal{L}'(-p)) = h^0(\mathcal{L}')$. Applying the Riemann–Roch formula, and using $h^1(\mathcal{L}') = 0$, we see that $h^0(\mathcal{L}'(-p)) = (d-1) - g + 1 + h^1(\mathcal{L}'(-p)) = d - g + 1$; that is, $h^1(\mathcal{L}'(-p)) = 1$. It thus suffices to show that the set U'' of line bundles $\mathcal{L}''$ of degree $d - 1 \geq g$ with $h^1(\mathcal{L}'') \neq 0$ has dimension $\leq g - 2$.

Let $e = 2g - 2 - (d - 1)$. Under the isomorphism

$$\text{Pic}_{d-1}(X) \to \text{Pic}_e(X); \quad \mathcal{L}'' \mapsto \omega_X \otimes \mathcal{L}''^{-1}$$

the set U'' is carried into the set of bundles with a nonzero global section, the image of the Abel–Jacobi map $X^{(e)} \to \text{Pic}_e(X)$. This image has dimension at most $\dim X^{(e)} = e = 2g - 2 - (d-1) \leq 2g - 2 - g = g - 2$ as required.

Part 6: If $d \geq 2g + 1$ then $d - g - 1 \geq g$, so any line bundle of degree $d - g - 1$ can be written as $\mathcal{O}_X(D)$ for some effective divisor. Thus if $\mathcal{L}$ has degree d, and D is a general effective divisor of degree $d - g - 1$, then $\mathcal{L}'' := \mathcal{O}_X(D)$ is a general line bundle of degree $d - g - 1$, and $\mathcal{L}' = \mathcal{L} \otimes \mathcal{L}''^{-1}$ is a general line bundle of degree $g + 1$. The assertions of part 6 thus follow from those of part 3. $\qquad \square$

Returning to the proof of Theorem 8.1 and its notation, we suppose that D is a general divisor of degree $d - g - 1$, the sum of $d - g - 1$ general points. Since $\mathcal{L}|_D$ is a coherent sheaf with finite support, it is generated by global sections. The line bundle $\mathcal{L}$ is generated by global sections too, as already noted, and by Theorem 8.5, part 6, the same goes for $\mathcal{L}(-D)$. Thus all three rows of diagram $(*)$ are exact. The exactness of the right-hand column is immediate, while the exactness of the middle column follows from the fact that $H^1 \mathcal{L}(-D) = 0$. By the Snake Lemma, it follows that the left-hand column of $(*)$ is exact.

To understand $M_{\mathcal{L}(-D)}$, we use part 6 of Theorem 8.5 again. Let σ_1, σ_2 be a basis of the vector space $H^0(\mathcal{L}(-D))$. We can form a sort of Koszul complex

$$\mathbf{K} : 0 \to \mathcal{L}^{-1}(D) \xrightarrow{\begin{pmatrix} \sigma_2 \\ -\sigma_1 \end{pmatrix}} \mathcal{O}_X^2 \xrightarrow{(\sigma_1 \quad \sigma_2)} \mathcal{L}(-D) \to 0$$

whose right-hand map $\mathcal{O}_X^2 \xrightarrow{(\sigma_1 \quad \sigma_2)} \mathcal{L}(-D) \to 0$ is the map $H^0 \mathcal{L}(-D) \otimes_{\mathbf{K}} \mathcal{O}_X \to \mathcal{L}(-D)$ in the sequence $\epsilon_{\mathcal{L}(-D)}$. If $U = \text{Spec } R \subset X$ is an open set where $\mathcal{L}$ is trivial, then we may identify $\mathcal{L}|_U$ with R, and σ_1, σ_2 as a pair of elements generating the unit ideal. Thus $\mathbf{K}|_U$ is exact, and since X is covered by such open sets U, the complex $\mathbf{K}$ is exact. It follows that $M_{\mathcal{L}(-D)} = \mathcal{L}^{-1}(D)$.

Finally, to understand $M_{\mathcal{L}|_D}$ we choose an isomorphism $\mathcal{L}|_D = \mathcal{O}_D$. Writing $D = \sum_1^{d-g-1} p_i$, the defining sequence $\epsilon_{\mathcal{O}_D}$ becomes

$$0 \to M_{\mathcal{O}_D} \to \sum_1^{d-g-1} \mathcal{O}_X \to \sum_1^{d-g-1} \mathcal{O}_{p_i} \to 0,$$

and we deduce that $\mathcal{M}_{\mathcal{O}_D} = \sum_1^{d-g-1} \mathcal{O}_X(-p_i)$.

The left-hand column of diagram $(*)$ is thus an exact sequence

$$0 \to \mathcal{L}^{-1}(D) \to \mathcal{M}_X \to \sum_1^{d-g-1} \mathcal{O}_X(-p_i) \to 0.$$

Tensoring with $\mathcal{L}^n$ and taking cohomology, we get an exact sequence

$$H^1(\mathcal{L}^{n-1}(D)) \to H^1(\mathcal{L}^n \otimes \mathcal{M}_X) \to \sum H^1(\mathcal{L}^n(-p_i)).$$

As D is general of degree $d - g - 1 > g - 1$, part 1 of Theorem 8.5 gives $H^1(\mathcal{L}^{n-1}(D)) = 0$ for all $n \geq 1$. Since $\mathcal{L}^n(-p_i)$ has degree at least $n(2g+1)-1 \geq 2g$, its first cohomology also vanishes, whence $H^1(\mathcal{L}^n \otimes \mathcal{M}_X) = 0$ as required for the proof of Theorem 8.1. □

8B   Strands of the Resolution

Consider again the case of a curve X of genus g embedded in $\mathbb{P}^r$ by a complete linear series $|\mathcal{L}|$ of "high" degree $d = 2g+1+p \geq 2g+1$ (so that by Riemann–Roch we have $r = d-g$.) By Theorem 8.1 and Corollary 8.2 the resolution of S_X has the form

	0	1	2	$\cdots$	$\cdots$	$r-2$	$r-1$
0	1	-	-	$\cdots$	$\cdots$	-	-
1	-	$\beta_{1,2}$	$\beta_{2,3}$	$\cdots$	$\cdots$	$\beta_{r-2,r-1}$	$\beta_{r-1,r}$
2	-	$\beta_{1,3}$	$\beta_{2,4}$	$\cdots$	$\cdots$	$\beta_{r-2,r}$	$\beta_{r-1,r+1}$

where $\beta_{i,j}$ is the vector space dimension of $\mathrm{Tor}_i^S(S_X, \mathbb{K})_j$. The goal of this section is to explain what is known about the $\beta_{i,j}$. We will call the strand of the resolution corresponding to the $\beta_{i,i+1}$ the *quadratic strand*; the $\beta_{i,i+2}$ correspond to the *cubic strand*. (The names arise because $\beta_{1,2}$ is the number of quadratic generators required for the ideal of X, while $\beta_{1,3}$ is the number of cubic equations.)

Since I_X contains no linear forms, the number of generators of degree 2 is

$$\beta_{1,2} = \dim(I_X)_2 = \dim S_2 - \dim(S_X)_2 = \binom{r+2}{2} - (2d-g+1) = \binom{d-g-1}{2},$$

where the penultimate equality comes from Corollary 8.4 and the Riemann–Roch theorem. This argument extends a little. By Corollary 1.10, the formula in Corollary 8.4 determines the numbers $\beta_{i,i+1} - \beta_{i-1,i+1}$ for all i in terms of the genus g and degree d of $X \subset \mathbb{P}_0^r$.

We have already given a similar argument computing the "last" graded Betti number, $\beta_{r-1,r+1}(S_X)$ (Corollary 8.4). Now we will give a conceptual argument yielding much more.

Proposition 8.6. *With notation as above, $\beta_{r-1,r+1} = g$. In fact, if $\mathbf{F}$ is the minimal free resolution of S_X as an S-module, and ω_X is the canonical sheaf of X, then the twisted dual, $\mathrm{Hom}_S(\mathbf{F}, S(-r-1))$, of $\mathbf{F}$, is the minimal free resolution of the S-module $w_X := \bigoplus_n \mathrm{H}^0\omega_X(n)$.*

Proof. The first statement of the Proposition follows from the second because w_X is 0 in negative degrees, while $(w_X)_0 = \mathrm{H}^0\omega_X$ is a vector space of dimension g.

Since S_X is Cohen–Macaulay and of codimension $r-1$ we have

$$\mathrm{Ext}_S^i(S_X, S(-r-1)) = 0 \qquad \text{for } i \neq r-1.$$

In other words, the cohomology of the twisted dual $\mathrm{Hom}(\mathbf{F}, S(-r-1))$ is zero except at the end, so it is a free resolution of the module $\mathrm{Ext}_S^{r-1}(S_X, S(-r-1))$. It is minimal because it is the dual of a minimal complex. Because the resolution is of length $r-1$, the module $\mathrm{Ext}_S^{r-1}(S_X, S(-r-1))$ is Cohen–Macaulay, and it follows from Corollary A1.12 that $\mathrm{Ext}_S^{r-1}(S_X, S(-r-1)) = \bigoplus_n \mathrm{H}^0\omega_X(n)$. In particular, we see that

$$\beta_{r-1,r+1}(S_X) = \beta_{0,0}\big(\mathrm{Ext}_S^{r-1}(S_X, S(-r-1))\big) = \dim_{\mathbb{K}} \mathrm{H}^0\omega_X = h^0\omega_X.$$

From Serre duality we have $h^0\omega_X = h^1\mathcal{O}_X = g$, as required by the last formula. $\square$

In terms of Betti diagrams, Proposition 8.6 means that the Betti diagram of w_X is obtained by "reversing" that of S_X left-right and top-to-bottom. Taking account of what we know so far, it has the form:

	0	1	2	$\cdots$	$\cdots$	$r-2$	$r-1$
0	g	$\beta_{r-2,r}$	$\cdots$	$\cdots$		$\beta_{2,4}$	$\beta_{1,3}$	-
1	$\beta_{r-1,r}$	$\beta_{r-2,r-1}$	$\cdots$	$\cdots$		$\beta_{2,3}$	$\beta_{1,2}$	-
2	-	-	$\cdots$	$\cdots$		-	-	1

It would be fascinating to know what the value of each individual Betti number says about the geometry of the curve, but this is far beyond current knowledge. A cruder question is, "Which of the $\beta_{i,j}$ are actually nonzero?" In fact, there is just one block of nonzero entries in each row:

Proposition 8.7. *If $I \subset S$ is a homogeneous ideal that does not contain any linear forms, and if S/I is Cohen–Macaulay of regularity 3, then*

$$\beta_{i,i+1} = 0 \Rightarrow \beta_{j,j+1} = 0 \text{ for } j \geq i,$$
$$\beta_{i,i+2} = 0 \Rightarrow \beta_{j,j+2} = 0 \text{ for } j \leq i.$$

Proof. Using Proposition 1.9, applied to the resolution of S_X, gives the first conclusion. By Proposition 8.6 the dual complex is also a resolution; applying Proposition 1.9 to it, we get the second conclusion. $\square$

Because the projective dimension of S_X is $r-1$, at least one of $\beta_{i,i+1}$ and $\beta_{i,i+2}$ must be nonzero for $i = 1, \ldots, r-1$. Thus the nonzero entries in the Betti diagram of S_X are determined by two numbers $a = a(X)$ and $b = b(X)$ with $0 \leq a < b \leq r$ which may be defined informally from the diagram

	0	1	$\cdots$	a	$a+1$	$\cdots$	$b-1$	b	$\cdots$	$r-1$
0	1	$-$	$\cdots$	$-$	$-$	$\cdots$	$-$	$-$	$\cdots$	$-$
1	$-$	$*$	$\cdots$	$*$	$*$	$\cdots$	$*$	$-$	$\cdots$	$-$
2	$-$	$-$	$\cdots$	$-$	$*$	$\cdots$	$*$	$*$	$\cdots$	g

where "$-$" denotes a zero entry and "$*$" denotes a nonzero entry (we admit the possibilities $a = 0, b = r$, and $b = a+1$.) More formally, $0 \leq a(X) < b(X) \leq r$ are defined by letting $a(X)$ be the greatest number such that $\beta_{i,i+2}(S_X) = 0$ for all $i \leq a(X)$ and letting $b(X)$ be the least number such that $\beta_{i,i+1}(S_X) = 0$ for all $i \geq b(X)$.

Note that when $b \leq a+2$ Corollaries 8.4 and 1.10 determine all of the numbers $\beta_{i,j}$. However if $b \geq a+3$ there could be examples with the same genus and degree but with different graded Betti numbers.

8B.1 The Cubic Strand

What does the number a tell us? It is closely related to an important geometric invariant of the embedding $X \subset \mathbb{P}^r$, the dimension of the smallest degenerate secant plane. To understand this notion, recall that q general points span a projective $q-1$-plane. A plane in $\mathbb{P}^r$ is a *degenerate q-secant plane* to X if it has dimension at most $q-2$ and meets X in at least q points, or more generally if it meets X in a scheme of length at least q. We use $\lfloor x \rfloor$ and $\lceil x \rceil$ to denote the *floor* and *ceiling* of x, the largest integer $\leq x$ and the smallest integer $\geq x$ respectively.

Theorem 8.8. *Suppose that $X \subset \mathbb{P}^r$ is a curve embedded by a complete linear series of degree $2g+1+p$, with $p \geq 0$.*

1. *$p \leq a(X)$.*
2. *If X has a degenerate q-secant plane, then $a(X) \leq q-3$.*
3. *X always has a degenerate q-secant plane for $q = p+3+\max(0, \lceil \frac{g-p-3}{2} \rceil)$. Thus*

$$p \leq a(X) \leq p+\max\left(0, \left\lceil \frac{g-3-p}{2} \right\rceil\right).$$

When $p \geq g-3$, or in other words $d = 2g+1+p \geq 3g-2$, Parts 1 and 2 show that $a(X)$ determines the size of the smallest degenerate secant plane precisely. For smaller p, and special X other phenomena can occur. See the example and discussion in Section 8C.

Part 1 of Theorem 8.8, along with Theorem 8.1, is usually stated by saying that a linearly normal curve $X \subset \mathbb{P}^r$ of degree $2g+1+p$ *satisfies condition N_p*;

here N_0 is taken to mean that S_X is Cohen–Macaulay; N_1 means N_0 and the condition that I_X is generated by quadrics; N_2 means in addition that I_X is linearly presented; and so on.

Proof of Theorem 8.8.1. Let $\mathbf{F}$ be the minimal free resolution of S_X. By Proposition 8.6 the complex $\operatorname{Hom}_S(\mathbf{F}, S(-r-1))$ is the minimal free resolution of $w_X = \oplus_n H^0(\omega_X(n))$. We have $\mathrm{h}^0(\omega_X) = g$, while $h^0 \omega_X(n) = 0$ for $n < 0$ since $\deg \mathscr{O}_X(1) > \deg \omega_X = 2g - 2$. Thus we may apply Corollary 7.4, and we see that the linear strand of the free resolution $\operatorname{Hom}_S(\mathbf{F}, S(-r-1))$ has length $r - a - 2 \leq g - 1$, as required. $\qquad\square$

Theorem 8.8.2 is a special case (see Exercise 8.12) of a more general geometric result:

Theorem 8.9. *If a variety (or scheme) $X \subset \mathbb{P}^r$ intersects a plane Λ of dimension e in a finite scheme of length at least $e + 2$, the graded Betti number $\beta_{e,e+2}(S_X)$ is nonzero. In particular, $a \leq e - 1$.*

The idea is that by Theorem 4.1 and Proposition 4.12, the homogeneous coordinate ring of a set of dependent points in $\mathbb{P}^e$ cannot be 1-regular, and the cubic strand of its resolution begins by the e-th step. In general, the regularity of a subset $Y \subset X$ need not be bounded by the regularity of X, but in our setting the high degree syzygy in the e-th place of the resolution of the coordinate ring of the point somehow forces a high degree syzygy in the same place in the resolution of the coordinate ring of X. The proof we will give is indirect; we bound the local cohomology instead of the syzygies. Here is a general algebraic version, from which Theorem 8.9 will follow easily. The reader will recognize the idea used here from the proof of the Gruson–Lazarsfeld–Peskine Theorem 5.1: if the homology of a free complex has low dimension, then the complex can be used to compute regularity as if it were a resolution.

Theorem 8.9 follows at once from a still more general result of Eisenbud, Huneke and Ulrich [Eisenbud et al. 2004, Theorem 2.1]. We give here the special case of the result that is needed:

Theorem 8.10. *Let M be a finitely generated gradedmodule over a polynomial ring $S = \mathbf{K}[x_0, \ldots, x_r]$. Set $\bar{S} = \mathbf{K}[x_0, \ldots, x_p]$ be the quotient of S by an ideal generated by $r - p$ linear forms, and $\bar{M} = M \otimes_S \bar{S}$. If $\dim \bar{M} \leq 1$ then $\operatorname{reg} \mathrm{H}^1_{\mathfrak{m}}(\bar{M}) + 1 \leq \operatorname{reg} \operatorname{Tor}_p(M, \mathbb{K}) - p$.*

Proof. Let $\mathbf{F} : \cdots \to F_1 \to F_0 \to M \to 0$ be the minimal free resolution of M as an S-module, and write $\bar{F}_i$ for $\bar{S} \otimes F_i$. Let $K_i = \ker \bar{F}_i \to \bar{F}_{i-1}$ be the module of i-cycles, and let $B_i = \operatorname{im} \bar{F}_{i+1} \to \bar{F}_i$ be the module of boundaries, so that there are exact sequences

$$(E_i) \qquad\qquad 0 \to B_i \to K_i \to \mathrm{H}_i(\bar{S} \otimes \mathbf{F}) \to 0,$$
$$(G_i) \qquad\qquad 0 \to K_i \to \bar{F}_i \to B_{i-1} \to 0,$$
$$(G_0) \qquad\qquad 0 \to K_0 \to \bar{F}_0 \to \bar{M} \to 0.$$

The objects that play a role in the proof appear in the diagram

$$H^1_m \bar{M} \xrightarrow{s_0} H^2_m K_0 \xleftarrow{t_0} H^2_m B_0 \xrightarrow{s_1} H^3_m K_1 \xleftarrow{t_1} H^3_m B_1 \xrightarrow{s_2} H^4_m K_2 \xleftarrow{t_2} \cdots$$

$$\Big\uparrow{u_1} \qquad\qquad\qquad \Big\uparrow{u_2} \qquad\qquad \cdots$$

$$H^2_m \bar{F}_1 \qquad\qquad\qquad H^3_m \bar{F}_2 \qquad\qquad \cdots$$

where the map t_i is induced by the inclusion $B_i \subset K_i$, the map s_i is the connecting homomorphism coming from the sequence G_i and the map u_i comes from the surjection $\bar{F}_i \to B_{i-1}$. We will prove:

1. Each t_i is an isomorphism;
2. For $i < p$ the map s_i is a monomorphism;
3. For $i = p$ the map u_i is a surjection.

It follows from items 1–3 that $H^1_m(\bar{M})$ is a subquotient of $H^{p+1}_m(\bar{F}_p)$. In particular, since both of these are Artinian modules, $\operatorname{reg} H^1_m(\bar{M}) \leq \operatorname{reg} H^{p+1}_m(\bar{F}_p)$. By Lemma A1.6,

$$\operatorname{reg} H^{p+1}_m(\bar{F}_p) + p + 1$$

is the maximum degree of a generator of $\bar{F}_p$ or, equivalently, of F_p; this number is also equal to $\operatorname{reg} \operatorname{Tor}_p(M, \mathbf{K})$. Putting this together we get $\operatorname{reg} H^1_m(\bar{M}) + 1 \leq \operatorname{reg} \operatorname{Tor}_p(M, \mathbf{K}) - p$ as required.

The map t_i is an isomorphism for $i = 0$ simply because $B_0 = K_0$. For $i > 0$, we first note that $H_i(\bar{S} \otimes F) = \operatorname{Tor}_i(\bar{S}, M)$. Since $\bar{M} = \bar{S} \otimes M$ has dimension ≤ 1, the annihilator of M plus the annihilator of $\bar{S}$ is an ideal of dimension ≤ 1. This ideal also annihilates $\operatorname{Tor}_i(\bar{S}, M)$, so $\dim \operatorname{Tor}_i(\bar{S}, M) \leq 1$ also. It follows that $H^j_m(H_i(\bar{S} \otimes F)) = 0$ for all $j \geq 2$ and all i. The short exact sequence (E_i) gives rise to a long exact sequence containing

$$H^{i+1}_m(H_i(\bar{S} \otimes F)) \longrightarrow H^{i+2}_m(B_i) \xrightarrow{t_i} H^{i+2}_m(K_i) \longrightarrow H^{i+2}_m(H_i(\bar{S} \otimes F))$$

and we have just shown that for $i \geq 1$ the two outer terms are 0. Thus t_i is an isomorphism, proving the statement in item 1.

For items 2 and 3 we use the long exact sequence

$$\cdots \longrightarrow H^{i+1}_m \bar{F}_i \longrightarrow H^{i+1}_m \bar{B}_{i-1} \xrightarrow{s_i} H^{i+2}_m \bar{K}_i \longrightarrow \cdots$$

corresponding to the short exact sequence (G_i). For $i < p$ we have $H^{i+1}_m \bar{F}_i = 0$, giving the conclusion of item 2. Finally, $\dim \bar{S} = p+1$, so $H^{p+2}_m K_i = 0$. This gives the statement of item 3. $\qquad\square$

Conclusion of the proof of Theorem 8.8. It remains to prove part 3, and for this it is enough to produce a degenerate q-secant plane with

$$q = p + 3 + \max\big(0, \lceil (g-p-3)/2 \rceil\big),$$

to which to apply Theorem 8.9.

To do this we will focus not on the q-plane but on the subscheme D in which it meets X. We don't need to know about schemes for this: in our case D is an effective divisor on X. Thus we want to know when an effective divisor spans "too small" a plane.

The hyperplanes in $\mathbb{P}^r$ correspond to the global sections of $\mathscr{L} := \mathcal{O}_X(1)$, so the hyperplanes containing D correspond to the global sections of $\mathscr{L}(-D)$. Thus the number of independent sections of $\mathscr{L}(-D)$ is the codimension of the span of D. That is, D spans a projective plane of dimension $e = r - \mathrm{h}^0(\mathscr{L}(-D)) = \mathrm{h}^0(\mathscr{L}) - 1 - \mathrm{h}^0(\mathscr{L}(-D))$.

The Riemann–Roch formula applied to $\mathscr{L}$ and to $\mathscr{L}(-D)$ shows that

$$
\begin{aligned}
e &= (\deg \mathscr{L} - g + 1 - \mathrm{h}^1\mathscr{L}) - 1 - (\deg \mathscr{L} - \deg D - g + 1 - \mathrm{h}^1\mathscr{L}(-D)) \\
&= \deg D + \mathrm{h}^1\mathscr{L} - \mathrm{h}^1\mathscr{L}(-D) - 1 \\
&= \deg D - \mathrm{h}^1\mathscr{L}(-D) - 1,
\end{aligned}
$$

since $\mathrm{h}^1\mathscr{L} = 0$. From this we see that the points of D are linearly dependent, that is, $e \le \deg D - 2$, if and only if

$$
\mathrm{h}^1\mathscr{L}(-D) = \mathrm{h}^0\big(\omega_X \otimes \mathscr{L}^{-1}(D)\big) \ne 0.
$$

This nonvanishing means $\omega_X \otimes \mathscr{L}^{-1}(D) = \mathcal{O}_X(D')$, or equivalently that $\mathscr{L} \otimes \omega_X^{-1} = \mathcal{O}_X(D - D')$, for some effective divisor D'.

The degree of $\mathscr{L} \otimes \omega_X^{-1}$ is $2g + 1 + p - (2g - 2) = p + 3$, but we know nothing else about it. If $p \ge g - 3$, then $\deg \mathscr{L} \otimes \omega_X^{-1} \ge g$. By Theorem 8.5, Part 8.5, there is an effective divisor D such that $\mathscr{L} \otimes \omega_X^{-1} = \mathcal{O}_X(D)$, and taking $D' = 0$ we see that the span of D is a degenerate $p + 3$-secant plane, as required in this case.

On the other hand, if $p < g - 3$, the subset of $\mathrm{Pic}_{p+3}(X)$ that consists of line bundles of degree $p + 3$ that can be written in the form $\mathcal{O}_X(D)$ is the image of X^{p+3}, so it has at most dimension $p + 3 < g$. Thus it cannot be all of the variety $\mathrm{Pic}_{p+3}(X)$, and we will not in general be able to take $D' = 0$. From this argument it is clear that we may have to take the degree q of D large enough so that the sum of the degrees of D and D' is at least g. Moreover this condition suffices: if q and q' are integers with $q + q' = g$ then the map

$$
X^q \times X^{q'} \to \mathrm{Pic}_{q-q'}(X)
$$

$$
((a_1, \ldots, a_q), (b_1, \ldots b_{q'})) \mapsto \mathcal{O}_X\left(\sum_1^q a_i - \sum_1^{q'} b_j\right)
$$

is surjective (see [Arbarello et al. 1985, V.D.1]).

With this motivation we take

$$
q = p + 3 + \left\lceil \frac{g - p - 3}{2} \right\rceil = \left\lceil \frac{g + p + 3}{2} \right\rceil,
$$

$$
q' = \left\lfloor \frac{g - p - 3}{2} \right\rfloor.
$$

We get $q - q' = p + 3$ and $q + q' = g$, so by the result above we may write the line bundle $\mathscr{L} \otimes \omega_X^{-1}$ in the form $\mathscr{O}_X(D - D')$ for effective divisors D and D' of degrees q and q', and the span of D will be a degenerate q-secant plane as required. $\qquad\square$

Some of the uncertainty in the value of $a(X)$ left by Theorem 8.8 can be explained in terms of the quadratic strand; see Example 8.21 and Theorem 8.23.

8B.2   The Quadratic Strand

We turn to the invariant of X given by $b(X) = \min\{i \geq 1 \mid \beta_{i,i+1}(X) = 0\}$. Theorem 8.8 shows that some $\beta_{i,i+2}$ is nonzero when X contains certain "interesting" subschemes. By contrast, we will show that some $\beta_{i,i+1}$ is nonzero by showing that X is contained in a variety Y with $\beta_{i,i+1}(S_Y) \neq 0$. To do this we compare the resolution of I_X with that of its submodule I_Y.

Proposition 8.11. *Suppose that $M' \subset M$ are graded S-modules. If $M_n = 0$ for $n < e$, then $\beta_{i,i+e}(M') \leq \beta_{i,i+e}(M)$ for all i.*

Proof. If $M'_e = 0$ then $\beta_{0,e}(M') = 0$, and since the differential in a minimal resolution maps each module into $\mathfrak{m}$ times the next one, it follows by induction that $\beta_{i,i+e}(M') = 0$ for every i. Thus we may assume that $M'_e \subset M_e$ are both nonzero. Under this hypothesis, we will show that any map $\phi : \mathbf{F}' \to \mathbf{F}$ from the minimal free resolution of M' to that of M that lifts the inclusion $M' \subset M$ must induce an inclusion of the linear strands. To simplify the notation we may shift both M and M' so that $e = 0$.

Let $\mathbf{G} \subset \mathbf{F}$ be the linear strand, so that the i-th free module G_i in G is a direct sum of copies of $S(-i)$, and similarly for $\mathbf{G}' \subset \mathbf{F}'$. To prove that $\phi_i|_{G_i} : G_i' \to F_i$ is an inclusion, we do induction on i, starting with $i = 0$.

Because the resolution is minimal, we have $F_0/\mathfrak{m}F_0 = M/\mathfrak{m}M$. In particular $G_0/\mathfrak{m}G_0 = M_0$, and similarly $G_0'/\mathfrak{m}G_0' = M_0'$, which is a subspace of M_0. Thus the map $\phi_0|_{G_0'}$ has kernel contained in $\mathfrak{m}G_0'$. Since G_0' and G_0 are free modules generated in the same degree, and $\phi_0|_{G_0'}$ is a monomorphism in the degree of the generators, $\phi_0|_{G_0'}$ is a monomorphism (even a split monomorphism.)

For the inductive step, suppose that we have shown $\phi_i|_{G_i'}$ is a monomorphism for some i. Since $\mathbf{F}'$ is a minimal resolution, the kernel of the differential $d : F_{i+1}' \to F_i'$ is contained in $\mathfrak{m}F_{i+1}'$. Since $d(G_{i+1}') \subset G_i'$, and G_{i+1}' is a summand of F_{i+1}', the composite map $\phi_i|_{G_{i+1}'} \circ d$ has kernel contained in $\mathfrak{m}G_{i+1}'$. From the commutativity of the diagram

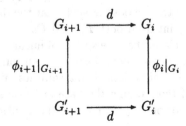

we see that the kernel of $\phi_{i+1}|_{G_{i+1}}$ must also be contained in $\mathfrak{m}G'_{i+1}$. Once again, $\phi_{i+1}|_{G_{i+1}}$ is a map of free modules generated in the same degree that is a monomorphism in the degree of the generators, so it is a (split) monomorphism. □

To apply Proposition 8.11 we need an ideal generated by quadrics that is contained in I_X. We will use an ideal of 2×2 minors of a 1-generic matrix, as described in Chapter 6. Recall that the integer $b(X)$ was defined as the smallest integer such that $\beta_{i,i+1}(S_X) = 0$ for all $i \geq b(X)$.

Theorem 8.12. *Suppose that $X \subset \mathbb{P}^r$ is a curve embedded by a complete linear series $|\mathcal{L}|$. Suppose a divisor $D \subset X$ has has $\mathrm{h}^0\mathcal{O}_X(D) = s+1 \geq 2$. If $\mathrm{h}^0\mathcal{L}(-D) = t+1 \geq 2$, then $\beta_{s+t-1,s+t}(S_X) \neq 0$. In particular $b(X) \geq s+t$.*

Proof. After picking bases for $\mathrm{H}^0\mathcal{O}_X(D)$ and $\mathrm{H}^0\mathcal{L}(-D)$, the multiplication map

$$\mathrm{H}^0\mathcal{O}_X(D) \otimes \mathrm{H}^0\mathcal{L}(-D) \to \mathrm{H}^0\mathcal{L}$$

corresponds, as in Proposition 6.10, to a 1-generic $(s+1) \times (t+1)$ matrix A of linear forms on $\mathbb{P}^r$ whose 2×2 minors lie in I_X.

Since I_X contains no linear forms we may apply Proposition 8.11, and it suffices to show that the ideal $I = I_2(A) \subset I_X$ has $\beta_{s+t-2,s+t}(I) \neq 0$.

If $s = 1$, we can get the result from the Eagon–Northcott complex as follows. By Theorem 6.4 the maximal minors of A generate an ideal I of codimension $(t+1) - (s+1) + 1 = t$ whose minimal free resolution is given by the Eagon–Northcott complex (see Section A2H). Examining this complex, we see that $\beta_{t-1,t+1}(I) \neq 0$. A similar argument holds when $t = 1$.

If $s > 2$ and $t > 2$ we use a different technique, which also covers the previous case and is in some ways simpler. Since the matrix A is 1-generic, the elements of the first row are linearly independent, and the same goes for the first column. We first show that by choosing bases that are sufficiently general, we can ensure that the $s+t+1$ elements in the union of the first row and the first column are linearly independent.

Choose bases $\sigma_0, \ldots, \sigma_s$ and $\tau_0, \ldots, \tau_t$ for $\mathrm{H}^0\mathcal{O}_X(D)$ and $\mathrm{H}^0\mathcal{L}(-D)$ respectively, so that the (i,j)-th element of the matrix A is the linear form corresponding to $\sigma_i\tau_j \in \mathrm{H}^0\mathcal{L} = S_1$. Let B_σ and B_τ be the base divisors of the linear series $|\mathcal{O}_X(D)| = (\mathcal{O}_X(D), \langle \sigma_0, \ldots, \sigma_{s-1} \rangle)$ and $|\mathcal{L}(-D)| = (\mathcal{L}(-D), \langle \tau_0, \ldots, \tau_{t-1} \rangle)$ respectively. Since the linear series $|\mathcal{O}_X(D - B_\sigma)|$ is basepoint-free, we may choose the basis $\{\sigma_i\}$ so that the divisor corresponding to σ_0 is $B_\sigma + D_0$, and D_0 is disjoint from B_τ. We may then choose τ_0 such that the divisor corresponding to τ_0 is $B_\tau E_0$ and E_0 is disjoint from both B_σ and D_0.

With these choices, we claim that the spaces of linear forms $\langle \sigma_0\tau_0, \ldots, \sigma_0\tau_{t-1} \rangle$ and $\langle \sigma_0\tau_0, \ldots, \sigma_{s-1}\tau_0 \rangle$ intersect only in the one-dimensional space $\langle \sigma_0\tau_0 \rangle$. Indeed, if a linear form ℓ is in the intersection, then ℓ vanishes on both D_0 and E_0, so it vanishes on $D_0 + E_0$ and thus, taking the base loci into account, on $B_\sigma + \mathcal{B}_\tau + D_0 + E_0$. This is the divisor of $\sigma_0\tau_0$, so ℓ is a scalar multiple of $\sigma_0\tau_0$ as required.

It follows that the linear forms that appear in the first row and column of A, that is the $s+t+1$ elements

$$\sigma_0 \tau_0 \quad \cdots \quad \sigma_0 \tau_t$$
$$\vdots$$
$$\sigma_s \tau_0$$

are linearly independent.

The following more general result now concludes the proof of Theorem 8.12.

□

Theorem 8.13. *Let $A = (\ell_{i,j})_{0 \leq i \leq s, 0 \leq j \leq t}$ be an $s+1 \times t+1$ matrix of linear forms. If the first row and column of A consist of $s+t+1$ linearly independent elements, then $\beta_{s+t-1,s+t}(S/I_2(A)) \neq 0$.*

A weaker version of Theorem 8.13 was proved by Green and Lazarsfeld to verify one inequality of Green's conjecture, as explained below. A similar theorem holds for the 4×4 pfaffians of a suitably conditioned skew-symmetric matrix of linear forms, and in fact this represents a natural generalization of the result above. See [Koh and Stillman 1989] for details.

Example 8.14. Consider the matrix

$$A = \begin{pmatrix} x_0 & x_1 & x_2 & \cdots & x_t \\ x_{1+t} & 0 & 0 & \cdots & 0 \\ \vdots & \vdots & \vdots & \cdots & \vdots \\ x_{s+t} & 0 & 0 & \cdots & 0 \end{pmatrix} \qquad (*)$$

where $x_0, \ldots, x_{s+t}$ are indeterminates. To simplify the notation, let

$$P = (x_1, \ldots, x_t) \quad \text{and} \quad Q = (x_{1+t}, \ldots, x_{s+t})$$

be the ideals of S corresponding to the first row and the first column of A, respectively. It is easy to see that $I_2(A) = PQ = P \cap Q$. Consider the exact sequence

$$0 \to S/P \cap Q \to S/P \oplus S/Q \to S/P + Q \to 0.$$

The corresponding long exact sequence in Tor includes

$$\mathrm{Tor}_{s+t}(S/P \oplus S/Q, \mathbb{K}) \to \mathrm{Tor}_{s+t}(S/(P+Q), \mathbb{K}) \to \mathrm{Tor}_{s+t-1}(S/(P \cap Q), \mathbb{K}).$$

The free resolutions of S/P, S/Q and $S/(P+Q)$ are all given by Koszul complexes, and we see that the left-hand term is 0 while the middle term is $\mathbb{K}$ in degree $s+t$, so

$$\beta_{s+t-1,s+t}(S/I_2(A)) = \dim \mathrm{Tor}_{s+t-1}(S/(P \cap Q)) \geq 1$$

as required.

Note that x_0 actually played no role in this example — we could have replaced it by 0. Thus the conclusion of Theorem 8.13 holds in slightly more generality than we have formulated it. But some condition is necessary: see Exercise 8.14.

The proof of Theorem 8.13 uses the Koszul complex through the following result.

Theorem 8.15. *Let $I \subset S$ be a homogeneous ideal containing no linear form, and let δ be the differential of the Koszul complex $\mathbf{K}(x_0, \ldots, x_r)$. The graded Betti number $\beta_{i,i+1}(S/I)$ is nonzero if and only if there is an element $u \in \bigwedge^i S^{r+1}(-i)$ of degree $i+1$, such that $\delta(u) \in I \bigwedge^{i-1} S^{r+1}(-i+1)$ and $\delta(u) \neq 0$.*

Given an element $u \in \bigwedge^i S^{r+1}(-i)$ of degree $i+1$ with $\delta(u) \neq 0$, there is a smallest ideal I such that $\delta(u) \in I \bigwedge^{i-1} S^{r+1}(-i+1)$; it is the ideal generated by the coefficients of $\delta(u)$ with respect to some basis of $\bigwedge^{i-1} S^{r+1}(-i+1)$, and is thus generated by quadrics. This ideal I is called the *syzygy ideal* of u, and by Theorem 8.15 we have $\beta_{i,i+1}(S/I) \neq 0$.

Proof. Suppose first that $\beta_{i,i+1}(S/I) = \dim_{\mathbb{K}} \mathrm{Tor}_i(S/I, \mathbb{K})_{i+1} \neq 0$, so we can choose a nonzero element $t \in \mathrm{Tor}_i(S/I, \mathbb{K})_{i+1}$. Since $\mathrm{Tor}_i(S/I, \mathbb{K})$ is the i-th homology of $S/I \otimes \mathbf{K}(x_0, \ldots, x_r)$, we may represent t as the class of a cycle $1 \otimes u$ with $u \in \bigwedge^i S^{r+1}(-i)$ and $\deg u = i+1$. Thus $\delta(u) \in I \bigwedge^{i-1} S^{r+1}(-i+1)$. If $\delta(u) = 0$, then u would be a boundary in $\mathbf{K}(x_0, \ldots, x_r)$, and thus also a boundary in $S/I \otimes \mathbf{K}(x_0, \ldots, x_r)$, so that $t = 0$, contradicting our hypothesis.

Conversely, let $u \in \bigwedge^i S^{r+1}(-i)$ be an element with $\deg u = i+1$ and $\delta(u) \neq 0$. If $\delta(u) \in I \bigwedge^{i-1} S^{r+1}(-i+1)$ then the element $1 \otimes u$ is a cycle in

$$S/I \otimes \mathbf{K}(x_0, \ldots, x_r).$$

We show by contradiction that $1 \otimes u$ is not a boundary. The generators of $\bigwedge^i S^{r+1}(-i)$ are all in degree exactly i. Since I contains no linear forms, the degree $i+1$ part of $S/I \otimes \bigwedge^i S^{r+1}(-i)$ may be identified with the degree $i+1$ part of $\bigwedge^i S^{r+1}(-i)$. If $1 \otimes u$ were a boundary in $S/I \otimes \mathbf{K}(x_0, \ldots, x_r)$, then u would be a boundary in $\mathbf{K}(x_0, \ldots, x_r)$ itself. But then $\delta(u) = 0$, contradicting our hypothesis.

Since $1 \otimes u$ is not a boundary, $\mathrm{Tor}_i(S/I, \mathbb{K})_{i+1} \neq 0$, so $\beta_{i,i+1}(S/I) \neq 0$. $\square$

The hypothesis that I contain no linear forms is necessary in Theorem 8.15. For example, if $I = \mathfrak{m}$, then $\delta(u) \in I \bigwedge^{i-1} S^{r+1}(-i+1)$ for any u, but $\beta_{i,i+1} S/\mathfrak{m} = 0$ for all i.

As an application, using Theorem 8.15 we can easily describe all the ideals $I \subset S$ such that I contains no linear form but $\beta_{r+1,r+2}(S/I) \neq 0$. (For the case of $\beta_{r,r+1}(S/I) \neq 0$ see the next theorem.) Since $\bigwedge^{r+1} S^{r+1} \cong S$, an element of degree $r+2$ in $\bigwedge^{r+1} S^{r+1}(-r-1)$ may be written as a linear form ℓ times the generator. Applying δ gives an element whose coefficients are $\pm x_i \ell$. By Theorem 8.15, if I is a homogeneous ideal that contains no linear forms, then $\beta_{r+1,r+2}(S/I) \neq 0$ if and only if I contains the ideal $\ell(x_0, \ldots, x_r)$ for some linear form ℓ.

Proof of Theorem 8.13. To simplify notation, set $I = I_2(A)$. We must show that the vector space $\mathrm{Tor}_{s+t-1}(S/I, \mathbb{K})_{s+t}$ is nonzero, and we use the free resolution

K of $\mathbb{K}$ to compute it. We may take **K** to be the Koszul complex

$$\mathbf{K}: 0 \longrightarrow \bigwedge^{r+1} S^{r+1}(-r-1) \xrightarrow{\delta} \bigwedge^r S^r(-r) \xrightarrow{\delta} \cdots \xrightarrow{\delta} S,$$

Thus it suffices to give a cycle of degree $s+t$ in

$$S/I \otimes \mathbf{K}_{s+t-1} = S/I \otimes \overset{s+t-1}{\bigwedge} S^{r+1}(-s-t+1)$$

that is not a boundary. The trick is to find an element α, of degree $s+t$ in $\mathbf{K}_{s+t-1}$, such that

1. $\delta(\alpha) \neq 0 \in \mathbf{K}_{s+t-2}$; and
2. $\delta(\alpha)$ goes to zero in $S/I \otimes \mathbf{K}_{s+t-2}$.

Having such an element will suffice to prove the Theorem: From condition 2 it follows that the image of α in $S/I \otimes \mathbf{K}$ is a cycle. On the other hand, the generators of $\mathbf{K}_{s+t-1}$ have degree $s+t-1$, and the elements of I are all of degree 2 or more. Thus the degree $s+t$ part of $\mathbf{K}_{s+t-1}$ coincides with that of $S/I \otimes \mathbf{K}_{s+t-1}$. If the image of α were a boundary in $S/I \otimes \mathbf{K}$, then α would also be a boundary in **K**, and $\delta(\alpha)$ would be zero, contradicting condition 1.

To write down α, let $x_0, \ldots x_t$ be the elements of the first row of A, and let $x_{1+t}, \ldots, x_{s+t}$ be the elements of the first column, starting from the position below the upper left corner, as in equation $(*)$ in the example above. Thus if $0 \leq j \leq t$ then $\ell_{0,j} = x_j$, while if $1 \leq i \leq s$ then $\ell_{i,0} = x_{i+t}$. Complete the sequence $x_0, \ldots, x_{s+t}$ to a basis of the linear forms in S by adjoining some linear forms $x_{s+t+1}, \ldots, x_r$. Let $\{e_i\}$ be a basis of $S^{r+1}(-1)$ such that $\delta(e_i) = x_i$ in the Koszul complex.

The free module $\bigwedge^{s+t-1} S^{r+1}(-s-t+1)$ has a basis consisting of the products of $s+t-1$ of the e_i. If $0 \leq j \leq t$ and $1 \leq i \leq s$, then we denote by $e_{[i+t,j]}$ the product of all the $e_1, \ldots, e_{s+t}$ except e_j and e_{i+t}, in the natural order, which is such a basis element. With this notation, set

$$\alpha = \sum_{\substack{1 \leq i \leq s \\ 0 \leq j \leq t}} (-1)^{i+j} \ell_{i,j} e_{[i+t,j]}.$$

If $0 \leq k \leq s+t$ and $k \neq i+t$, $k \neq j$ then we write $e_{[k,i+t,j]}$ for the product of all the $e_1, \ldots, e_{s+t}$ except for e_{i+t}, e_k and e_j, as always in the natural order. These elements are among the free generators of $\bigwedge^{s+t-2} S^{r+1}(-s-t+2)$. The formula for the differential of the Koszul complex gives

$$\delta(e_{[i+t,j]}) = \sum_{0 \leq k < j} (-1)^k e_{[k,i+t,j]} + \sum_{j < k < i+t} (-1)^{k-1} e_{[k,i+t,j]}$$

$$+ \sum_{i+t < k \leq s+t} (-1)^k e_{[k,i+t,j]}.$$

Write $(p, q \mid u, v)$ for the 2×2 minor of A involving rows p, q and columns u, v. Straightforward computation shows that the coefficient of $e_{[k,i+t,j]}$ in $\delta(\alpha)$ is

$$\begin{cases} \pm(0, i \mid j, k)e_{[k,i+t,j]} & \text{if } 0 \leq k \leq t, \\ \pm(i, k-t \mid 0, j)e_{[k,i+t,j]} & \text{if } 1+j \leq k = \leq s+t. \end{cases}$$

In particular, the coefficients of the $e_{[k,i+t,j]}$ in $\delta(\alpha)$ are all in I.

Consider a 2×2 minor of A involving the upper left corner, say

$$(0, 1 \mid 0, 1) = \det \begin{pmatrix} \ell_{0,0} & \ell_{0,1} \\ \ell_{1,0} & \ell_{1,1} \end{pmatrix} = \ell_{0,0}\ell_{1,1} - \ell_{0,1}\ell_{1,0}.$$

Since $\ell_{0,0}, \ell_{0,1}$, and $\ell_{1,0}$ are distinct prime elements of S, and S is factorial, this element is nonzero. Thus the coefficients of $\delta(\alpha)$ are not all 0, so α satisfies conditions 1 and 2 as required. □

One way to get a divisor to which to apply Theorem 8.12 is to choose D to be a general divisor of degree $g+1$. By Theorem 8.5.1, we have $h^0\mathcal{O}_X(D) = 2$. Since $\mathcal{L}(-D)$ is a general line bundle of degree $2g+1+p-g-1 = g+p$ the bundle $\mathcal{L}(-D)$ will be nonspecial, whence $h^0(\mathcal{L}(-D)) = p+1$ by the Riemann–Roch formula. Thus $b(X) \geq p+1$. However, we could already have deduced this from the fact that $b(X) > a(X)$ and $a(X) \geq p$ by Theorem 8.8.

To do better, we need to invoke a much deeper result, the Brill–Noether Theorem. The statement first appears in [Brill and Noether 1873], but it was realized fairly soon that the proof given by Brill and Noether was incomplete. The first complete proofs are found in [Kempf 1972; Kleiman and Laksov 1972; 1974] (see [Arbarello et al. 1985, Chapter V] for an exposition and history). The application to high degree curves was first noted in the thesis of Schreyer [1983].

Theorem 8.16 (Brill–Noether). *If X is a curve of genus g, then the set J_d^r of line bundles $\mathscr{F} \in \mathrm{Pic}_d(X)$ with $h^0\mathscr{F} \geq s+1$ is an algebraic subset with dimension*

$$\dim J_d^r \geq \rho(d, s) = g - (s+1)(g-d+s).$$

In particular, if if $d \geq 1 + \lceil g/2 \rceil$ then X has a line bundle of degree d with at least 2 independent sections.

It is known that the Brill–Noether theorem is sharp for a general curve (that is, for the curves in on open dense set of the moduli space of curves of genus g.) See [Gieseker 1982] or, for a simpler proof, [Eisenbud and Harris 1983].

Idea of the proof. The formula is easy to understand, even though it is hard to prove. Take an arbitrary divisor E that is the sum of a large number e of distinct points of X. The divisor E corresponds to a section of the line bundle $\mathcal{O}_X(E)$ from which we get a short exact sequence

$$0 \longrightarrow \mathcal{O}_X \longrightarrow \mathcal{O}_X(E) \longrightarrow \mathcal{O}_E \longrightarrow 0.$$

Let $\mathscr{F}$ be a line bundle of degree d on X. We tensor the exact sequence above with $\mathscr{F}$. Since E is a finite set of points we may identify $\mathscr{F} \otimes \mathcal{O}_E$ with $\mathcal{O}_E$. Taking cohomology, we get a left exact sequence

$$0 \longrightarrow H^0 \mathscr{F} \longrightarrow H^0 \mathscr{F}(E) \xrightarrow{\alpha_{\mathscr{F}}} H^0 \mathcal{O}_E.$$

Now $H^0 \mathcal{O}_E \cong \mathbb{K}^e$ is the e-dimensional vector space of functions on E. If we choose e so large that $\deg \mathscr{F}(E) = d + e > 2g - 2$, then by the Riemann–Roch formula

$$h^0 \mathscr{F}(E) = (d+e) - g + 1 + h^1 \mathscr{F}(E) = d + e - g + 1.$$

Thus the dimension of $H^0 \mathscr{F}(E)$ does not vary as $\mathscr{F}$ runs over $\mathrm{Pic}_d(X)$. Locally on $\mathrm{Pic}_d(X)$ we may think of $\alpha_{\mathscr{F}}$ as a varying map between a fixed pair of vector spaces (globally it is a map between a certain pair of vector bundles). The set of $\mathscr{F}$ with $h^0 \mathscr{F} \geq s + 1$ is the set of $\mathscr{F}$ with $\mathrm{rank}\, \alpha_{\mathscr{F}} \leq d + e - g + 1 - (s+1) = d + e - g - s$, so, locally, $J_d^r(X)$ is defined by the $(d+e-g-s-1) \times (d+e-g-s-1)$ minors of a $(d+e-g+1) \times e$ matrix. Macaulay's formula, Theorem A2.54 shows that *if the set J_d^r is nonempty* then its codimension is at most $(s+1)(g-d+s)$, so the dimension is at least $g - (s+1)(g-d+s)$ as required. The argument we have given is essentially the original argument of Brill and Noether. Its main problem is that is does not prove that the locus $J_d^r(X)$ is nonempty — the very fact we were interested in.

One way to address this point is to identify $\alpha_{\mathscr{F}}$ as the map on fibers of a map of explicitly given vector bundles, $\alpha : \mathscr{E}_1 \to \mathscr{E}_2$. To see what might be required, replace $\mathrm{Pic}_d(X)$ by a projective space $\mathbb{P}^r$, and α by a map

$$\alpha : \mathscr{E}_1 = \mathcal{O}_{\mathbb{P}^r}(a) \to \mathscr{E}_2 = \mathcal{O}_{\mathbb{P}^r}(b).$$

Let Y be the locus of points $y \in \mathbb{P}^r$ such that the fiber of α at y has rank 0. There are three cases:

- If $b - a < 0$ then $\alpha = 0$ and $Y = \mathbb{P}^r$.
- If $b - a = 0$ then either $Y = \mathbb{P}^r$ or $Y = \varnothing$.
- If $b - a > 0$ then Y is always nonempty, and has codimension ≤ 1 by Macaulay's formula, Theorem A2.54, or just the Principal Ideal Theorem, of which Macaulay's Theorem is a generalization.

Thus the case in which Macaulay's formula is relevant is the case where $\mathscr{E}_1^* \otimes \mathscr{E}_2 = \mathcal{O}_{\mathbb{P}^r}(b-a)$ with $b - a > 0$. This suggests the general case: by [Fulton and Lazarsfeld 1983, Prop. 3.5], the determinantal loci are really nonempty if $\mathscr{E}_1^* \otimes \mathscr{E}_2$ is ample in the vector bundle sense. This turns out to be true for the bundles that appear in the Brill–Noether theorem, completing the proof. □

As promised, we can use the Brill–Noether theorem to give a lower bound for the number $b(X)$ that is better than $p+1$:

Theorem 8.17 (Schreyer). *If $X \subset \mathbb{P}^r$ is a curve ,embedded by a complete linear series of degree $2g+1+p$, with $p \geq 0$, then*

$$b(X) \geq p+1+\left\lfloor \frac{g}{2} \right\rfloor.$$

Proof. Theorem 8.16 tells us that X must have a line bundle $\mathscr{F}$ of degree $1 + \lceil g/2 \rceil$ with $h^0 \mathscr{F} \geq 2$. Let D be the divisor corresponding to a global section of $\mathscr{F}$. As before, set $\mathscr{L} = \mathscr{O}_X(1)$. The codimension of the span of D in $\mathbb{P}^r$ is number of independent hyperplanes containing D, that is $h^0 \mathscr{L}(-D)$. By the Riemann–Roch formula,

$$h^0 \mathscr{L}(-D) \geq \deg \mathscr{L} - \deg D - g + 1 = 2g+1+p-\left\lceil \frac{g}{2} \right\rceil -1-g+1 = p+1+\left\lfloor \frac{g}{2} \right\rfloor,$$

and the desired result follows from Theorem 8.12. $\qquad\qquad\square$

With this lower bound for $b(X)$ in hand, we turn to the question of an upper bound. When $X \subset \mathbb{P}^r$ is the rational normal curve, then the Eagon–Northcott construction (Theorem A2.60) shows that the quadratic strand is the whole resolution. Thus $b(X) = 1 + \operatorname{pd} S_X = r$. However, $b(X) \leq r-1$ for curves of higher genus. To derive this bound we use Koszul homology, which enables us to go directly from information about the $\beta_{i,i+1}(X)$ to information about quadrics in the ideal of X.

Theorem 8.18. *Suppose that $\mathbb{K}$ is algebraically closed. If $I \subset S$ is a homogeneous ideal not containing any linear form, then $\beta_{r,r+1}(S/I)$ is nonzero if and only if, after a linear change of variables, I contains the ideal of 2×2 minors of a matrix of the form*

$$\begin{pmatrix} x_0 & \cdots & x_s & x_{s+1} & \cdots & x_r \\ \ell_0 & \cdots & \ell_s & 0 & \cdots & 0 \end{pmatrix}$$

where $0 \leq s < r$ and $\ell_0, \ldots, \ell_s$ are linearly independent linear forms.

See Exercise 8.10 for an example in the non–algebraically closed case.

Proof. Consider the Koszul complex

$$\mathbf{K}(x_0, \ldots, x_r): \quad 0 \longrightarrow \bigwedge^{r+1} S^{r+1}(-r-1) \xrightarrow{\delta} \cdots \xrightarrow{\delta} S^{r+1}(-1) \xrightarrow{\delta} S.$$

By Theorem 8.15 it suffices to show that if $u \in \bigwedge^r S^{r+1}(-r)$ is an element of degree $r+1$ such that $\delta(u) \neq 0$, then the syzygy ideal of u has the given determinantal form.

Let $e_0, \ldots, e_r$ be the basis of S^{r+1} such that $\delta(e_i) = x_i$. There is a basis for $\bigwedge^{r-1} S^{r+1}$ consisting of all products of "all but one" of the e_j; we shall write

$$_i e = e_0 \wedge \cdots \wedge e_{i-1} \wedge e_{i+1} \wedge \cdots \wedge e_r$$

for such a product. Similarly, we write $_{ij}e$ for the product of all but the i-th and j-th basis vectors, so the $_{ij}e$ form a basis of $\bigwedge^{r-1} S^{r+1}$.

Suppose that $u = \sum_i m_{i\ i}e$. Since $\deg u = i+1$, the m_i are linear forms. Now

$$\delta(_ie) = \sum_{j<i}(-1)^j x_j{}_{ij}e + \sum_{j>i}(-1)^{j-1} x_j{}_{ij}e,$$

so

$$\delta(u) = \sum_{g<h}\left((-1)^g x_g m_h + (-1)^{h-1} x_h m_g\right){}_{ij}e$$

$$= \sum_{g<h}\det\begin{pmatrix}(-1)^g x_g & (-1)^h x_h \\ m_g & m_h\end{pmatrix}{}_{ij}e$$

$$= \sum_{g<h}(-1)^{g+h}\det\begin{pmatrix}x_g & x_h \\ (-1)^g m_g & (-1)^h m_h\end{pmatrix}{}_{ij}e.$$

Setting $\ell'_i = (-1)^i m_i$, it follows that the syzygy ideal of u is the ideal of 2×2 minors of the matrix

$$M = \begin{pmatrix} x_0 & x_1 & \cdots & x_r \\ \ell'_0 & \ell'_1 & \cdots & \ell'_r \end{pmatrix}.$$

If we set $e = e_0 \wedge \cdots \wedge e_r$, then $\delta(e) = \sum(-1)^i x_i{}_ie$. Moreover, the Koszul complex is exact, so the hypothesis $\delta(u) \neq 0$ translates into the hypothesis that u is not a scalar multiple of $\delta(e)$. It follows in particular that the two rows R_1, R_2 of the matrix M, regarded as vectors of linear forms, are linearly independent, so no scalar linear combination of R_1 and R_2 can be 0. If the elements ℓ'_i are linearly dependent, then after a column transformation and a linear change of variables, the matrix M will have the desired form. Furthermore, we could replace the second row R_2 by $\lambda R_1 + R_2$ for any $\lambda \in \mathbb{K}$ without changing the situation, so it is enough to show that the linear forms in the vector $\lambda R_1 + R_2$ are linearly dependent for some λ.

Each vector $\ell_0, \ldots, \ell_r$ of $r+1$ linear forms corresponds to a linear transformation of the space of linear forms sending x_i to ℓ_i. Because R_2 is not a scalar multiple of R_1, the set of vectors $\lambda R_1 + R_2$ correspond to a line in the projective space of matrices modulo scalars. In this projective space, any line must meet the hypersurface of matrices of vanishing determinant, so some row $\lambda R_1 + R_2$ consists of linearly dependent forms, and we are done. (This last argument is a special case of a general fact about 1-generic matrices, for which see [Eisenbud 1988, Proposition 1.3].) $\qquad\square$

Putting these ideas together, we can characterize rational normal curves in terms of syzygies.

Corollary 8.19. *Suppose that $X \subset \mathbb{P}^r$ is an irreducible nondegenerate curve such that S_X is Cohen–Macaulay and some hyperplane section $H \cap X$ of X consists of simple points in linearly general position. X is a rational normal curve if and only if $b(X) = r$.*

Proof. The points of a general hyperplane section of the rational normal curve $\mathbb{P}^1 \simeq X \subset \mathbb{P}^r$ correspond to the roots of a general polynomial of degree r, so they are distinct, simple and independent. We already know that $\beta_{r-1,r}(X) \neq 0$ and $\beta_{r,r+1}(X) = 0$, so $b(X) = r$.

Conversely, the hypothesis $b(X) = r$ means that $\beta_{r-1,r} \neq 0$. Let Y be a hyperplane section $Y = X \cap H$. After a change of variable, we may suppose that the ideal of H is generated by the last variable, x_r. Since S_X is Cohen–Macaulay, the minimal free resolution of S_Y as an $\bar{S} = S/(x_r)$-module is obtained by reducing the resolution of S_X modulo x_r.

We consider Y as a subset of $H = \mathbb{P}^{r-1}$. Write $\beta_{r-1,r}(S_Y, \bar{S})$ for the graded Betti number of this $\bar{S}$-free resolution. We have $\beta_{r-1,r}(S_Y, \bar{S}) \neq 0$ and $\bar{S}$ has only r variables, so we may apply Theorem 8.18. In particular, the ideal of Y contains a product $(\ell_0, \ldots, \ell_s)(x_{s+1}, \ldots, x_{r-1})$ with $0 \leq s < r-1$. Since Y is reduced, it is contained in the union of the linear subspaces L and L' in $\mathbb{P}^{r-1}$ defined by the ideals $(\ell_0, \ldots, \ell_s)$ and $(x_{s+1}, \ldots, x_{r-1})$ respectively. The dimensions of L and L' are $r-1-(s+1) < r-1$ and $s < r-1$. Since the points of Y are in linearly general position, at most $(r-1-(s+1))+1$ points of Y can be contained in L and at most $s+1$ points of Y can be contained in L', so the cardinality of Y, which is the degree of X, is at most

$$\deg X \leq (r-1-(s+1)+1) + (s+1) = r.$$

By Theorem 6.8, X is a rational normal curve.   □

Corollary 8.20. *If $X \subset \mathbb{P}^r$ is a curve embedded by a complete linear series of degree $2g+1+p$, with $p \geq 0$, and X is not a rational normal curve, then $b(X) < r$. In particular, the graded S-module $w_X = \oplus H^0(\omega_X(n))$ is generated by $H^0(\omega_X)$.*

The method explained at the end of Section 2B can be used to derive the value of the second-to-last Betti number in the cubic strand from this; see Exercise 8.9.

Proof. The hypothesis of Corollary 8.19 holds for all smooth curves X embedded by linear series of high degree. The Cohen–Macaulay property is proved in Theorem 8.1 and the general position property is proved in the case char $\mathbb{K} = 0$ in Exercises 8.18–8.20. A general proof may be found in [Rathmann 1987].

Because $\operatorname{pd} S_X \leq r-1$ we have $b(X) \leq r$. By Corollary 8.19, except possibly $b(X) < r$.

To prove the second statement we must show that $\beta_{0,m}(w_X) = 0$ for $m \neq 0$. Since S_X is Cohen–Macaulay, the dual of its free resolution, twisted by $-r-1$, is a free resolution of the canonical module $w_X = \operatorname{Ext}^{r-1}(S_X, S(-r-1))$. Thus $\beta_{0,m}(w_X) = \beta_{r-1,r+1-m}(S_X)$. When $m \geq 2$ this is zero because I_X is 0 in degrees ≤ 1, and when $m < 0$ we have $H^0(\omega_X(m)) = 0$ because then $\omega_X(m)$ has negative degree. Thus only $\beta_{0,0}(w_X) = \beta_{r-1,r}(S_X)$ and $\beta_{0,1}(w_X)$ could be nonzero. But $\beta_{r-1,r}(S_X) = 0$ by the first part of the Corollary.   □

8C   Conjectures and Problems

Again, let X be a (smooth irreducible) curve embedded in $\mathbb{P}^r$ as a curve of degree $d = 2g+1+p \geq 2g+1$ by a complete linear series $|\mathcal{L}|$. We return to the diagram at the end of the introduction to Section 8B:

	0	1	$\cdots$	a	$a+1$	$\cdots$	$b-1$	b	$\cdots$	$r-1=g+p$
0	1	$-$	$\cdots$	$-$	$-$	$\cdots$	$-$	$-$	$\cdots$	$-$
1	$-$	$\binom{d-g-1}{2}$	$\cdots$	$*$	$*$	$\cdots$	$*$	$-$	$\cdots$	$-$
2	$-$	$-$	$\cdots$	$-$	$*$	$\cdots$	$*$	$*$	$\cdots$	g

We have shown that

$$p \leq a \leq p + \max\left(0,\ \left\lceil \frac{g-p-3}{2} \right\rceil\right)$$

and

$$p+1+\left\lfloor \frac{g}{2} \right\rfloor \leq b \leq r-1.$$

Using Proposition 2.7 and Corollary 8.4 we can compute some of the nonzero graded Betti numbers, namely $\beta_{i,i+1}$ for $i \leq a+1$ and $\beta_{i,i+2}$ for $i \geq b-1$ in terms of g and d. When $b(X) \leq a(X)+2$ (and this includes all cases with $g \leq 3$) we get the values of all the graded Betti numbers. However, in the opposite case, for example when $g \geq 4, p \geq 2$, we will have both $\beta_{a,a+2} \neq 0$ and $\beta_{a+1,a+2} \neq 0$, so $\beta_{a,a+2}$ cannot be determined this way. In such cases the remaining values, and their significance, are mostly unknown.

We can probe a little deeper into the question of vanishing in the cubic strand, that is, the value of $a(X)$. Part 3 of Theorem 8.8 shows that, when the degree d is at least $3g-2$, the value of $a(X)$ is accounted for by degenerate secant planes to X. But when $2g+1 \leq d < 3g-2$, other phenomena may intervene, as the next example shows.

Example 8.21. When does a high degree curve require equations of degree 3?
Suppose that $X \subset \mathbb{P}^r$ is a curve embedded by a complete linear series of degree $d = 2g+1+p$. By Corollary 8.2, the ideal I_X of X is generated by forms of degrees ≤ 3. We know that I_X contains exactly $\binom{g+2}{2}$ quadrics. But these quadrics might not generate the ideal of X. For example, if X has a trisecant line L, every quadric containing X vanishes at three points L, and thus vanishes on all of L. This shows that I_X is not generated by quadrics. (This is an simple special case of Theorem 8.8.)

Another way to see that I_X is not generated by quadrics is to show that the quadrics it contains have "too many" linear relations. By Corollary 8.4, we may choose linear forms $x, y \in S$ that form a regular sequence on S_X, and the nonzero values of the Hilbert function of $S_X/(x,y)S_X$ are $1, g, g$. Using Proposition 2.7 we see that

$$\beta_{1,3}(S_X) - \beta_{2,3}(S_X) = g^2 - g\binom{g}{2} + \binom{g}{3} = g^2 - 2\binom{g+1}{3}.$$

From this it follows that if $\beta_{2,3}(S_X) > 2\binom{g+1}{3} - g^2$, then $\beta_{1,3}(S_X) \neq 0$. (A similar argument shows that if any Betti number $\beta_{j-1,j}(S_X)$ in the quadratic strand is unusually large, then the Betti number $\beta_{j-2,j}$ in the cubic strand is nonzero, so $a(X) \leq j - 3$.)

One geometric reason for the quadratic strand to be large would be the presence of a special variety containing X (Theorem 8.12). The most extreme examples come from two-dimensional scrolls, defined by the 2×2 minors of a 1-generic matrix of linear forms on $\mathbb{P}^r$. Such scrolls appear, for example, when X is *hyperelliptic* in the sense that $g \geq 2$ and there is a degree 2 map $X \to \mathbb{P}^1$. Let D be a fiber of this map, so that $\deg D = 2$ and $h^0(\mathscr{O}_X(D)) = 2$. The multiplication matrix $H^0(\mathscr{O}_X(D)) \otimes H^0(\mathscr{L}(-D)) \to H^0\mathscr{L} = S_1$ corresponds to a $2 \times h^0(\mathscr{L}(-D))$ matrix of linear forms. Since $\mathscr{L}$ has degree $2g+1$, the line bundle $\mathscr{L}(-D)$ is nonspecial, so the Riemann–Roch theorem yields $h^0\mathscr{L}(-D) = g+p-1 = r-2$. By Theorem 6.4 the variety Y defined by the 2×2 minors is irreducible and has the "generic" codimension for a variety defined by such matrices, namely $r - 2$, so it is a surface, the union of the lines spanned by divisors linearly equivalent to D (see the equality (*) on page 100 and [Eisenbud 1988]). Moreover $X \subset Y$ by Proposition 6.10.

The minimal free resolution of S/I_Y is an Eagon–Northcott complex, and it follows that $\beta_{2,3}(S_Y) = 2\binom{r-2}{3}$. By Theorem 8.12

$$\beta_{2,3}(S_X) \geq 2\binom{r-2}{3} = 2\binom{g-1}{3}.$$

But $2\binom{g-1}{3} > 2\binom{g+1}{3} - g^2$ for every $g \geq 1$. This proves the first statement of the following Proposition.

Proposition 8.22. *If $X \subset \mathbb{P}^r$ is a hyperelliptic curve embedded by a complete linear series $|\mathscr{L}|$ of degree $2g+1$, then I_X is not generated by quadrics (so $a(X) = 0$). Moreover, if $g \geq 4$ and $\mathscr{L}$ is general in $\mathrm{Pic}_{2g+1}(X)$, then X has no trisecant.*

Using the same method one can show that $a(X) = p$ whenever $X \subset \mathbb{P}^r$ is hyperelliptic of degree $2g+1+p$ (Exercise 8.15).

Proof. We can characterize a trisecant as an effective divisor D of degree 3 on X lying on $r - 2$ independent hyperplanes, which means $h^0\mathscr{L}(-D) = r - 2$. Since $\deg \mathscr{L}(-D) = 2g+1-3 = 2g-2$, the Riemann–Roch theorem yields $h^0\mathscr{L}(-D) = 2g-2+h^1\mathscr{L}(-D) = r-1+h^0(\omega_X \otimes \mathscr{L}^{-1}(D))$. Since $\omega_X \otimes \mathscr{L}^{-1}(D)$ is a line bundle of degree 0, it cannot have sections unless it is trivial. Unwinding this, we see that there exists a trisecant D to X if and only if the line bundle $\mathscr{L} = \mathscr{O}_X(1)$ can be written as $\mathscr{O}_X(1) = \omega_X(D)$ for some effective divisor D of degree 3. When $g \leq 3$ this is always the case — that is, there is always a trisecant — by Theorem 8.5.3. But when $g \geq 4$ most line bundles of degree 3 are ineffective, so when $\mathscr{L}$ is general X has no trisecant.   □

It turns out that hyperellipticity is the only reason other than a degenerate secant plane for having $a(X) = p$.

Theorem 8.23. [Green and Lazarsfeld 1988] *Suppose that X is a curve of genus g. If $X \subset \mathbb{P}^r$ is embedded by a complete linear series of degree $2g+1+p$, with $p \geq 0$, and $a(X) = p$, then either $X \subset \mathbb{P}^r$ has a degenerate $(p+3)$-secant plane (that is, $\mathcal{O}_X(1) = \omega_X(D)$ for some effective divisor D) or X is hyperelliptic.*

Hyperelliptic curves are special in other ways too; for example $b(X)$ takes on its maximal value $r - 1$ for hyperelliptic curves: if X is hyperelliptic, then the scroll $Y \supset X$ constructed above has $\beta_{r-2,r-1}(S_Y) \neq 0$ because the free resolution is given by the Eagon–Northcott complex of 2×2 minors of a $2 \times (r-1)$ matrix. Thus $\beta_{r-2,r-1}(S_X) \neq 0$ by Theorem 8.12.

More generally, we say that a curve X is δ-*gonal* if there is a nonconstant map $\phi : X \to \mathbb{P}^1$ of degree δ. The *gonality* of X is then the minimal δ such that X is δ-gonal. (The name came from the habit of calling a curve with a three-to-one map to $\mathbb{P}^1$ "trigonal".) Suppose that $X \subset \mathbb{P}^r$ is a δ-gonal curve in a high degree embedding, and set $\mathscr{L} = \mathcal{O}_X(1)$. Let D be a fiber of a map $\phi : X \to \mathbb{P}^1$ of degree δ. By the same arguments as before, X is contained in the variety Y defined by the 2×2 minors of the matrix $M(\mathcal{O}_X(D), \mathscr{L}(-D))$. This matrix has size at least $2 \times (r+1-\delta)$, so the Eagon–Northcott complex resolving S_Y has length at least $r - \delta$, and $b(X) \geq r - \delta + 1$ by Theorem 8.12. For embedding of very high degree, this may be the only factor:

Gonality Conjecture. [Green and Lazarsfeld 1985, Conjecture 3.7] *If $d \gg g$ and X is a δ-gonal curve of genus g embedded by a complete linear series of degree d in $\mathbb{P}^r$, then $b(X) = r - \delta + 1$.*

This conjecture was recently verified for generic curves and in some further cases by Aprodu and Voisin. See [Aprodu 2004].

8D   Exercises

1. Suppose that $X \subset \mathbb{P}^r$ is a curve of genus > 0. Use the sheaf cohomology description of regularity to prove that the regularity of S_X is at least 2.

2. Show that if $X \subset \mathbb{P}^r$ is any scheme with S_X Cohen–Macaulay of regularity 1, then X has degree at most $1 + \operatorname{codim} X$ (this gives another approach to Exercise 8.1 in the arithmetically Cohen–Macaulay case.)

3. Let X be a reduced curve in $\mathbb{P}^r$. Show that S_X is Cohen–Macaulay if and only if the space of forms of degree n in $\mathbb{P}^r$ vanishing on X has dimension at most (equivalently: exactly)

$$\dim(I_X)_n = \binom{r+n}{r} - h^0 \mathcal{O}_X(n),$$

or equivalently $\dim(S_X)_n = h^0 \mathcal{O}_X(n)$ for every $n \geq 0$.

4. Suppose that $X \subset \mathbb{P}^r$ is a hyperelliptic curve of genus g. Show that if S_X is Cohen–Macaulay then $\deg X \geq 2g + 1$, by using Exercise 8.3 and the 2×2 minors of the matrix $M(\mathcal{L}', \mathcal{L} \otimes \mathcal{L}'^{-1})$ as defined on page 99, where $\mathcal{L}'$ is the line bundle of degree 2 defining the two-to-one map from $X \to \mathbb{P}^1$.

5. Show that if $X \subset \mathbb{P}^r$ is any variety (or even any scheme) of dimension d, and $\nu_d : X \to \mathbb{P}^N$ is the d-th Veronese embedding (the embedding by the complete linear series $|\mathcal{O}_X(d)|$) then for $d \gg 0$ the image $\nu_d(X)$ is $(1 + \dim X)$-regular. (This can be proved using just Serre's and Grothendieck's vanishing theorems [Hartshorne 1977, Theorems III.2.7 and III.5.2].)

6. Suppose that X is an irreducible algebraic variety of dimension ≥ 1 and that $\mathcal{L} \not\cong \mathcal{O}_X$ is a line bundle on X with $H^0 \mathcal{L} \neq 0$. Show that $H^0 \mathcal{L}^{-1} = 0$. (Hint: A nonzero section of $\mathcal{L}$ must vanish somewhere...).

7. Suppose that X is a smooth projective hyperelliptic curve of genus g, and let $\mathcal{L}_0$ be the line bundle that is the pull-back of $\mathcal{O}_{\mathbb{P}^1}(1)$ under the two-to-one map $X \to \mathbb{P}^1$. Show that if L is any line bundle on X that is special (which means $h^1(\mathcal{L}) \neq 0$) then $\mathcal{L} = \mathcal{L}_0^a \mathcal{L}_1$ where $\mathcal{L}_1$ is a special bundle satisfying $h^0 \mathcal{L}_1 = 1$ and $a \geq 0$. Show under these circumstances that $h^0 \mathcal{L} = g + 1$. Deduce that any very ample line bundle on X is nonspecial.

8. Compute all the $\beta_{i,j}$ for a curve of genus 2, embedded by a complete linear series of degree 5.

9. Let $X \subset \mathbb{P}^r$ be a curve of degree $2g + 1 + p$ embedded by a complete linear series in $\mathbb{P}^r$. Use Corollary 8.20 and the method of Section 2B to show that $\beta_{r-2,r}(X) = g(g+p-1)$ (the case $g = 2, p = 0$ may look familiar.)

10. Let $r = 1$, and let

$$Q = \det \begin{pmatrix} x_0 & x_1 \\ -x_1 & x_0 \end{pmatrix}; \qquad I = (Q) \subset \mathbb{R}[x_0, x_1].$$

Show that $\beta_{r,r+1}(\mathbb{R}[x_0, x_1]/I) \neq 0$, but that I does not satisfy the conclusion of Theorem 8.18. Show directly that I does satisfy Theorem 8.18 if we extend the scalars to be the complex numbers.

11. Prove parts 4 and 5 of Theorem 8.5.

12. Complete the proof of the second statement of Theorem 8.8 by showing that there are effective divisors D and E such that $\mathcal{L}^{-1} \otimes \omega_X(D) = \mathcal{O}_X(E)$ with

$$\deg D \leq 2 + \max(p+1, \lceil (g+p-1)/2 \rceil).$$

Hint: the numbers are chosen to make $\deg D + \deg E \geq g$.

13. Theorem 8.8 implies that a smooth irreducible curve X of genus g, embedded in $\mathbb{P}^r$ by a complete linear series of degree $2g+1+p$, cannot have a degenerate q-secant plane for $q < p+3$. Give a direct proof using just the Riemann–Roch Theorem.

14. Find a $2 \times t + 1$ matrix of linear forms

$$\begin{pmatrix} \ell_{0,0} & \cdots & \ell_{0,t} \\ \ell_{1,0} & \cdots & \ell_{1,t} \end{pmatrix}$$

such that the $1+t$ elements $\ell_{1,0}, \ell_{0,1}, \ell_{0,2}, \ldots, \ell_{0,t}$ are linearly independent, but all the 2×2 minors are 0.

15. Let $X \subset \mathbb{P}^r$ be a hyperelliptic curve embedded by a complete linear series of degree $2g + 1 + p$ with $p \geq 0$. Show by the method of Section 8C that $a(X) \leq p$, and thus $a(X) = p$ by Theorem 8.8.

Many deep properties of projective curves can be proved by Harris' "Uniform Position Principle" [Harris 1979], which says that, in characteristic 0, two subsets of points of a general hyperplane section are geometrically indistinguishable if they have the same cardinality. A consequence is that the points of a general hyperplane section always lie in linearly general position. It turns out that Theorem 8.1 (in characteristic 0) can easily be deduced from this. The following exercises sketch a general approach to the arithmetic Cohen–Macaulay property for "nonspecial" curves — that is, curves embedded by linear series whose line bundle has vanishing first cohomology — that includes this result.

16. Suppose that $X \subset \mathbb{P}^r$ is a (reduced, irreducible) curve. Use Exercise 8.2 to show that if X is linearly normal and the points of some hyperplane section of X impose independent conditions on quadrics, then S_X is Cohen–Macaulay. If $h^1 \mathcal{O}_X(1) = 0$, show that the converse is also true.

17. Suppose that X is a curve of genus g, embedded in $\mathbb{P}^r$ by a complete linear series of degree $d \geq 2g + 1$. Show that $d \leq 2(r-1) + 1$. Deduce from Exercise 2.9 that if the points of the hyperplane section $H \cap X$ are in linearly general position, then they impose independent conditions on quadrics. By Exercise 8.16, this statement implies Theorem 8.1 for any curve of high degree whose general hyperplane section consists of points in linearly general position.

Here are two sharp forms of the uniform position principle, from [Harris 1979]. The exercises below sketch a proof of the first, and suggest one of its simplest corollaries.

Theorem 8.24. *Let $X \subset \mathbb{P}^r_\mathbb{C}$ be an irreducible reduced complex projective curve. If $U \subset \check{\mathbb{P}}^r_\mathbb{C}$ is the set of hyperplanes H that meet X transversely then the fundamental group of U acts by monodromy as the full symmetric group on the hyperplane section $H \cap X$.*

In other words, as we move the hyperplane H around a loop in U and follow the points of intersection $H \cap X$ (which we can do, since the intersection remains transverse) we can achieve *any* permutation of the set $H \cap X$. The result can be restated in a purely algebraic form, which makes sense over any field, and is true in somewhat more generality.

Theorem 8.25. [Rathmann 1987] *Let* $S = \mathbb{K}[x_0, \ldots, x_r]$ *be the homogeneous coordinate ring of* $\mathbb{P}^r$, *and let* $X \subset \mathbb{P}^r_\mathbb{K}$ *be an irreducible reduced curve. Assume that* $\mathbb{K}$ *is algebraically closed, and that either* $\mathbb{K}$ *has characteristic* 0 *or that* X *is smooth. Let* H *be the universal hyperplane, defined over the field of rational functions* $\mathbb{K}(u_0, \ldots, u_r)$, *with equation* $\sum u_i x_i = 0$. *The intersection* $H \cap X$ *is an irreducible variety and the natural map* $H \cap X \to X$ *is a finite covering with Galois group equal to the full symmetric group on* $\deg X$ *letters.*

Theorem 8.25 can be stated as the same way as Theorem 8.24 by using the étale fundamental group. It remains true for singular curves in $\mathbb{P}^5$ or higher-dimensional spaces. Amazingly, it really can fail for singular curves in $\mathbb{P}^3$: [Rathmann 1987] contains examples where the general hyperplane section looks like the set of points of a finite projective plane (with many collinear points, for example).

Theorem 8.24 may be proved by following the steps in Exercises 8.19–8.20. But first, here is an application.

18. Use Theorem 8.24 to show that if $X \subset \mathbb{P}^r_\mathbb{C}$ is an irreducible curve, then the general hyperplane section $\Gamma = H \cap X$ consists of points in linearly general position (If a point $p \in H \cap X$ lies in the span of $p_1, \ldots, p_k \in H \cap X$, use a permutation to show that every point of $H \cap X$ lies in this span.) Use Exercise 8.17 to deduce Theorem 8.1 for projective curves over $\mathbb{C}$.

19. Let $X \subset \mathbb{P}^r_\mathbb{C}$ be a reduced, irreducible, complex projective curve. Show that a general tangent line to X is simply tangent, and only tangent at 1 point of X as follows.

    (a) Reduce to the case $r = 2$ by showing that $X \subset \mathbb{P}^r_\mathbb{C}$ can be projected birationally into $\mathbb{P}^2$ (Show that if $r > 2$ then there is a point of $\mathbb{P}^r$ on only finitely many (or no) secant lines to X at smooth points. Sard's Theorem implies that projection from such a point is generically an isomorphism. For a version that works in any characteristic see [Hartshorne 1977, Proposition IV.3.7]).

    (b) Assume that $r = 2$. Show that the family of tangent lines to X is irreducible and one-dimensional, and that not all the tangent lines pass through a point. (For the second part, you can use Sard's theorem on the projection from the point.) Thus the general tangent line does not pass through any singular point of the curve.

    (c) Let U be an open subset of $\mathbb{C}$. Show that the general point of any analytic map $v : U \to \mathbb{C}^2$, is uninflected. (This means that there are points $p \in U$ such that the derivatives $v'(p)$ and $v''(p)$ are linearly independent.) Deduce that the general tangent line is at worst simply tangent at several smooth points of X.

    (d) Let $p \in X \subset \mathbb{P}^2_\mathbb{C}$ be an uninflected point. Show that in suitable analytic coordinates there is a local parametrization at p of the form

$$v(x) = p + v_0(x) \quad \text{with} \quad v_0(x) = (x, x^2).$$

Deduce that as p moves only X the motion of the tangent line is approximated to first order by "rolling" on the point p.

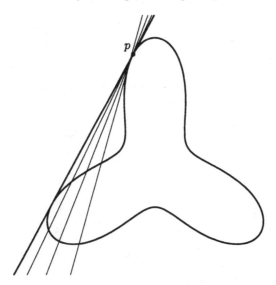

(e) Conclude that there are only finitely many lines that are simply tangent to X at more than one point. Thus the general tangent line to X is tangent only at a single, smooth point.

20. Complete the proof of Theorem 8.24 as follows.

(a) Use Exercise 8.19 to prove that the general tangent hyperplane to X is tangent at only one point, and is simply tangent there.

(b) Suppose that H meets X at an isolated point p, at which H is simply tangent to X. Show that a general hyperplane H' near H meets X in two points near p, and that these two points are exchanged as H' moves along a small loop around the divisor of planes near H that are tangent to X near p. That is, the local monodromy of $H' \cap X$ is the transposition interchanging these two points.

(c) Show that the incidence correspondence

$$I := \big\{(p_1, p_2, H) \in X^2 \times \check{\mathbb{P}}^r :$$
$$p_1 \neq p_2, \ p_1, p_2 \in H \text{ and } H \text{ meets } X \text{ transversely}\big\}$$

is an irreducible quasiprojective variety, and is thus connected (this depends on the complex numbers: over the real numbers, an irreducible variety minus a proper closed set may be disconnected).

(d) Deduce that the monodromy action in Theorem 8.24 is doubly transitive. Show that a doubly transitive permutation group that contains a transposition is the full symmetric group.

/>

9

Clifford Index and Canonical Embedding

If $X \subset \mathbb{P}^r$ is a curve embedded by a complete linear series $|\mathscr{L}|$ of high degree, as studied in Chapter 8, then properties of the homogeneous coordinate ring S_X, such as its graded betti numbers, depend both on the curve and on $\mathscr{L}$. But there is a distinguished linear series on X, called the *canonical series*. It is the complete linear series $|\omega_X|$ associated to the to the *canonical bundle* ω_X, the cotangent bundle of the curve. For most curves the canonical series gives an embedding, and properties of the homogeneous coordinate ring of a curve X in this embedding are intrinsic properties of X alone. We will call the image of X under the map defined by $|\omega_X|$ the *canonical model* of X, and refer to it as a canonical curve.

Green's Conjecture says that the invariants $a(X)$ and $b(X)$, studied in Chapter 8, measure, in the case of canonical curves, the *Clifford index* of X. In this chapter we briefly introduce the Clifford index, canonical curves, and Green's conjecture. As this book is being completed there are dramatic advances in our knowledge, and we will finish the chapter with some pointers to this literature.

9A The Cohen–Macaulay Property and the Clifford Index

To introduce the Clifford index, we will consider the question: when is the homogeneous coordinate ring of a curve $X \subset \mathbb{P}^r$ Cohen–Macaulay? To avoid technicalities, we will restrict to the case of smooth curves. By Proposition 8.3 this is the case if and only if the linear series of hypersurface sections of degree d on

X is complete for every d. In particular, the linear series of hyperplane sections must be the complete. It makes no difference to restrict our attention to the linear span of X, and thus to assume that X is nondegenerate in $\mathbb{P}^r$. With this restriction, to say that the linear series of hyperplane sections is complete is to say that X is embedded by the complete series $|\mathscr{L}|$, where $\mathscr{L}$ is the line bundle $\mathscr{O}_X(1)$. Thus we may restate the original question: for which line bundles $\mathscr{L}$ on a curve X is it the case that the homogeneous coordinate ring of the curve embedded by $|\mathscr{L}|$ is Cohen–Macaulay? Theorem 8.1 asserts that this is the case whenever $\deg \mathscr{L} \geq 2g+1$. What about bundles of lower degree?

For at least one sort of curve, there are no such bundles of lower degree: Recall that X is *hyperelliptic* if X has genus ≥ 2 and X admits a map of degree 2 onto $\mathbb{P}^1$, or, equivalently, X has a line bundle of degree 2 with 2 independent global sections (such a line bundle is then unique.) Exercise 8.4 shows that if $X \subset \mathbb{P}^r$ is a hyperelliptic curve with S_X Cohen–Macaulay, then X must have degree $\geq 2g+1$, so Theorem 8.1 is sharp in this sense. However, among curves of genus ≥ 2, hyperelliptic curves are the only curves for which Theorem 8.1 is sharp! To give a general statement we will define the *Clifford index*, which measures how far a curve is from being hyperelliptic. It may be thought of as a refinement of the *gonality*, the lowest degree of a nonconstant morphism from the curve to the projective line; the Clifford index is the gonality minus 2 in many cases.

The name "Clifford index" comes from Clifford's Theorem (see, for example, [Hartshorne 1977, Theorem IV.5.4]). Recall that a line bundle $\mathscr{L}$ on a curve X is called *special* if $h^1(\mathscr{L}) \neq 0$. Clifford's classic result gives an upper bound on the number of sections of a special line bundle $\mathscr{L}$. If we set $r(\mathscr{L}) = h^0(\mathscr{L}) - 1$, the dimension of the projective space to which $|\mathscr{L}|$ maps X, Clifford's Theorem asserts that a special line bundle $\mathscr{L}$ satisfies $r(\mathscr{L}) \leq (\deg \mathscr{L})/2$, with equality if and only if $\mathscr{L} = \mathscr{O}_X$ or $\mathscr{L} = \omega_X$ or X is hyperelliptic and $\mathscr{L} = \mathscr{L}_0^k$, where $\mathscr{L}_0$ is the line bundle of degree 2 with two independent sections and $0 \leq k \leq g-1$.

With this in mind, we define the *Clifford index* of (any) line bundle $\mathscr{L}$ on a curve X to be

$$\text{Cliff } \mathscr{L} = \deg \mathscr{L} - 2(h^0(\mathscr{L}) - 1) = g + 1 - h^0(\mathscr{L}) - h^1(\mathscr{L}),$$

where g is the genus of X and the two formulas are related by the Riemann–Roch Theorem.

For example, if $\mathscr{L}$ is nonspecial (that is, $h^1 \mathscr{L} = 0$) then $\text{Cliff } \mathscr{L} = 2g - \deg L$ depends only on the degree of $\mathscr{L}$, and is negative when $\deg L \geq 2g+1$. By Serre duality,

$$\text{Cliff } \mathscr{L} = \text{Cliff}(\mathscr{L}^{-1} \otimes \omega_X).$$

Finally, the *Clifford index of the curve* X of genus $g \geq 4$ is defined by taking the minimum of the Clifford indices of all "relevant" line bundles on X:

$$\text{Cliff } X = \min\{\text{Cliff } \mathscr{L} \mid h^0 \mathscr{L} \geq 2 \text{ and } h^1 \mathscr{L} \geq 2\}.$$

If $g \leq 3$ (in which case there are no line bundles $\mathscr{L}$ with $h^0 \mathscr{L} \geq 2$ and $h^1 \mathscr{L} \geq 2$) we instead make the convention that a nonhyperelliptic curve of genus 3 has

Clifford index 1, while any hyperelliptic curve or curve of genus ≤ 2 has Clifford index 0.

Thus, by Clifford's Theorem, $\text{Cliff}\,X \geq 0$ and $\text{Cliff}\,X = 0$ if and only if X is hyperelliptic (or $g \leq 1$). If $\mathscr{L}$ is a line bundle defining a map of degree δ from X to $\mathbb{P}^1$, and $h^1(\mathscr{L}) \geq 2$ (the Brill–Noether Theorem 8.16 shows that this is always the case if X if X is δ-gonal (see page 171), then $h^0(\mathscr{L}) \geq 2$, so $\text{Cliff}\,\mathscr{L} \leq \delta - 2$. On the other hand, the Brill–Noether Theorem also shows the gonality of any curve is at most $\lceil \tfrac{1}{2}(g+2) \rceil$, and it follows that

$$0 \leq \text{Cliff}\,X \leq \lceil \tfrac{1}{2}(g-2) \rceil.$$

The sharpness of the Brill–Noether Theorem for general curves implies that for a general curve of genus g we actually have $\text{Cliff}\,X = \lceil \tfrac{1}{2}(g-2) \rceil$, and that (for $g \geq 4$) the "relevant" line bundles achieving this low Clifford index are exactly those defining the lowest degree maps to $\mathbb{P}^1$.

For a first example where the gonality does not determine the Clifford index, let X be a smooth plane quintic curve. The line bundle $\mathscr{L}$ embedding X in the plane as a quintic has

$$g = 6, \quad \deg \mathscr{L} = 5, \quad h^0\mathscr{L} = 3,$$

whence

$$h^1\mathscr{L} = 3, \quad \text{Cliff}\,\mathscr{L} = 1 \quad \text{and} \quad \text{Cliff}\,X \leq 1.$$

Any smooth plane quintic X is in fact 4-gonal: the lowest degree maps $X \to \mathbb{P}^1$ are projections from points on X, as indicated in the drawing.

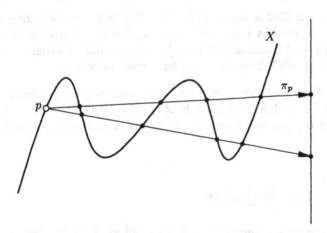

In general, one can show that $\text{Cliff}\,X = 1$ if and only if X is either trigonal or X can be represented as a smooth plane quintic. This sort of analysis can be carried much farther; see for example [Eisenbud et al. 1989], joint work with Herbert Lange and Frank-Olaf Schreyer.

Using the Clifford index we are able to state a strong result about the Cohen–Macaulay property:

Theorem 9.1. *Suppose that $X \subset \mathbb{P}^r$ is a smooth curve over an algebraically closed field of characteristic 0, embedded by a complete linear series. If*

$$\text{Cliff } \mathscr{O}_X(1) < \text{Cliff } X,$$

then S_X is Cohen–Macaulay.

This was first proved in [Green and Lazarsfeld 1985] over the complex numbers. See [Koh and Stillman 1989] for a proof in all characteristics, along lines developed in this book. Theorem 9.1 includes Theorem 8.1 and some other classical assertions:

Corollary 9.2. *Let $X \subset \mathbb{P}^r$ be a smooth nondegenerate curve of degree d and genus $g \geq 2$, embedded by a complete linear series, and let $\mathscr{L} = \mathscr{O}_X(1)$. The homogeneous coordinate ring S_X is Cohen–Macaulay if any of the following conditions are satisfied:*

1. *(Castelnuovo) $d \geq 2g+1$.*
2. *(Max Noether) X is nonhyperelliptic and $\mathscr{L} = \omega_X$.*
3. *(Arbarello, Cornalba, Griffiths, Harris) X is a general curve, $\mathscr{L}$ is a general bundle on X, and $d \geq \lfloor \frac{3}{2}g \rfloor + 2$.*

Proof. 1. If $d \geq 2g+1$ then $\mathscr{L}$ is nonspecial so $\text{Cliff } \mathscr{L} = 2g-d < 0$ while $\text{Cliff } X \geq 0$.

2. $\text{Cliff } \omega_X = 0$, and by Clifford's theorem $\text{Cliff } X = 0$ only if X is hyperelliptic.

3. If X is general, $\text{Cliff } X$ equals $\lceil \frac{1}{2}(g-2) \rceil$. If $\mathscr{L}$ is general of degree at least $\frac{3}{2}g$, then $\mathscr{L}$ is nonspecial by Lemma 8.5, so $\text{Cliff } \mathscr{L} = 2g-d$. Arithmetic shows that $2g-d < \lceil \frac{1}{2}(g-2) \rceil$ if and only if $d \geq \lfloor \frac{3}{2}g \rfloor + 2$ (compare Theorem 8.16). See and [Arbarello et al. 1985, Exercises V.C] for further information. $\square$

Because of the way $\text{Cliff } X$ is defined, the only very ample bundles that can have $\text{Cliff } \mathscr{L} < \text{Cliff } X$ must have $\text{h}^1 \mathscr{L} \leq 1$. It would be interesting to know what is true beyond this range. Some results of this sort appear in [Yau and Chen 1996].

9B   Green's Conjecture

The homogeneous coordinate ring of a canonical curve has been an object of study in algebraic geometry and commutative algebra since the work of Petri [1922] in the early part of the twentieth century. It was my own path of entry from commutative algebra to algebraic geometry, and it still contains plenty of mysteries! In this last section we will concentrate on one of the major conjectures, relating the the free resolution of the homogeneous coordinate ring of a canonical curve with the Clifford index of the curve.

The Homogeneous Coordinate Ring of a Canonical Curve

Let X be a smooth projective curve. If X has genus 0 — and since we are working over an algebraically closed field, this just means $X \cong \mathbb{P}^1$ — the canonical bundle has only the 0 section. For a curve of genus $g > 0$, however, the canonical series is basepoint-free. If X has genus 1, the canonical line bundle is $\mathcal{O}_X$, and the canonical model is a point. For a curve of genus 2, there are 2 sections, so the canonical model is $\mathbb{P}^1$. In these cases the canonical series is not very ample. But for $g \geq 3$, the canonical series is very ample on most curves of genus g.

Theorem 9.3. [Hartshorne 1977, Proposition IV.5.2] *Let X be a smooth curve of genus $g \geq 2$. If X is hyperelliptic, then the canonical series maps X two-to-one onto $\overline{X}$, which is a rational normal curve of degree $g-1$ in $\mathbb{P}^{g-1}$. Otherwise, the canonical series is very ample and embeds $X = \overline{X}$ as a curve of degree $2g-2$ in $\mathbb{P}^{g-1}$.* □

Since the hyperelliptic case is so simple we will normally exclude it from consideration, and we will discuss only canonical models $X \subset \mathbb{P}^{g-1}$ of smooth, non-hyperelliptic curves of genus $g \geq 3$. By Part 2 of Corollary 9.2, the homogeneous coordinate ring S_X of X in its canonical embedding is then Cohen–Macaulay.

For example, it follows from Exercise 9.2 or from the adjunction formula [Hartshorne 1977, Example II.8.20.3] that any smooth plane curve of degree $4 = 2 \cdot 3 - 2$ is the canonical model of a smooth nonhyperelliptic curve of genus 3, and conversely; see Exercise 9.1. The Betti diagram is

	0	1
0	1	–
1	–	–
2	–	–
3	–	1

$g = 3$:

For a nonhyperelliptic curve X of genus $g = 4$, we see from the Hilbert function that the canonical model $X \subset \mathbb{P}^3$ has degree 6 and lies on a unique quadric. In fact, X is a complete intersection of the quadric and a cubic (see Exercise 9.3). Conversely, the adjunction formula shows that every such complete intersection is the canonical model of a curve of genus 4.

	0	1	2
0	1	–	–
1	–	1	–
2	–	1	–
3	–	–	1

$g = 4$:

Finally, we shall see in Exercise 9.4 that two Betti diagrams are possible for the homogeneous coordinate ring of the canonical model of a curve of genus 5:

$g = 5:$

	0	1	2	3
0	1	-	-	-
1	-	3	-	-
2	-	-	3	-
3	-	-	-	1

or

	0	1	2	3
0	1	-	-	-
1	-	3	2	-
2	-	2	3	-
3	-	-	-	1

In all these examples we see that S_X has regularity 3. This is typical:

Corollary 9.4. *If $X \subset \mathbb{P}^{g-1}$ is the canonical model of a nonhyperelliptic curve of genus $g \geq 3$, the Hilbert function of S_X is given by*

$$H_{S_X}(n) = \begin{cases} 0 & \text{if } n < 0, \\ 1 & \text{if } n = 0, \\ g & \text{if } n = 1, \\ (2g-2)n - g + 1 = (2n-1)(g-1) & \text{if } n > 1. \end{cases}$$

In particular, $\beta_{1,2}(S_X)$, the dimension of the space of quadratic forms in the ideal of X, is $\binom{g-1}{2}$ and the Castelnuovo–Mumford regularity of S_X is 3.

Proof. Because S_X is Cohen–Macaulay, its n-th homogeneous component $(S_X)_n$ is isomorphic to $H^0(\mathcal{O}_X(n)) = H^0(\omega_X^n)$. Given this, the Hilbert function values follow at once from the Riemann–Roch theorem.

Because S_X is Cohen–Macaulay we can find a regular sequence on X consisting of 2 linear forms ℓ_1, ℓ_2. The regularity of S_X is the same as that of $S_X/(\ell_1, \ell_2)$. The Hilbert function of this last module has values $1, g-2, g-2, 1$, and thus $\operatorname{reg} S_X/(\ell_1, \ell_2) = 3$. (See also Theorem 4.2.) □

The question addressed by Green's conjecture is: which $\beta_{i,j}$ are nonzero? Since the regularity is 3 rather than 2 as in the case of a curve of high degree, one might think that many invariants would be required to determine this. But in fact things are simpler than in the high degree case, and a unique invariant suffices. The simplification comes from a self-duality of the resolution of S_X, equivalent to the statement that S_X is a Gorenstein ring. See [Eisenbud 1995, Chapter 20] for an introduction to the rich theory of Gorenstein rings, as well as [Huneke 1999] and [Eisenbud and Popescu 2000] for some manifestations.

As in the previous chapter, we write $a(X)$ for the largest integer a such that $\beta_{i,i+2}(S_X)$ vanishes for all $i \leq a(X)$, and $b(X)$ for the smallest integer such that $\beta_{i,i+1}(S_X) = 0$ for all $i \geq b(X)$. The next result shows that, for a canonical curve, $b(X) = g - 2 - a(X)$.

Proposition 9.5. *If $X \subset \mathbb{P}^{g-1}$ is the canonical model of a nonhyperelliptic curve of genus $g \geq 3$, then $w_X = \operatorname{Ext}^{g-2}(S_X, S(-g)) \cong S_X(1)$, so the minimal free resolution of S_X is, up to shift, self-dual, with*

$$\beta_{i,j}(S_X) = \beta_{g-2-i, g+1-j}(S_X).$$

Setting $\beta_i := \beta_{i,i+1}$ the Betti diagram of S_X has the form

	0	1	$\cdots$	a	$a+1$	$\cdots$	$b-1$	b	$\cdots$	$g-3$	$g-2$
0	1	$-$	$\cdots$	$-$	$-$	$\cdots$	$-$	$-$	$\cdots$	$-$	$-$
1	$-$	β_1	$\cdots$	β_a	β_{a+1}	$\cdots$	β_{g-3-a}	$-$	$\cdots$	$-$	$-$
2	$-$	$-$	$\cdots$	$-$	β_{g-3-a}	$\cdots$	β_{a+1}	β_a	$\cdots$	β_1	$-$
3	$-$	$-$	$\cdots$	$-$	$-$	$\cdots$	$-$	$-$	$\cdots$	$-$	1

where the terms marked "$-$" are zero, the β_i are nonzero, and $\beta_1 = \binom{g-2}{2}$.

Proof. By Theorem 9.3, local duality (Theorem A1.9), and Corollary 9.2 we have

$$S_X = \bigoplus \mathrm{H}^0 \mathscr{O}_X(n) = \bigoplus \mathrm{H}^0(\omega_X^n) = \bigoplus \mathrm{H}^0(\omega_X(n-1)) = w_X(-1).$$

The rest of the statements follow.   □

Here is Green's Conjecture, which stands at the center of much current work on the topics of this book.

Conjecture 9.6. [Green 1984b] *Let $X \subset \mathbb{P}^{g-1}$ be a smooth nonhyperelliptic curve over a field of characteristic 0 in its canonical embedding. The invariant $a(X)$ of the free resolution of S_X is equal to* Cliff $X - 1$.

The first case in which Green's conjecture is nontrivial is that of a nonhyperelliptic curve X of genus 5. In this case X has Clifford index 1 if and only if X has a degree 3 divisor that "moves" in the sense that $h^0 \mathscr{O}_X(D) = 2$; otherwise X has Clifford index 2. If the Clifford index of X is 2, then the canonical model $X \subset \mathbb{P}^4$ is a complete intersection of 3 quadrics, with Betti diagram

	0	1	2	3
0	1	$-$	$-$	$-$
1	$-$	3	$-$	$-$
2	$-$	$-$	3	$-$
3	$-$	$-$	$-$	1

$g = 5$, Cliff $X = 2$:

On the other hand, if X has Clifford index 1 then the Betti diagram of X is

	0	1	2	3
0	1	$-$	$-$	$-$
1	$-$	3	2	$-$
2	$-$	2	3	$-$
3	$-$	$-$	$-$	1

$g = 5$, Cliff $X = 1$:

(Exercise 9.4). In the case $g = 6$ one encounters for the first time a case in which the Clifford index itself, and not just the gonality of X enters the picture. If X is a smooth plane curve of degree d, then by the adjunction formula [Hartshorne 1977, Example 8.20.3] the canonical series is the restriction of $\mathscr{O}_{\mathbb{P}^2}(d-3) = \mathscr{O}_{\mathbb{P}^2}(2)$ to X. Thus in the case $d = 5$ the canonical model of X in $\mathbb{P}^5$ is the image of

$X \subset \mathbb{P}^2$ under the quadratic Veronese map $\nu_2 : \mathbb{P}^2 \to \mathbb{P}^5$. The Veronese surface $V := \nu_2(\mathbb{P}^2)$ has degree 4, and thus its hyperplane section is a rational normal curve. Since S_V is Cohen–Macaulay (A2.44), the graded Betti numbers of S_V are the same as those for the rational normal quartic, namely

	0	1	2	3
0	1	–	–	–
1	–	6	8	3

It follows from Theorem 8.12 that $\beta_{3,4}(S/I_X) \neq 0$, so $a(X) = 0$ in this case, just as it would if X admitted a line bundle $\mathscr{L}$ of degree 3 with $h^0 \mathscr{L} = 2$. This corresponds to the fact that $\mathrm{Cliff}\, X = 1$ in both cases.

Green and Lazarsfeld proved one inequality of the Conjecture, using the same technique that we have used above to give a lower bound for $b(X)$ [Green 1984b, Appendix].

Corollary 9.7. *With hypothesis as in Green's Conjecture,*

$$a(X) \leq \mathrm{Cliff}\, X - 1.$$

Proof. Let D be a divisor on X with $h^0 \mathcal{O}_X(D) \geq 2$ and $h^1 \mathcal{O}_X(D) \geq 2$. Theorem 8.12 shows that $b(X)$ is bounded below by

$$h^0 \mathcal{O}_X(D) - 1 + h^0 \omega_X(-D) - 1 = h^0 \mathcal{O}_X(D) + h^1 \mathcal{O}_X(D) - 2 = g - 1 - \mathrm{Cliff}\, \mathcal{O}_X(D).$$

By virtue of the duality above, this bound can also be viewed as an upper bound

$$a(X) = g - 2 - b(X) \leq g - 2 - (g - 1 - \mathrm{Cliff}\, \mathcal{O}_X(D)) = \mathrm{Cliff}\, \mathcal{O}_X(D) - 1. \qquad \square$$

Thus, to prove Green's Conjecture for a particular curve X we need to show that $a(X) \geq \mathrm{Cliff}(X) - 1$. As of this writing the result is known for all curves of genus up to 9 (see [Schreyer 1986] for genus up to 8, and [Hirschowitz and Ramanan 1998a], [Mukai 1995] and [Schreyer 1989] for different subcases of genus 9). It is known for all curves of Clifford index at most 4 [Voisin 1988, Schreyer 1991], and it is also known for some special classes of curves, such as those that can be represented as smooth plane curves [Loose 1989].

On the other hand, The obvious extension of Green's conjecture to positive characteristic is known to fail in characteristic 2 for curves of genus 7 [Schreyer 1986] and 9 [Mukai 1995] and there is strong probabilistic evidence that it fails in various other cases of positive characteristic. For this and a very interesting group of conjectures about the possible Betti diagrams of canonical curves of genus up to 14 in any characteristic, see [Schreyer 2003, Section 6].

As of this writing, a series of spectacular papers [Voisin 2002, Voisin 2003, Teixidor I Bigas 2002] has greatly advanced our knowledge: roughly speaking, we now know that the conjecture holds for the generic curves of each genus and Clifford index. By [Hirschowitz and Ramanan 1998b] this implies that the conjecture is true for every curve of odd genus g that has the maximal possible Clifford index, $\frac{1}{2}(g - 1)$.

As always in mathematics, when one approaches the frontier one begins to realize that the unknown is far larger and than the known. But the recent progress in Green's conjecture offers plenty of hope for further breakthroughs. Perhaps the reader will take the next step!

9C   Exercises

1. Show that a smooth plane curve is a canonical model if and only if it is a plane quartic (you might use the Adjunction Formula of [Hartshorne 1977, Example 8.20.3].

2. Suppose that $X \subset \mathbb{P}^{g-1}$ is a nondegenerate curve such that S_X is Cohen–Macaulay. Show that X is the canonical model of the abstract curve X if and only if
$$\beta_{g-2,n} = \begin{cases} 1 & \text{if } n = g, \\ 0 & \text{otherwise.} \end{cases}$$

3. Prove that a curve in $\mathbb{P}^3$ is a canonical model if and only if it is a complete intersection of a quadric and a cubic. (You might use Exercise 9.2.)

4. Let $X \subset \mathbb{P}^4$ be a nondegenerate smooth irreducible curve. If X is the complete intersection of three quadrics, show that X is a canonical model. In this case $a(X) = 1$.

    Now let $X \subset \mathbb{P}^4$ be a canonical model with $a(X) = 0$; that is, suppose that I_X is not generated by quadrics. Show that the quadratic forms in I_X form a three-dimensional vector space, and that each of them is irreducible. Show that they define a two-dimensional irreducible nondegenerate variety Y of degree 3. This is the minimal possible degree for a nondegenerate surface in $\mathbb{P}^4$ [Hartshorne 1977, Exercise I.7.8]. By the classification of such surfaces (see for example [Eisenbud and Harris 1987]) Y is a scroll. Using the Adjunction formula ([Hartshorne 1977, Proposition V.5.5]) show that the curve meets each line of the ruling of Y in 3 points. The divisor defined by these three points moves in a one-dimensional linear series by Theorem 9.8, and thus the Clifford index of X is 1, as required by Green's Theorem.

5. Suppose that $X \subset \mathbb{P}^{g-1}$ is a smooth, irreducible, nondegenerate curve of degree $2g-2$ where $g \geq 3$ is the genus of X. Using Clifford's Theorem (page 178), show that $\mathcal{O}_X(1) = \omega_X$, so X is a canonical model.

6. Let $X \subset \mathbb{P}^{g-1}$ be the canonical model of a smooth irreducible curve of genus $g \geq 3$. Assume that for a general hyperplane $H \subset \mathbb{P}^{g-1}$ the hyperplane section $\Gamma = H \cap X$ consists of points in linearly general position. Show that Γ fails by at most 1 to impose independent conditions on quadrics in H, and imposes independent conditions on hypersurfaces of degree n for $n > 2$. Deduce that the linear series of hypersurfaces of degree n is complete for every n, and thus that S_X is Cohen–Macaulay.

7. Reinterpret the Riemann–Roch theorem to prove the following:

Theorem 9.8 (Geometric Riemann–Roch). *Let $X \subset \mathbb{P}^{g-1}$ be a canonically embedded nonhyperelliptic curve. If D is an effective divisor on X and L is the smallest linear space in $\mathbb{P}^{g-1}$ containing D, then $h^0 \mathcal{O}_X(D) = \deg D - \dim L$.*

Otherwise stated: The (projective) dimension, $h^0(\mathcal{O}_X(D)) - 1$, of the linear series D equals the amount by which the points of D fail to be linearly independent. (Some care is necessary when the points of D are not distinct. In the statement of the theorem, we must insist that L cut X with multiplicity at least as great as that of D at each point. And "the amount by which the points of D fail to be linearly independent" requires us to think of the span of a multiple point as the dimension of the smallest linear space that contains it, in the sense just given.)

8. Use Theorem 8.9, Corollary 9.7, and Theorem 9.8 to show that for a canonically embedded, nonhyperelliptic curve $X \subset \mathbb{P}^{g-1}$, with genus $g \geq 4$,

$$a(X) \leq \mathrm{Cliff}\, \mathcal{O}_X(D) - 1 \leq d - 3.$$

9. Work through the document "Canonical Embeddings of Plane Curves and Gonality" found at www.math.uiuc.edu/Macaulay2/Manual/1617.html as part of the Macaulay manual.

Appendix 1

Introduction to Local Cohomology

This appendix is an introduction to local cohomology, including the results used in the text and the connection with the cohomology of sheaves on projective space. For another version, see [Hartshorne/Grothendieck 1967]; for more results, and a very detailed and careful treatment, see [Brodmann and Sharp 1998]. A partial idea of recent work in the subject can be had from the survey [Lyubeznik 2002].

We will work over a Noetherian ring, with a few comments along the way about the differences in the non-Noetherian case. (I am grateful to Arthur Ogus and Daniel Schepler for straightening out my ideas about this.)

A1A   Definitions and Tools

First of all, the definition: If R is a Noetherian ring, $Q \subset R$ is an ideal, and M is an R-module, then the *zeroth local cohomology module of M* is

$$\mathrm{H}^0_Q(M) := \{m \in M \mid Q^d m = 0 \text{ for some } d\}.$$

H^0_Q is a functor in an obvious way: if $\varphi : M \to N$ is a map, the induced map $\mathrm{H}^0_Q(\varphi)$ is the restriction of φ to $\mathrm{H}^0_Q(M)$. One sees immediately from this that the functor H^0_Q is left exact, so it is natural to study its derived functors, which we call H^i_Q.

For example, suppose that R is a local ring and Q is its maximal ideal. If M is a finitely generated R-module then we claim that $\mathrm{H}^0_Q(M)$ is the (unique) largest submodule of M with finite length. On one hand, Nakayama's Lemma shows that any finite length submodule is annihilated by some power of Q, and thus is contained in $\mathrm{H}^0_Q(M)$. On the other hand, since R is Noetherian and M is finitely generated, $\mathrm{H}^0_Q(M)$ is generated by finitely many elements. Some power of Q^d annihilates each of them, and thus $\mathrm{H}^0_Q(M)$ is a finitely generated module over the ring R/Q^n, which has finite length, completing the argument. The same proof works in the case where R is a graded algebra, generated over a field $\mathbb{K}$ by elements of positive degree, the ideal Q is the homogeneous maximal ideal, and M is a finitely generated graded R-module.

Local Cohomology and Ext

We can relate the local cohomology to the more familiar derived functor Ext.

Proposition A1.1. *There is a canonical isomorphism*

$$\mathrm{H}^i_Q(M) \cong \varinjlim \mathrm{Ext}^i_R(R/Q^d, M),$$

where the limit is taken over the maps $\mathrm{Ext}^i_R(R/Q^d, M) \to \mathrm{Ext}^i_R(R/Q^e, M)$ induced by the natural epimorphisms $R/Q^e \longrightarrow R/Q^d$ for $e \geq d$.

Proof. There is a natural injection

$$\mathrm{Ext}^0_R(R/Q^d, M) = \mathrm{Hom}(R/Q^d, M) \longrightarrow M$$
$$\phi \longmapsto \phi(1)$$

whose image is $\{m \in M \mid Q^d m = 0\}$. Thus the direct limit $\varinjlim \mathrm{Ext}^0_R(R/Q^d, M) = \varinjlim \mathrm{Hom}(R/Q^d, M)$ may be identified with the union

$$\bigcup_d \{m \in M \mid Q^d m = 0\} = \mathrm{H}^0_Q(M).$$

The functor $\mathrm{Ext}^i_R(R/Q^d, -)$ is the i-th derived functor of $\mathrm{Hom}_R(R/Q^d, -)$. Taking filtered direct limits commutes with taking derived functors because of the exactness of the filtered direct limit functor [Eisenbud 1995, Proposition A6.4]. □

Corollary A1.2. *Any element of $\mathrm{H}^i_Q(M)$ is annihilated by some power of Q.*

Proof. Any element is in the image of some $\mathrm{Ext}^i_R(R/Q^d, M)$, which is itself annihilated by Q^d. □

Local Cohomology and Čech Cohomology

Another useful expression for the local cohomology is obtained from a Čech *complex:* Suppose that Q is generated by elements $(x_1, \ldots, x_t)$. We write $[t] = \{1, \ldots, t\}$ for the set of integers from 1 to t, and for any subset $J \subset [t]$ we let $x_J = \prod_{j \in J} x_j$. We denote by $M[x_J^{-1}]$ the localization of M by inverting x_J.

Theorem A1.3. *Suppose that R is a Noetherian ring and $Q = (x_1, \ldots, x_t)$. For any R-module M the local cohomology $\mathrm{H}^i_Q(M)$ is the i-th cohomology of the complex*

$$C(x_1, \ldots, x_t; M) : 0 \longrightarrow M \xrightarrow{d} \overset{t}{\underset{1}{\bigoplus}} M[x_i^{-1}] \xrightarrow{d} \cdots$$

$$\longrightarrow \underset{\#J=s}{\bigoplus} M[x_J^{-1}] \xrightarrow{d} \cdots \longrightarrow M[x_{\{1,\ldots,t\}}^{-1}] \longrightarrow 0,$$

whose differential takes an element

$$m_J \in M[x_J^{-1}] \subset \underset{\#J=s}{\bigoplus} M[x_J^{-1}]$$

to the element

$$d(m_J) = \sum_{k \notin J} (-1)^{o_J(k)} m_{J \cup \{k\}},$$

where $o_J(k)$ denotes the number of elements of J less than k, and $m_{J \cup \{k\}}$ denotes the image of m_J in the further localization $M[(x_{J \cup \{k\}})^{-1}] = M[x_J^{-1}][x_k^{-1}]$.

Here the terms of the Čech complex are numbered from left to right, counting M as the 0-th term, and we write $C^s(M) = \bigoplus_{\#J=s} M[x_J^{-1}]$ for the term of cohomological degree s. If R is non-Noetherian, then the Čech complex as defined here does *not* always compute the derived functors in the category of R-modules of $\mathrm{H}^0_I()$ as defined above, even for finitely generated I; see Exercise A1.7. Rather, it computes the derived functors in the category of (not necessarily quasi-coherent) sheaves of $\mathcal{O}_{\mathrm{Spec}\,R}$ modules. For this and other reasons, the general definition of the local cohomology modules should probably be made in this larger category. As we have no use for this refinement, we will not pursue it further. See [Hartshorne/Grothendieck 1967] for a treatment in this setting.

Proof. An element $m \in M$ goes to zero under $d : M \to \bigoplus_j M[x_j^{-1}]$ if and only if m is annihilated by some power of each of the x_i. This is true if and only if m is annihilated by a sufficiently big power of Q, so $\mathrm{H}^0(C(M)) = \mathrm{H}^0_Q(M)$ as required.

The complex $C(x_1, \ldots, x_t; M)$ is obviously functorial in M. Since localization is exact, a short exact sequence of modules gives rise to a short exact sequence of complexes, and thus to a long exact sequence in the homology functors $\mathrm{H}^i(C(M))$. To prove that $\mathrm{H}^i(C(M)) = \mathrm{H}^i_Q(M)$ we must show it is the derived functor of $\mathrm{H}^0_Q(M) = \mathrm{H}^0(C(M))$. For this it is enough to show that $\mathrm{H}^i(C(M)) = 0$ when M is an injective module and $i > 0$ (see for example [Eisenbud 1995, Proposition A3.17 and Exercise A3.15]). We need two properties of injective modules over Noetherian rings:

Lemma A1.4. *Suppose that R is a Noetherian ring, and M is an injective R-module.*

1. *For any ideal $Q \subset R$, the submodule $\mathrm{H}^0_Q(M)$ is also an injective module.*
2. *For any $x \in R$, the localization map $M \to M[x^{-1}]$ is surjective.*

Proof. 1. We must show that if $I \subset R$ is an ideal and $\phi : I \to \mathrm{H}^0_Q(M)$ is a map, then ϕ extends to a map $R \to \mathrm{H}^0_Q(M)$. We first extend ϕ to an ideal containing a power of Q: Since I is finitely generated, and each generator goes to an element annihilated by a power of Q, we see that for sufficiently large d the ideal $Q^d I$ is in the kernel of ϕ. By the Artin–Rees Lemma [Eisenbud 1995, Lemma 5.1], the ideal $Q^d I$ contains an ideal of the form $Q^e \cap I$. It follows that the map $(\phi, 0) : I \oplus Q^e \to \mathrm{H}^0_Q(M)$ factors through the ideal $I + Q^e \subset R$. Changing notation, we may assume that $I \supset Q^e$ from the outset.

By the injectivity of M we may extend ϕ to a map $\phi' : R \to M$. Since $\phi'(Q^e) = \phi(Q^e) \subset \mathrm{H}^0_Q(M)$, it follows that some power of Q annihilates $Q^e \phi'(1)$, and thus some power of Q annihilates $\phi'(1)$; that is, $\phi'(1) \in \mathrm{H}^0_Q(M)$, so ϕ' is the desired extension.

2. Given $m \in M$ and natural number d, we want to show that m/x^d is in the image of M in $M[x^{-1}]$. Since R is Noetherian, the annihilator of x^e in R is equal to the annihilator of x^{d+e} in R when e is large enough. Thus the annihilator of x^{d+e} is contained in the annihilator of $x^e m$. It follows that there is a map from the principal ideal (x^{d+e}) to M sending x^{d+e} to $x^e m$. Since M is injective, this map extends to a map $R \to M$; write $m' \in M$ for the image of 1, so that $x^{e+d} m' = x^e m$. Since $x^e(x^d m' - m) = 0$, the element m' goes, under the localization map, to $m/x^d \in M[x^{-1}]$, as required. $\qquad\square$

To complete the proof of Theorem A1.3 we prove that $\mathrm{H}^i(C(x_1, \ldots, x_t; M)) = 0$ for all $i > 0$, when M is an injective module. We apply induction on t, the case $t = 0$ being obvious. For the case $t = 1$ we must show that, for any injective R-module M and any $x \in R$, the localization map $M \to M[x^{-1}]$ is surjective, and this is the content of part 2 of Lemma A1.4.

When $t > 1$, we observe that the submodules

$$\bigoplus_{\substack{t \in J \\ \#J = s}} M[x_J^{-1}] \subset C^s(x_1, \ldots, x_t; M)$$

for a subcomplex isomorphic to $C(x_1, \ldots, x_t; M)[x_J^{-1}][1]$, where the [1] indicates that the cohomological degree is shifted by 1. Since the quotient involves no terms where x_t is inverted, we get a short exact sequence of complexes

$$0 \longrightarrow C(x_1, \ldots, x_{t-1}; M)[x_t^{-1}][1] \longrightarrow C(x_1, \ldots, x_t; M) \longrightarrow$$
$$\longrightarrow C(x_1, \ldots, x_{t-1}; M) \longrightarrow 0.$$

The associated long exact sequence contains the terms

$$\mathrm{H}^{i-1}(C(x_1, \ldots, x_{t-1}; M)) \xrightarrow{\delta_i} \mathrm{H}^{i-1}(C(x_1, \ldots, x_{t-1}; M)[x_t^{-1}]) \longrightarrow$$
$$\longrightarrow \mathrm{H}^i(C(x_1, \ldots, x_t; M)) \longrightarrow \mathrm{H}^i(C(x_1, \ldots, x_{t-1}; M)).$$

It is easy to check from the definitions that the connecting homomorphism δ is simply the localization map. If M is injective and $i > 1$ we derive

$$\mathrm{H}^i\big(C(x_1, \ldots, x_t; M)\big) = 0$$

by induction. For the case $i = 1$ we use Lemma A1.4, which implies that $\delta_0 :$ $M \to M[x_t^{-1}]$ is surjective. By induction, $\mathrm{H}^1(C(x_1, \ldots, x_t; M)) = 0$, and the result follows. $\qquad\square$

Corollary A1.5. *If M is a graded S-module of finite length, then $\mathrm{H}_Q^0(M) = M$, while $\mathrm{H}_Q^i(M) = 0$ for $i > 0$.*

Note the contrast with the case of $\mathrm{Ext}_S^i(S/Q^j, M)$; for example, when M is the module $\mathbb{K}$, of length 1, the value is nonzero for all j and all $0 \le i \le r$. The corollary says that in the limit everything goes to zero except when $i = 0$!

Proof. The first assertion is the definition of $\mathrm{H}_Q^0(M)$. Since a power of each x_i annihilates M, we have $M[x_i^{-1}] = 0$ for each i. Thus the complex $C(x_1, \ldots, x_t; M)$ reduces to $0 \to M \to 0$, so the the second assertion follows from Theorem A1.3. $\qquad\square$

Theorem A1.3 also allows us to compute the $\mathrm{H}_{(x_0, \ldots, x_r)}^i(S)$ explicitly. For any finitely generated graded S-module M let $M^\vee$ be the graded vector space dual $\bigoplus_d(\mathrm{Hom}_\mathbb{K}(M_d, \mathbb{K}))$, regarded as an S-module by the rule $(s\phi)(m) := \phi(sm)$ for $s \in S$, $\phi \in \oplus \mathrm{Hom}_\mathbb{K}(M_d, \mathbb{K})$ and $m \in M$. As usual we set $\omega_S = S(-r-1)$, called the canonical module of S.

Corollary A1.6. *If $S = \mathbb{K}[x_0, \ldots, x_r]$ is the polynomial ring in $r+1$ variables, and $Q = (x_0, \ldots, x_r)$ is the ideal generated by the variables, then $\mathrm{H}_Q^i(S) = 0$ for $i < r+1$ while*

$$\mathrm{H}_Q^{r+1}(S) = (\omega_S)^\vee$$

functorially as S-modules. That is, the functor on free S-modules that takes F to $\mathrm{H}_Q^{r+1}(F)$ is naturally isomorphic to the functor $F \mapsto (\mathrm{Hom}_S(F, \omega_S))^\vee$.

Proof. To show that $\mathrm{H}_Q^i(S) = 0$ for $i < r+1$ it suffices, since the complex $C(x_0, \ldots, x_r; S)$ is multigraded, to work on one multi-degree $\alpha \in \mathbb{Z}^{r+1}$ at a time. Let J be the set of those indices $j \in \{0, \ldots, r+1\}$ such that $\alpha_j < 0$. The summand $S[x_I^{-1}]$ contains the monomial x^α if and only if $I \supset J$, so

$$S[x_I^{-1}]_\alpha = \begin{cases} \mathbb{K} & \text{if } I \subset J; \\ 0 & \text{otherwise.} \end{cases}$$

Consider the simplex Δ whose faces are the subsets of $K := \{0, \ldots, r\} \setminus J$. Examining the maps of the complex, we see that the the degree α part of $C(x_0, \ldots, x_r; S)$ is the reduced chain complex of Δ, where the face with index set L corresponds to the monomial x^α in the component $S[x_{J \cup L}^{-1}]$ of $C(x_0, \ldots, x_r; S)$. Since the reduced homology of a simplex is 0 unless the simplex is empty, we are done.

The same argument shows that $H^r_Q(S)_\alpha$ is $\mathbb{K}$ if every component of α is negative, and zero otherwise, which agrees as a vector space with $(\omega_S)^\vee = (\mathrm{Hom}_S(S, \omega_S))^\vee$. To say that the identification is functorial means that if $f : S(d) \to S(e)$ is a map, so that $f \in S_{e-d}$, then the induced map $f^\vee : \omega_S(d)^\vee \to \omega_S(d)^\vee$ is the map $f : H^{r+1}_Q(S(d)) \to H^{r+1}_Q(S(e))$ induced by multiplication. For this we must simply show that the module structure of $H^{r+1}_Q(S)$ agrees with that of $(\omega_S)^\vee$.

As an S-module $H^{r+1}_Q(S)$ is, by definition, the cokernel of the natural map

$$\bigoplus_{\#I=r} S[x_I^{-1}] \longrightarrow S[(x_0 \cdots x_r)^{-1}].$$

The image is the vector space spanned by those monomials x^α such that one of the components α_i of the multi-index $\alpha = (\alpha_0, \ldots, \alpha_r)$ is non-negative. Thus the cokernel, $H^{r+1}_Q(S)$, may be identified with the vector space

$$(x_0 \cdots x_r)^{-1} \mathbb{K}[x_0^{-1}, \cdots, x_r^{-1}].$$

The S-module structure on $H^{r+1}_Q(S)$ induced from $S[(x_0 \cdots x_r)^{-1}]$ may be described, with this identification, by saying that for $x^\beta \in S$ and $x^\alpha \in H^{r+1}_Q(S)$ we have

$$x^\beta x^\alpha = \begin{cases} x^{\alpha+\beta} & \text{if all components of } \alpha+\beta \text{ are negative,} \\ 0 & \text{otherwise.} \end{cases}$$

Thus the map $H^{r+1}_Q \to \bigoplus_d \mathrm{Hom}_{\mathbb{K}}(S_d, \mathbb{K})$ sending x^α to the dual basis vector of $x^{-\alpha}$ is an isomorphism. $\qquad\square$

One of the most important applications of local cohomology depends on the following easy consequence.

Corollary A1.7. *Suppose $Q = (x_1, \ldots, x_t)$. If M is an R-module, $H^i_Q(M) = 0$ for $i > t$.*

Proof. The length of the Čech complex $C(x_1, \ldots, x_t; M)$ is t. $\qquad\square$

We say that an algebraic set X is defined set-theoretically by n equations if there is an ideal Q with n generators whose radical is $I(X)$. Corollary A1.7 is a powerful tool for testing whether this holds. Since the local cohomology $H^i_I(M)$ depends only on the radical of I, this implies $H^i_{I(X)}(M) = H^i_Q(M) = 0$ for all $i > n$ and all modules M. See [Schmitt and Vogel 1979] and [Stückrad and Vogel 1982] for examples of use of this technique, and [Lyubeznik 2002] for a recent survey including many pointers to the literature.

By far the most famous open question of this type is whether each irreducible curve in $\mathbb{P}^3_{\mathbb{K}}$ can be defined set-theoretically by just two equations; it is not even known whether this is the case for the smooth rational quartic curve X in $\mathbb{P}^3_{\mathbb{K}}$ defined as the image of the map

$$\mathbb{P}^1_{\mathbb{K}} \ni (s, t) \to (s^4, s^3 t, s t^3, t^4) \in \mathbb{P}^3_{\mathbb{K}}.$$

For this curve it is known that $H^i_{I(X)}(M) = 0$ for all $i > 2$ and all modules M [Hartshorne 1970, Chapter 3], so the local cohomology test is not useful here. To add to the fun, it is known that if we replace $\mathbb{K}$ by a field of characteristic $p > 0$ then this curve *is* set-theoretically the complete intersection of two surfaces [Hartshorne 1979]. See [Lyubeznik 1989] for an excellent review of this whole area.

Change of Rings

Suppose $\varphi : R \to R'$ is a homomorphism of rings, Q is an ideal of R, and M is an R'-module. Using the map φ we can also regard M as an R-module. For any given d, the relation between $\mathrm{Ext}^i_R(R/Q^d, M)$ and $\mathrm{Ext}^i_{R'}(R'/Q'^d, M)$, where $Q' = QR'$, is mysterious (there is a change of rings spectral sequence that helps a little). For some reason taking the limit, and thus passing to local cohomology, fixes this.

Corollary A1.8. *Suppose that* $\varphi : R \to R'$ *is a homomorphism of Noetherian rings. With notation as above, there is a canonical isomorphism* $H^i_Q(M) \cong H^i_{QR'}(M)$.

Proof. If $x \in R$ is any element, the localization $M[x^{-1}]$ is the same whether we think of M as an R-module or an R'-module: it is the set of ordered pairs (m, x^d) modulo the equivalence relation $(m, x^d) \sim (m', x^e)$ if $x^f(x^e m - x^d m') = 0$ for some f. Thus the Čech complex $C(x_1, \ldots, x_t; M)$ is the same whether we regard M as an R-module or an R'-module, and we are done by Theorem A1.3.   □

Corollary A1.8 fails in the non-Noetherian case even when $R = \mathbb{K}[t]$ and $I = t$; see Exercise A1.7.

Local Duality

Because it comes up so often in applications, we mention a convenient way to compute local cohomology with respect to the maximal ideal of a homogeneous polynomial ring. The same method works more generally over regular local rings, and, with some care, over arbitrary rings.

Theorem A1.9. *Let* $S = \mathbb{K}[x_0, \ldots, x_r]$ *be the polynomial ring, and let* $\mathfrak{m} = (x_0, \ldots, x_r)$ *be the homogeneous maximal ideal. If* M *is a finitely generated graded S-module then* $H^i_{\mathfrak{m}}(M)$ *is (as S-module) the graded $\mathbb{K}$-vectorspace dual of* $\mathrm{Ext}^{r+1-i}(M, S(-r-1))$.

Proof. Let $\mathbf{F} : \cdots F_1 \to F_0$ be a free resolution of M. Tensoring $\mathbf{F}$ with the complex $\mathbf{C} := C(x_0, \cdots, x_r; S)$ gives a double complex. If we think of the differentials from $\mathbf{F}$ as horizontal, and the differentials induced from $\mathbf{C}$ as vertical, then since localization is an exact functor the horizontal homology of the double complex is just the complex $C(x_0, \cdots, x_r; M)$. The i-th homology of $C(x_0, \cdots, x_r; M)$ is

$H^i_Q(M)$. Thus one spectral sequence of the double complex degenerates, and the homology of the total complex $\mathbf{F} \otimes \mathbf{C}$ is $H^i_Q(M)$.

By Corollary A1.6 $H^i_Q F_j = 0$ for $i < r+1$, so the columns of the double complex have homology only at the end, and the vertical homology of the double complex is the complex $H^{r+1}_Q(\mathbf{F}) \cong ((\mathrm{Hom}(\mathbf{F}, \omega_S))^\vee$. Since $A \mapsto A^\vee$ is an exact functor, it comutes with homology, and the j-th homology of $((\mathrm{Hom}(\mathbf{F}, \omega_S))^\vee$ is $(\mathrm{Ext}^{r+1-j}_S(M, \omega_S))^\vee$. Thus the other spectral sequence degenerates too, and the homology of the total complex $\mathbf{F} \otimes \mathbf{C}$ is $(\mathrm{Ext}^{r+1-j}_S(M, \omega_S))^\vee$, proving the desired equality. $\qquad \square$

Example A1.10. A simple example may serve to make all these computations clearer.

Let $S = \mathbb{K}[x, y]$, $\mathfrak{m} = (x, y)$, and consider the S-module $R = \mathbb{K}[x, y]/(x^2, xy)$. We will compute the local cohomology $H^i_\mathfrak{m}(R)$ (which is the same, by Theorem A1.8, as the local cohomology of R as a module over itself) in two ways:

From the Čech complex: The Čech complex of R is by definition

$$0 \longrightarrow R \xrightarrow{\binom{1}{1}} R[x^{-1}] \oplus R[y^{-1}] \xrightarrow{(1,-1)} R[(xy)^{-1}] \longrightarrow 0.$$

However, R is annihilated by x^2, and thus also by $(xy)^2$. Consequently, the Čech complex takes the simpler form

$$0 \to R \xrightarrow{(1)} R[y^{-1}] \longrightarrow 0,$$

where the map denoted (1) is the canonical map to the localization.

The kernel of this map is the 0-th homology of the Čech complex, and thus by Theorem A1.3 it is $H^0_\mathfrak{m}(R)$. As the kernel of the localization map $R \to R[y^{-1}]$, it is the set of elements of R annihilated by a power of y, which is the one-dimensional vector space

$$H^0_\mathfrak{m}(R) = (x^2, xy) : y^\infty/(x^2, xy) = (x)/(x^2, xy) = \mathbb{K} \cdot x = \mathbb{K}(-1).$$

Since the localization map kills x, we see that $R[y^{-1}] = S/(x)[y^{-1}]$, and the image of R in $R[y^{-1}]$ is the same as the image of $S/(x)$ in $S/(x)[y^{-1}]$. Thus the first homology of the Čech complex, which is equal by Theorem A1.3 to the first local cohomology of R, is

$$H^1_\mathfrak{m}(R) = S/(x)[y^{-1}]/(S/(x)) = \mathbb{K} \cdot y^{-1} \oplus \mathbb{K} \cdot y^{-2} \oplus \cdots = \mathbb{K}(1) \oplus \mathbb{K}(2) \oplus \cdots.$$

From local duality: Because (x^2, xy) is generated by just two elements it is easy to write down a free resolution of $R = S/(x^2, xy)$:

$$0 \longrightarrow S(-3) \xrightarrow{\binom{y}{-x}} S^2(-2) \xrightarrow{(x^2 \;\; xy)} S \longrightarrow R \longrightarrow 0.$$

The modules $\mathrm{Ext}_S^i(R, S(-2)) = \mathrm{Ext}_S(R, S(-2))$ are the homology of the dual complex, twisted by -2, which is

$$0 \longleftarrow S(1) \xleftarrow{(y \ \ -x)} S^2 \xleftarrow{\begin{pmatrix} x^2 \\ xy \end{pmatrix}} S(-2) \longleftarrow 0.$$

It is thus immediate that $\mathrm{Ext}_S^0(R, S(-2)) = 0$. We also see at once that

$$\mathrm{Ext}_S^2(R, S(-2)) = S(1)/(x, y) = \mathbb{K}(1),$$

the dual of $\mathbb{K}(-1) = \mathrm{H}_m^0(R)$, as claimed by Theorem A1.9.

To analyze $\mathrm{Ext}_S^1(R, S(-2)) = 0$ we note that the actual kernel of the map

$$S(1) \xrightarrow{(y \ \ -x)} S^2$$

is the image of the map

$$S^2 \xleftarrow{\begin{pmatrix} x \\ y \end{pmatrix}} S(-1),$$

so the desired homology is

$$\mathrm{Ext}_S^1(R, S(-2)) = S \cdot \begin{pmatrix} x \\ y \end{pmatrix} / S \cdot \begin{pmatrix} x^2 \\ xy \end{pmatrix} = S/(x)(-1) = \mathbb{K}(-1) \oplus \mathbb{K}(-2) \oplus \cdots,$$

which is indeed the dual of the local cohomology module $\mathrm{H}_m^1(R)$, as computed above.

A1B   Local Cohomology and Sheaf Cohomology

If M is any module over a Noetherian ring R and $Q = (x_1, \ldots, x_t) \subset R$ is an ideal, then M gives rise by restriction to a sheaf $\mathscr{F}_M$ on the scheme $\mathrm{Spec}\, R \backslash V(Q)$. The i-th Zariski cohomology $\mathrm{H}^i(\mathscr{F}_M)$ may be defined as the i-th cohomology of the Čech complex

$\check{\mathrm{C}}\mathrm{ech}(x_1, \ldots, x_t; M)$:

$$0 \longrightarrow \bigoplus_1^t M[x_i^{-1}] \xrightarrow{d} \cdots \bigoplus_{\#J=s} M[x_J^{-1}] \xrightarrow{d} \cdots \longrightarrow M[x_{\{1,\ldots,t\}}^{-1}] \longrightarrow 0,$$

whose differential is defined as in Theorem A1.3. The reader who has not yet studied schemes and their cohomology should think of $\mathrm{H}^i(\mathscr{F}_M)$ as a functor of M without worrying about the nature of $\mathscr{F}_M$. The definition is actually independent of the choice of generators $x_1, \ldots, x_t$ for Q; one can show that $\mathrm{H}^0(\mathscr{F}_M) = \varinjlim_d \mathrm{Hom}(Q^d, M)$. This module is sometimes called the *ideal transform of M with respect to Q* (see Exercise A1.3). Further, $\mathrm{H}^i(M)$ is the i-th right derived functor of the ideal transform functor — this follows just as in the proof of Theorem A1.3. As a consequence, $H^i(\mathscr{F}_M) = \lim_n \text{Ext}^i(Q^n, M)$.

The local cohomology is related to Zariski cohomology in a simple way:

Proposition A1.11. *Suppose* $Q = (x_1, \ldots, x_t)$. *Then:*

1. *There is an exact sequence of R-modules*

$$0 \to H_Q^0(M) \to M \to H^0(\mathscr{F}_M) \to H_Q^1(M) \to 0.$$

2. *For every* $i \geq 2$,

$$H_Q^i(M) = H^{i-1}(\mathscr{F}_M).$$

Proof. $\text{Čech}(x_1, \ldots, x_t; M)$ is the subcomplex of the complex $C(x_1, \ldots, x_t; M)$ obtained by dropping the first term, M; so we get an exact sequence of complexes

$$0 \longrightarrow \text{Čech}(x_1, \ldots, x_t; M)[-1] \longrightarrow C(x_1, \ldots, x_t; M) \longrightarrow M \longrightarrow 0,$$

where M is regarded as a complex with just one term, in degree 0. Since this one-term complex has no higher cohomology, the long exact sequence in cohomology coming from this short exact sequence of complexes gives exactly statements 1 and 2. ☐

Henceforward we will restrict our attention to the case where R is a graded polynomial ring $S = \mathbb{K}[x_0, \ldots, x_r]$, *each variable* x_i *ahs degree 1, the ideal Q is the homogeneous maximal ideal* $Q = (x_0, \ldots, x_r)$, *and the module M is finitely generated and graded.*

It follows that all the cohomology is graded too. Following our usual convention, we will write $H_Q^i(M)_d$ for the d-th graded component of $H_Q^i(M)$, and similarly for the Zariski cohomology of $\mathscr{F}_M$.

In this setting the Zariski cohomology has another interpretation: Any graded S-module M gives rise to a quasicoherent sheaf $\widetilde{M}$ on the projective space $\mathbb{P}^r$ (for the definition and properties of this construction see [Hartshorne 1977, II.5], for example). The Čech complex for $\widetilde{M}$ is the degree 0 part of the complex $\text{Čech}(x_0, \ldots, x_r; M)$. In particular, the i-th (Zariski) cohomology of the sheaf $\widetilde{M}$ is the degree 0 part of the cohomology of $\mathscr{F}_M$, that is $H^i(\widetilde{M}) = H^i(\mathscr{F}_M)_0$. If we shift the grading of M by d to get $M(d)$, then $\widetilde{M}(d)$ is the sheaf on $\mathbb{P}^r$ associated to $M(d)$, so in general $H^i(\widetilde{M}(d)) = H^i(\mathscr{F}_M)_d$. Thus Theorem A1.11 takes on the following form:

Corollary A1.12. *Let M be a graded S-module, and let* $\widetilde{M}$ *be the corresponding quasicoherent sheaf on* $\mathbb{P}^r$.

1. *There is an exact sequence of graded S-modules*

$$0 \longrightarrow H_Q^0(M) \longrightarrow M \longrightarrow \bigoplus_d H^0(\widetilde{M}(d)) \longrightarrow H_Q^1(M) \longrightarrow 0.$$

2. *For every* $i \geq 2$,

$$H_Q^i(M) = \bigoplus_d H^{i-1}(\tilde{M}(d)). \qquad \square$$

This corollary reduces the computation of the cohomology of line bundles on projective space to Corollary A1.6.

Corollary A1.13. *Let M be a finitely generated graded S-module. The natural map $M \to \bigoplus_d H^0(\tilde{M}(d))$ is an isomorphism if and only if depth $M \geq 2$.*

Proof. We have seen that depth $M \geq 2$ if and only if $H_Q^i M = 0$ for $i = 0, 1$; the first assertion now follows from the first assertion of Proposition A1.12. $\square$

Corollary A1.14. *Let $S = \mathbb{K}[x_0, \ldots, x_r]$, with $r \geq 1$. The line bundle $\mathcal{O}_{\mathbb{P}^r}(d)$ on $\mathbb{P}^r = \mathrm{Proj}(S)$ has cohomology*

$$H^i(\mathcal{O}_{\mathbb{P}^n}(d)) = \begin{cases} S_d & \text{if } i = 0, \\ 0 & \text{if } 0 < i < n, \\ (S_{r-1-d})^\vee & \text{if } i = r. \end{cases}$$

The final result of this section explains the gap between the Hilbert function and the Hilbert polynomial:

Corollary A1.15. *Let M be a finitely generated graded S-module. For every $d \in \mathbb{Z}$,*

$$P_M(d) = H_M(d) - \sum_{i \geq 0} (-1)^i \dim_{\mathbb{K}} H_Q^i(M)_d.$$

Proof. The Euler characteristic of the sheaf $\tilde{M}(d)$ is by definition

$$\chi(\tilde{M}(d)) = \sum_{i \geq 0} (-1)^i \dim_{\mathbb{K}} H^i \tilde{M}(d).$$

We first claim that $P_M(d) = \chi(\tilde{M}(d))$ for every d. Indeed, by Serre's Vanishing Theorem [Hartshorne 1977, Chapter 3], $H^i(\tilde{M}(d))$ vanishes for $i > 0$ when $d \gg 0$ so $\chi(\tilde{M}(d)) = \dim_{\mathbb{K}} H^0(\tilde{M}(d)) = M_d$ for large d. Thus for the claim it suffices to show that $\chi(\tilde{M}(d))$ is a polynomial function of d. This is done by induction: if x is a general linear form on $\mathbb{P}^r$ then from the exact sequence

$$0 \to \tilde{M}(-1) \xrightarrow{x} \tilde{M} \longrightarrow \widetilde{M/xM} \to 0$$

we derive a long exact sequence in cohomology which (since it has only finitely many terms) establishes the recursion formula

$$\chi(\tilde{M}(d)) - \chi(\tilde{M}(d-1)) = \chi(\widetilde{M/xM}(d)).$$

Since the support of $\widetilde{M/xM}$ is the hyperplane section of the support of $\tilde{M}$, we see by induction on the dimension of the support that

$$\chi(\widetilde{M/xM}(d))$$

is a polynomial, and thus $\chi(\widetilde{M}(d))$ is also.

By Corollary A1.12 we have, as required,

$$\chi(\widetilde{M}(d)) = \dim_{\mathbb{K}} H^0(\widetilde{M}(d)) - \sum_{i \geq 1}(-1)^i \dim_{\mathbb{K}} H^i(\widetilde{M}(d))$$

$$= \dim_{\mathbb{K}} M_d - \dim_{\mathbb{K}} H^0_Q(M)_d + \dim_{\mathbb{K}} H^1_Q(M)_d - \sum_{i \geq 2}(-1)^i \dim_{\mathbb{K}} H^i_Q(M)_d.$$

$\square$

A1C   Vanishing and Nonvanishing Theorems

In this section we maintain the hypothesis that $S = \mathbb{K}[x_0, \dots, x_r]$, the ideal Q is the homogeneous maximal ideal $Q = (x_0, \dots, x_r)$, and the module M is finitely generated and graded.

The converse of Corollary A1.5 is also true; it is a special case of the dimension assertion in the following result. The proofs of the next two results require slightly more sophisticated commutative algebra than what has gone before.

Proposition A1.16. *Let M be a finitely generated graded S-module.*

1. *If $i < \operatorname{depth} M$ or $i > \dim M$ then $H^i_Q(M) = 0$ for all $e \geq d$.*
2. *If $i = \operatorname{depth} M$ or $i = \dim M$ then $H^i_Q(M) \neq 0$.*
3. *There is an integer d (depending on M) such that $H^i_Q(M)_e = 0$*

In the context of sheaf cohomology the dimension statement of part 1 is called Grothendieck's Vanishing Theorem, and part 3 is called Serre's Vanishing Theorem. In between the depth and dimension the local cohomology modules can be zero and nonzero in any pattern; see [Evans and Griffith 1979].

Proof. We will use local duality (Theorem A1.9) and the Auslander–Buchsbaum Formula (Theorem A2.15). With these tools, the depth assertions of parts 1 and 2 are equivalent to the statements that $\operatorname{Ext}^j_S(M, S) = 0$ for $j > n - \operatorname{depth} M$, while $\operatorname{Ext}^j_S(M, S) \neq 0$ for $j = n - \operatorname{depth} M$. By the Auslander–Buchsbaum Formula, $n - \operatorname{depth} M$ is the projective dimension of M, and the depth statement in part 1 follows. When j is equal to the projective dimension of M, the module $\operatorname{Ext}^j_S(M, S)$ is the cokernel of the dual of the last map in a minimal free resolution of M. This cokernel is nonzero by Nakayama's Lemma, since the minimality of the resolution implies that the entries of a matrix representing the map are contained in the maximal ideal of S. This gives the depth statement in part 2.

The dimension assertion of part 1 is likewise equivalent to the statement that $\operatorname{Ext}^j_S(M, S)$ vanishes for $j < \operatorname{codim} M = \operatorname{codim} \operatorname{ann}_S M$. The polynomial ring S is Cohen–Macaulay, so the depth of $\operatorname{ann}_S M$ on S is equal to the codimension $\operatorname{codim} \operatorname{ann}_S M = \operatorname{codim} M$. We will show that $\operatorname{Ext}^j_S(M, N) = 0$ whenever $j < d := \operatorname{depth}(\operatorname{ann}_S M, N)$. For this we do not need the hypothesis that S is a polynomial ring — any Noetherian ring will do. We do induction on the this depth, the case $d = 0$ being trivial, since then $j < 0$. If $d > 0$, then by the definition of depth, the annihilator of M contains an element f that is a nonzerodivisor on N. The exact sequence

$$0 \longrightarrow N \xrightarrow{f} N \longrightarrow N/fF \longrightarrow 0$$

gives rise to a long exact sequence in $\mathrm{Ext}_S(M, -)$. The maps in this sequence corresponding to multiplication by f are zero since f annihilates M. Thus $\mathrm{Ext}_S^j(M, N) \cong \mathrm{Ext}_S^{j-1}(M, N/fN)$. Since $\operatorname{depth} N/fN = \operatorname{depth} N - 1$, we are done by induction.

To finish the proof of part 2 we must show that $\mathrm{Ext}_S^j(M, S) \neq 0$ for $j = \operatorname{codim} M$. Choose a codimension j prime P of S that is minimal over the annihilator of $\operatorname{ann}_S M$. Since the construction of Ext commutes with localization, it suffices to show that $\mathrm{Ext}_{S_P}^j(M_P, S_P)$ is nonzero when M_P is a module of finite length and $\dim S_P = j$. As any module of finite length also has depth 0, and as S_P is a Cohen-Macaulay ring, we may apply the nonvanishing result of part 1.

Finally, part 3 is equivalent to the statement that when d is sufficientlly negative the d-th graded component of the module $E := \mathrm{Ext}_S^j(M, S)$ is 0. This holds because E is a finitely generated module — just take d less than the degree of any generator of E. $\qquad \square$

A1D   Exercises

1. (Cofinality.) Let $R \supset J_1 \supset J_2 \supset \cdots$ and $R \supset K_1 \supset K_2 \supset \cdots$ be sequences of ideals in a ring R, and suppose that there exist functions $m(i)$ and $n(i)$ such that $J_i \supset K_{m(i)}$ and $K_i \supset J_{n(i)}$ for all i. Show that for any R-module M we have

$$\varinjlim_i \mathrm{Ext}_R^P(S/J_i, M) = \varinjlim_i \mathrm{Ext}_R^P(S/K_i, M).$$

2. Use Exercise A1.1 and the Artin–Rees Lemma to show that if R is a Noetherian ring containing ideals Q_1 and Q_2, and M is an R-module, then there is a long exact sequence

$$\cdots \longrightarrow H_{Q_1+Q_2}^i(M) \longrightarrow H_{Q_1}^i(M) \oplus H_{Q_2}^i(M) \longrightarrow$$
$$\longrightarrow H_{Q_1 \cap Q_2}^i(M) \longrightarrow H_{Q_1+Q_2}^{i+1}(M) \longrightarrow \cdots.$$

3. Let Q be an ideal in a Noetherian ring R. Let $\mathscr{F}$ be a coherent sheaf on $\operatorname{Spec} R \setminus V(Q)$. Prove that $H^0(\mathscr{F}_M) = \varinjlim \operatorname{Hom}(Q^d, M)$ by defining maps

$$\{m_i/x_i^d\} \mapsto [f : x_i^e \mapsto x_i^{e-d} m_i]$$

in both directions restricted to $Q^{(r+1)e} \subset (x_0^e, \ldots, x_r^e)$ for big e; and

$$[f : Q^d \to M] \mapsto \{f(x_i^d)/x_i^d\}.$$

4. Prove that, for any R-module M over any Noetherian ring,

$$\varinjlim_d \mathrm{Hom}((x^d), M) = M[x^{-1}].$$

5. Show that the complex $C(x_1, \ldots, x_t; M)$ is the direct limit of the Koszul complexes. Use this to give another proof of Theorem A1.3 in the case where $x_1, \ldots, x_t$ is a regular sequence in R.

6. The ring $R = k[x, y_1, y_2, \ldots]/(xy_1, x^2 y_2, \ldots)$ is non-Noetherian: the sequence of ideals $\mathrm{ann}(x^n)$ increases forever. Show that the formula in Exercise A1.4 fails over this ring for $M = R$.

7. Let R be any ring containing an element x such that the sequence of ideals $\mathrm{ann}(x^n)$ increases forever. If an R-module M contains R, show that the map $M \to M[x^{-1}]$ cannot be surjective; that is, the first homology of the Čech complex

$$0 \to M \to M[x^{-1}] \to 0$$

is nonzero. In particular, this is true for the injective envelope of R in the category of R-modules. Conclude that the cohomology of this Čech complex of M does not compute the derived functors of the functor H^0_{Rx}, and in particular that Corollary A1.8 fails for the map $\mathbb{Z}[t] \to R$ with $t \mapsto x$.

Appendix 2
A Jog Through Commutative Algebra

My goal in this appendix is to lead the reader on a brisk jog through the garden of commutative algebra. There won't be time to smell many flowers, but I hope to impart a sense of the landscape, at least of that part of the subject used in this book.

Each section focuses on a single topic. It begins with some motivation and the main definitions, and then lists some central results, often with illustrations of their use. Finally, there are some further, perhaps more subtle, examples. There are practically no proofs; these can be found, for example, in [Eisenbud 1995].

I assume that the reader is familiar with

- rings, ideals, and modules, and occasionally homological notions such as Hom and ⊗, Ext and Tor;
- prime ideals and the localizations of a ring; and
- the correspondence between affine rings and algebraic sets.

The few references to sheaves and schemes can be harmlessly skipped.

The topics treated in Sections A to H are:

A. associated primes
B. depth
C. projective dimension and regular local rings
D. normalization (resolution of singularities for curves)
E. the Cohen–Macaulay property
F. the Koszul complex
G. Fitting ideals
H. the Eagon–Northcott complex and scrolls

Throughout, $\mathbb{K}$ *denotes a field and* R *denotes a commutative Noetherian ring.*
You can think primarily of the cases where $R = \mathbb{K}[x_1, \ldots, x_n]/I$ for some ideal
I, or where R is the localization of such a ring at a prime ideal. Perhaps the most
interesting case of all is when R is a *homogeneous algebra* (or *standard graded
algebra*), by which is meant a graded ring of the form

$$R = \mathbb{K}[x_0, \ldots, x_r]/I,$$

where all the x_i have degree 1 and I is a homogeneous ideal (an ideal generated
by homogeneous polynomials, so that a polynomial f is in I if and only if each
homogeneous component of f is in I).

There is a fundamental similarity between the local and the homogeneous
cases. Many results for local rings depend on *Nakayama's Lemma*, which states
(in one version) that if M is a finitely generated module over a local ring R
with maximal ideal $\mathfrak{m}$ and $g_1, \ldots, g_n \in M$ are elements whose images in $M/\mathfrak{m}M$
generate $M/\mathfrak{m}M$, then $g_1, \ldots, g_n$ generate M. A closely analogous result is true
in the homogeneous situation: if M is a finitely generated graded module over
a homogeneous algebra R with maximal homogeneous ideal $\mathfrak{m} = \sum_{d>0} R_d$, and
if $g_1, \ldots, g_n \in M$ are homogeneous elements whose images in $M/\mathfrak{m}M$ generate
$M/\mathfrak{m}M$, then $g_1, \ldots, g_n$ generate M. These results can be unified: following
[Goto and Watanabe 1978a; 1978b], one can define a *generalized local ring* to be
a graded ring $R = R_0 \oplus R_1 \oplus \cdots$ such that R_0 is a local ring. If $\mathfrak{m}$ is the *maximal
homogeneous ideal*, that is, the sum of the maximal ideal of R_0 and the ideal of
elements of strictly positive degree, then Nakayama's Lemma holds for R and a
finitely generated graded R-module M just as before.

Similar homogeneous versions are possible for many results involving local
rings. Both the local and homogeneous cases are important, but rather than
spelling out two versions of every theorem, or passing to the generality of generalized local rings, we usually give only the local version.

A2A   Associated Primes and Primary Decomposition

Any integer admits a unique decomposition as a product of primes and a unit.
Attempts to generalize this result to rings of integers in number fields were
the number-theoretic origin of commutative algebra. With the work of Emanuel
Lasker (who was also a world chess champion) and Francis Macaulay around
1900 the theorems took something like their final form for the case of polynomial rings, the theory of *primary decomposition*. It was Emmy Noether's great
contribution to see that they followed relatively easily from just the ascending
chain condition on ideals. (Indeed, modern work has shown that most of the
important statements of the theory fail in the non-Noetherian case.) Though the
full strength of primary decomposition is rarely used, the concepts involved are
fundamental, and some of the simplest cases are pervasive.

The first step is to recast the unique factorization of an integer $n \in \mathbb{Z}$ into a unit and a product of powers of distinct primes p_i, say

$$n = \pm \prod_i p_i^{a_i},$$

as a result about intersections of ideals, namely

$$(n) = \bigcap_i (p_i^{a_i}).$$

In the general case we will again express an ideal as an intersection of ideals, called primary ideals, each connected to a particular prime ideal.

Recall that a proper ideal $I \subset R$ (that is, an ideal not equal to R) is *prime* if $xy \in I$ and $x \notin I$ implies $y \in I$. If M is a module then a prime ideal P is said to be *associated* to M if $P = \operatorname{ann} m$, the annihilator of some $m \in M$. We write Ass M for the set of associated primes of M. The module M is called P-*primary* if P is the only associated prime of M. The most important case occurs when $I \subset R$ is an ideal and $M = R/I$; then it is traditional to say that P is associated to I when P is associated to R/I, and to write Ass I in place of Ass R/I. We also say that I is P-primary if R/I is P-primary. (The potential confusion is seldom a problem, as the associated primes of I as a module are usually not very interesting.) The reader should check that the associated primes of an ideal $(n) \subset \mathbb{Z}$ are those ideals (p) generated by the prime divisors p of n. In particular, the (p)-primary ideals in $\mathbb{Z}$ are exactly those of the form (p^a).

For any ideal I we say that a prime P is *minimal over* I if P is is minimal among primes containing I. An important set of primes connected with a module M is the set Min M of primes *minimal over* the annihilator $I = \operatorname{ann} M$. These are called the *minimal primes of* M. Again we abuse the terminology, and when I is an ideal we define the *minimal primes of* I to be the minimal primes over I, or equivalently the minimal primes of the module R/I. We shall see below that all minimal primes of M are associated to M. The associated primes of M that are not minimal are called *embedded primes* of M.

Theorem A2.1. *Let M be a nonzero finitely generated R-module.*

1. *Min $M \subset$ Ass M, and both are nonempty finite sets.*
2. *The set of elements of R that are zerodivisors on M is the union of the associated primes of M.*

If M is a graded module over a homogeneous algebra R, all the associated primes of M are homogeneous.

Among the most useful corollaries is the following.

Corollary A2.2. *If I is an ideal of R and M is a finitely generated module such that every element of I annihilates some nonzero element of M, then there is a single nonzero element of M annihilated by all of I. In particular, any ideal of R that consists of zerodivisors is annihilated by a single element.*

The proof is immediate from Theorem A2.1 given the following result, often called "prime avoidance".

Lemma A2.3. *If an ideal I is contained in a finite union of prime ideals, then it is contained in one of them.*

It is easy to see that an element $f \in R$ is contained in an ideal I (equivalently: is zero in R/I) if and only if the image of f in the localization R_P is contained in I_P for all prime ideals, or even just for all maximal ideals P of R. Using Theorem A2.1 one can pinpoint the set of localizations it is necessary to test, and see that this set is finite.

Corollary A2.4. *If $f \in M$, then $f = 0$ if and only if the image of f is zero in M_P for each associated prime P of M. It even suffices that this condition be satisfied at each maximal associated prime of M.*

One reason for looking at associated primes for modules, and not only for ideals, is the following useful result, a component of the proof of Theorem A2.1.

Theorem A2.5. *Let $0 \to M' \to M \to M'' \to 0$ be a short exact sequence of finitely generated R-modules. Then*

$$\text{Ass}\, M' \subset \text{Ass}\, M \subset \text{Ass}\, M' \cup \text{Ass}\, M''$$

If $M = M' \oplus M''$ then the second inclusion becomes an equality.

Here is the primary decomposition result itself.

Theorem A2.6. *If I is an ideal of R then $\text{Ass}(R/I)$ is the unique minimal set of prime ideals S that can written as $I = \bigcap_{P \in S} Q_P$, where each Q_P is a P-primary ideal. (There is a similar result for modules.)*

In this decomposition the ideals Q_P with $P \in \text{Min}\, I$ are called *minimal components* and are unique. The others are called *embedded components* and are generally nonunique.

Example A2.7. Primary decomposition translates easily into geometry by means of Hilbert's Nullstellensatz [Eisenbud 1995, Theorem 1.6]. Here is a sample that contains a fundamental finiteness principle. Recall that the *radical* of an ideal I, written $\sqrt{I}$, is the ideal

$$\sqrt{I} = \{f \in R \mid f^m \in I \text{ for some } m\}.$$

We say that I is *radical* if $I = \sqrt{I}$. The primary decomposition of a radical ideal has the form

$$\sqrt{I} = \bigcap_{P \in \text{Min}\, I} P.$$

Any algebraic set X (say in affine n-space $\mathbb{A}^n_{\mathbb{K}}$ over an algebraically closed field $\mathbb{K}$, or in projective space) can be written uniquely as a finite union $X = \bigcup_i X_i$ of irreducible sets. The ideal $I = I(X)$ of functions vanishing on X is the intersection of the prime ideals $P_i = I(X_i)$. The expression $I = \bigcap_i P_i$ is the primary decomposition of I.

Example A2.8. For any ring R we write $K(R)$ for the result of localizing R by inverting all the nonzerodivisors of R. By Theorem A2.1, this is the localization of R at the complement of the union of the associated primes of R, and thus it is a ring with finitely many maximal ideals. Of course if R is a domain then $K(R)$ is simply its quotient field. The most useful case beyond is when R is reduced. Then $K(R) = K(R/P_1) \times \cdots \times K(R/P_m)$, the product of the quotient fields of R modulo its finitely many minimal primes.

Example A2.9. Let $R = \mathbb{K}[x,y]$ and let $I = (x^2, xy)$. The associated primes of I are (x) and (x,y), and a primary decomposition of I is $I = (x) \cap (x,y)^2$. This might be read geometrically as saying: for a function $f(x,y)$ to lie in I, the function must vanish on the line $x = 0$ in $\mathbb{K}^2$ and vanish to order 2 at the point $(0,0)$ (this last condition can be expressed by saying that the partial derivatives of f relative to x and y vanish at $(0,0)$). In this example, the (x,y)-primary component $(x,y)^2$ is not unique: we also have $I = (x) \cap (x^2, y)$. The corresponding geometric statement is that a function f lies in I if and only if f vanishes on the line $x = 0$ in $\mathbb{K}^2$ and $(\partial f/\partial x)(0,0) = 0$.

Example A2.10. If P is a prime ideal, the powers of P mayfail to be P-primary! In general, the P-primary component of P^m is called the m-th *symbolic power* of P, written $P^{(m)}$. In the special case where $R = \mathbb{K}[x_1, \ldots, x_n]$ and $\mathbb{K}$ is algebraically closed, a famous result of Zariski and Nagata (see for example [Eisenbud and Hochster 1979]) says that $P^{(m)}$ is the set of all functions vanishing to order $\geq m$ at each point of $V(P)$. For example, suppose that

$$A = \begin{pmatrix} x_{1,1} & x_{1,2} & x_{1,3} \\ x_{2,1} & x_{2,2} & x_{2,3} \\ x_{3,1} & x_{3,2} & x_{3,3} \end{pmatrix}$$

is a matrix of indeterminates. If P is the ideal $I_2(A)$ of 2×2 minors of A, then P is prime but, we claim, $P^{(2)} \neq P^2$. In fact, the partial derivatives of $\det A$ are the 2×2 minors of A, so $\det A$ vanishes to order 2 wherever the 2×2 minors vanish. Thus $\det A \in P^{(2)}$. On the other hand, $\det A \notin P^2$ because P^2 is generated by elements of degree 4, while $\det A$ only has degree 3.

A2B   Dimension and Depth

Perhaps the most fundamental definition in geometry is that of dimension. The dimension (also called Krull dimension) of a commutative ring plays a similarly central role. An arithmetic notion of dimension called depth is also important (the word "arithmetic" in this context refers to divisibility properties of elements in a ring). Later we shall see geometric examples of the difference between depth and dimension.

The *dimension* of R, written $\dim R$ is the supremum of lengths of chains of prime ideals of R. (Here a *chain* is a totally ordered set. The length of a chain of primes is, by definition, one less than the number of primes; that is $P_0 \subset P_1 \subset \cdots \subset P_n$ is a chain of length n.) If P is a prime ideal, the *codimension* of P, written $\operatorname{codim} P$, is the maximum of the lengths of chains of prime ideals $P \supset \cdots \supset P_0$ descending from P. If I is any ideal, the *codimension* of I is the minimum of the codimension of primes containing I. See [Eisenbud 1995, Ch. 8] for a discussion linking these very algebraic notions with geometry.) The generalization to modules doesn't involve anything new: we define the dimension $\dim M$ of an R-module M to be the dimension of the ring $R/\operatorname{ann} M$.

A sequence $\mathbf{x} = x_1, \ldots, x_n$ of elements of R is a *regular sequence* (or *R-sequence*) if $x_1, \ldots, x_n$ generate a proper ideal of R and if, for each i, the element x_i is a nonzerodivisor modulo $(x_1, \ldots, x_{i-1})$. Similarly, if M is an R-module, then $\mathbf{x}$ is a *regular sequence on M* (or *M-sequence*) if $(x_1, \ldots, x_n)M \neq M$ and, for each i, the element x_i is a nonzerodivisor on $M/(x_1, \ldots, x_{i-1})M$.

An ideal that can be generated by a regular sequence (or, in the geometric case, the variety it defines) is called a *complete intersection*.

If I is an ideal of R and M is a finitely generated module such that $IM \neq M$, then the *depth* of I on M, written $\operatorname{depth}(I, M)$, is the maximal length of a regular sequence on M contained in I. (If $IM = M$ we set $\operatorname{depth}(I, M) = \infty$.) The most interesting cases are the ones where R is a local or homogeneous algebra and I is the maximal (homogeneous) ideal. In these cases we write $\operatorname{depth} M$ in place of $\operatorname{depth}(I, M)$. We define the *grade* of I to be $\operatorname{grade} I = \operatorname{depth}(I, R)$. (Alas, terminology in this area is quite variable; see for example [Bruns and Herzog 1998, Section 1.2] for a different system.) We need one further notion of dimension, a homological one that will reappear in the next section. The *projective dimension* of an R-module is the minimum length of a projective resolution of M (or ∞ if there is no finite projective resolution.)

We will suppose for simplicity that R is local with maximal ideal $\mathfrak{m}$. Similar results hold in the homogeneous case. A fundamental geometric observation is that a variety over an algebraically closed field that is defined by one equation has codimension at most 1. The following is Krull's justly celebrated generalization.

Theorem A2.11 (Principal Ideal Theorem). *If I is an ideal that can be generated by n elements in a Noetherian ring R, then $\operatorname{grade} I \leq \operatorname{codim} I \leq n$. Moreover, any prime minimal among those containing I has codimension at most n. If M is a finitely generated R-module, then $\dim M/IM \geq \dim M - n$.*

For example, in

$$R = \mathbb{K}[x_1, \ldots, x_n] \quad \text{or} \quad R = \mathbb{K}[x_1, \ldots, x_n]_{(x_1, \ldots, x_n)} \quad \text{or} \quad R = \mathbb{K}[\![x_1, \ldots, x_n]\!],$$

the sequence $x_1, \ldots, x_n$ is a maximal regular sequence. It follows at once from Theorem A2.11 that in each of these cases the ideal $(x_1, \ldots, x_n)$ has codimension n, and for the local ring $R = \mathbb{K}[x_1, \ldots, x_n]_{(x_1, \ldots, x_n)}$ or $R = \mathbb{K}[\![x_1, \ldots, x_n]\!]$ this gives $\dim R = n$. For the polynomial ring R itself this argument gives only $\dim R \geq n$, but in fact it is not hard to show $\dim R = n$ in this case as well. This follows from a general result on affine rings.

Theorem A2.12. *If R is an integral domain with quotient field $K(R)$, and R is a finitely generated algebra over the field $\mathbb{K}$, then $\dim R$ is equal to the transcendence degree of $K(R)$ over $\mathbb{K}$. Geometrically: the dimension of an algebraic variety is the number of algebraically independent functions on it.*

The following is a generalization of Theorem A2.11 in which the ring R is replaced by an arbitrary module.

Theorem A2.13. *If M is a finitely generated R-module and $I \subset R$ is an ideal, then*
$$\operatorname{depth}(I, M) \leq \operatorname{codim}((I + \operatorname{ann} M)/\operatorname{ann} M) \leq \dim M.$$

A module is generally better behaved — more like a free module over a polynomial ring — if its depth is close to its dimension. See also Theorem A2.15.

Theorem A2.14. *Suppose R is a local ring and M is a finitely generated R-module.*

1. *All maximal regular sequences on M have the same length; this common length is equal to the depth of M. Any permutation of a regular sequence on M is again a regular sequence on M.*
2. *$\operatorname{depth} M = 0$ if and only if the maximal ideal of R is an associated prime of M (see Theorem A2.1.2).*
3. *For any ideal I, $\operatorname{depth}(I, M) = \inf\{i \mid \operatorname{Ext}^i_R(R/I, M) \neq 0\}$.*
4. *If $R = \mathbb{K}[x_0, \ldots, x_r]$ with the usual grading, M is a finitely generated graded R-module, and $\mathfrak{m} = (x_0, \ldots, x_r)$, then $\operatorname{depth} M = \inf\{i \mid \operatorname{H}^i_{\mathfrak{m}}(M) \neq 0\}$.*

Parts 3 and 4 of Theorem A2.14 are connected by what is usually called *local duality*; see Theorem A1.9.

Theorem A2.15 (Auslander–Buchsbaum formula). *If R is a local ring and M is a finitely generated R-module such that $\operatorname{pd} M$ (the projective dimension of M) is finite, then $\operatorname{depth} M = \operatorname{depth} R - \operatorname{pd} M$.*

The following results follow from Theorem A2.15 by localization.

Corollary A2.16. *Suppose that M is a finitely generated module over a local ring R.*

1. *If M has an associated prime of codimension n, then $\operatorname{pd} M \geq n$.*
2. *If M has finite projective dimension, then $\operatorname{pd} M \leq \operatorname{depth} R \leq \dim R$. If also $\operatorname{depth} M = \dim R$ then M is free.*
3. *If $\operatorname{pd} M = \dim R$ then R is Cohen–Macaulay and its maximal ideal is associated to M.*

Another homological characterization of depth, this time in terms of the Koszul complex, is given in Section A2G.

Example A2.17. Theorem A2.14 really requires the "local" hypothesis (or, of course, the analogous "graded" hypothesis). For example, in $\mathbb{K}[x] \times \mathbb{K}[y, z]$ the sequence $(1, y)$, $(0, z)$ of length two and the sequence $(x, 1)$ of length one are both maximal regular sequences. Similarly, in $R = \mathbb{K}[x, y, z]$ the sequence $x(1-x)$, $1 - x(1-y)$, xz is a regular sequence but its permutation $x(1-x)$, xz, $1 - x(1-y)$ is not. The ideas behind these examples are related: $R/(x(1-x)) = \mathbb{K}[y, z] \times \mathbb{K}[y, z]$ by the Chinese Remainder Theorem.

A2C   Projective Dimension and Regular Local Rings

After dimension, the next most fundamental geometric ideas may be those of smooth manifolds and tangent spaces. The analogues in commutative algebra are regular rings and Zariski tangent spaces, introduced by Krull [1937] and Zariski [1947]. Since the work of Auslander, Buchsbaum, and Serre in the 1950s this theory has been connected with the idea of projective dimension.

Let R be a local ring with maximal ideal $\mathfrak{m}$. The *Zariski cotangent space* of R is $\mathfrak{m}/\mathfrak{m}^2$, regarded as a vector space over $R/\mathfrak{m}$; the *Zariski tangent space* is the dual, $\mathrm{Hom}_{R/\mathfrak{m}}(\mathfrak{m}/\mathfrak{m}^2, R/\mathfrak{m})$. By Nakayama's Lemma, the vector space dimension of $\mathfrak{m}/\mathfrak{m}^2$ is the minimal number of generators of $\mathfrak{m}$. By the Principal Ideal Theorem A2.11 this is an upper bound for the Krull dimension $\dim R$. The ring R is called *regular* if $\dim R$ is equal to the vector space dimension of the Zariski tangent space; otherwise, R is *singular*. If R is a Noetherian ring that is not local, we say that R is regular if each localization at a maximal ideal is regular.

For example, the n-dimensional power series ring $\mathbb{K}[\![x_1, \ldots, x_n]\!]$ is regular because the maximal ideal $\mathfrak{m} = (x_1, \ldots, x_n)$ satisfies $\mathfrak{m}/\mathfrak{m}^2 = \bigoplus_1^n \mathbb{K}x_i$. The same goes for the localization of the polynomial ring $\mathbb{K}[x_1, \ldots, x_n]_{(x_1, \ldots, x_n)}$. Indeed any localization of one of these rings is also regular, though this is harder to prove; see Corollary A2.20.

Here is a first taste of the consequences of regularity.

Theorem A2.18. *Any regular local ring is a domain. A local ring is regular if and only if its maximal ideal is generated by a regular sequence.*

The following result initiated the whole homological study of rings.

Theorem A2.19 (Auslander–Buchsbaum–Serre). *A local ring R is regular if and only if the residue field of R has finite projective dimension if and only if every R-module has finite projective dimension.*

The abstract-looking characterization of regularity in Theorem A2.19 allowed a proof of two properties that had been known only in the "geometric" case (R a localization of a finitely generated algebra over a field). These were the first triumphs of representation theory in commutative algebra. Recall that a domain R is called *factorial* if every element of r can be factored into a product of prime elements, uniquely up to units and permutation of the factors.

Theorem A2.20. *Any localization of a regular local ring is regular. Every regular local ring is factorial (that is, has unique factorization of elements into prime elements.)*

The first of these statements is, in the geometric case, a weak version of the statement that the singular locus of a variety is a closed subset. The second plays an important role in the theory of divisors.

Example A2.21. The rings

$$\mathbb{K}[x_1,\ldots,x_n], \quad \mathbb{K}[x_1,\ldots,x_n,x_1^{-1},\ldots,x_n^{-1}], \quad \text{and} \quad \mathbb{K}[\![x_1,\ldots,x_n]\!]$$

are regular, and the same is true if $\mathbb{K}$ is replaced by the ring of integers $\mathbb{Z}$.

Example A2.22. A regular local ring R of dimension 1 is called a *discrete valuation ring*. By the definition, together with Nakayama's Lemma, the maximal ideal of R must be principal; let π be a generator. By Theorem A2.18, R is a domain. Conversely, any one-dimensional local domain with maximal ideal that is principal (and nonzero!) is a discrete valuation ring. Every nonzero element f of the quotient field $K(R)$ can be written uniquely in the form $u \cdot \pi^k$ for some unit $u \in R$ and some integer $k \in \mathbb{Z}$. The name "discrete valuation ring" comes from the fact that the mapping $\nu : K(R)^* \to \mathbb{Z}$ taking f to k satisfies the definition of a valuation on R and has as value group the discrete group $\mathbb{Z}$.

Example A2.23. A ring of the form $A = \mathbb{K}[\![x_1,\ldots,x_n]\!]/(f)$ is regular if and only if the leading term of f has degree ≤ 1 (if the degree is 0, of course A is the zero ring!) In case the degree is 1, the ring A is isomorphic to the ring of power series in $n-1$ variables. If $R = \mathbb{K}[\![x_1,\ldots,x_n]\!]/I$ is nonzero then R is regular if and only if I can be generated by some elements $f_1,\ldots,f_m$ with leading terms that are of degree 1 and linearly independent; in this case $R \cong \mathbb{K}[\![x_1,\ldots,x_{n-m}]\!]$. Indeed, Cohen's Structure Theorem says that any complete regular local ring containing a field is isomorphic to a power series ring (possibly over a larger field.)

This result suggests that all regular local rings, or perhaps at least all regular local rings of the same dimension and characteristic, look much alike, but this is only true in the complete case (things like power series rings). Example A2.30 shows how much structure even a discrete valuation ring can carry.

Example A2.24. Nakayama's Lemma implies that amodule over a local ring has projective dimension 0 if and only if it is free. It follows that an ideal of projective dimension 0 in a local ring is principal, generated by a nonzerodivisor. An ideal has projective dimension 1 (as a module) if and only if it is isomorphic to the ideal J of $n \times n$ minors of an $(n+1) \times n$ matrix with entries in the ring, and this ideal of minors has depth 2 (that is, $\mathrm{depth}(J,R) = 2$), the largest possible number. This is the Hilbert–Burch Theorem, described in detail in Chapter 3.

A2D   Normalization: Resolution of Singularities for Curves

If $R \subset S$ are rings, an element $f \in S$ is *integral* over R if f satisfies a monic polynomial equation

$$f^n + a_1 f^{n-1} + \cdots + a_n = 0$$

with coefficients in R. The *integral closure* of R in S is the set of all elements of S integral over R; it turns out to be a subring of S (Theorem A2.25). The ring R is *integrally closed* in S if all elements of S that are integral over R actually belong to R. The ring R is *normal* if it is reduced and integrally closed in the ring obtained from R by inverting all nonzerodivisors.

These ideas go back to the beginning of algebraic number theory: the integral closure of $\mathbb{Z}$ in a finite field extension $\mathbb{K}$ of $\mathbb{Q}$, defined to be the set of elements of $\mathbb{K}$ satisfying monic polynomial equations over $\mathbb{Z}$, is called the *ring of integers of* $\mathbb{K}$, and is in many ways the nicest subring of $\mathbb{K}$. For example, when studying the field $\mathbb{Q}[x]/(x^2-5) \cong \mathbb{Q}(\sqrt{5})$ it is tempting to look at the ring $R = \mathbb{Z}[x]/(x^2-5) \cong \mathbb{Z}[\sqrt{5}]$. But the slightly larger (and at first more complicated-looking) ring

$$\bar{R} = \frac{\mathbb{Z}[y]}{(y^2 - y - 1)} \cong \mathbb{Z}\left[\frac{1-\sqrt{5}}{2}\right]$$

is nicer in many ways: for example, the localization of R at the prime $P = (2, x-1) \subset R$ is not regular, since R_P is one-dimensional but P/P^2 is a two-dimensional vector space generated by 2 and $x - 1$. Since $x^2 - x - 1$ has no solution modulo 2, the ideal $P' = P\bar{R} = (2)\bar{R}$ is prime and $\bar{R}_{P'}$ is regular. In fact $\bar{R}$ itself is regular. This phenomenon is typical for one-dimensional rings.

In general, the first case of importance is the normalization of a reduced ring R in its quotient ring $K(R)$. In addition to the number-theoretic case above, this has a beautiful geometric interpretation. Let R be the coordinate ring of an affine algebraic set $X \subset \mathbb{C}^n$ in complex n-space. The normalization of R in $K(R)$ is then the ring of rational functions that are locally bounded on X.

For example, suppose that X is the union of two lines meeting in the origin in $\mathbb{C}^2$, with coordinates x, y, defined by the equation $xy = 0$. The function $f(x,y) = x/(x-y)$ is a rational function on X that is well-defined away from the point $(0,0)$. Away from this point, f takes the value 1 on the line $y = 0$ and 0 on the line $x = 0$, so although it is bounded near the origin, it does not extend to a continuous function at the origin. Algebraically this is reflected in the fact that f (regarded either as a function on X or as an element of the ring obtained from the coordinate ring $R = \mathbb{K}[x,y]/(xy)$ of X by inverting the nonzerodivisor $x-y$) satisfies the monic equation $f^2 - f = 0$, as the reader will easily verify. On the disjoint union $\bar{X}$ of the two lines, which is a smooth space mapping to X, the pull back of f extends to be a regular function everywhere: it has constant value 1 on one of the lines and constant value 0 on the other.

Another significance of the normalization is that it gives a *resolution of singularities in codimension 1*; we will make this statement precise in Example A2.32.

Theorem A2.25. *Let $R \subset S$ be rings. If $s, t \in S$ are integral over R, then $s+t$ and st are integral over R. Thus the of elements of S integral over R form a subring of S, called the normalization of R in S. If S is normal (for example if S is the quotient field of R), the integral closure of R in S is normal.*

The following result says that the normalization of the coordinate ring of an affine variety is again the coordinate ring of an affine variety.

Theorem A2.26. *If R is a domain that is a finitely generated algebra over a field $\mathbb{K}$, then the normalization of R (in its quotient field) is a finitely generated R-module; in particular it is again a finitely generated algebra over $\mathbb{K}$.*

It is possible to define the normalization of any abstract variety X (of finite type over a field $\mathbb{K}$), a construction that was first made and exploited by Zariski. Let $X = \bigcup X_i$ be a covering of X by open affine subsets, such that $X_i \cap X_j$ is also affine, and let $\overline{X}_i$ be the affine variety corresponding to the normalization of the coordinate ring of X_i. We need to show that the $\overline{X}_i$ patch together well, along the normalizations of the sets $X_i \cap X_j$. This is the essential content of the next result.

Theorem A2.27. *Normalization commutes with localization in the following sense. Let $R \subset S$ be rings and let $\overline{R}$ be the subring of S consisting of elements integral over R. If U is a multiplicatively closed subset of R, then the localization $\overline{R}[U^{-1}]$ is the normalization of $R[U^{-1}]$ in $S[U^{-1}]$.*

Here is what good properties we can expect when we have normalized a variety.

Theorem A2.28 (Serre's Criterion). *Any normal one-dimensionsonal ring is regular (that is, discrete valuation rings are precisely the normal one-dimensional rings). More generally, we have Serre's Criterion: A ring R is a finite direct product of normal domains if and only if*

1. *R_P is regular for all primes P of codimension ≤ 1; and*
2. *$\operatorname{depth}(P_P, R_P) \geq 2$ for all primes P of codimension ≥ 2.*

When R is a homogeneous algebra, it is only necessary to test conditions 1 and 2 at homogeneous primes.

When R is the coordinate ring of an affine variety X over an algebraically closed field, condition 1 has the geometric meaning one would hope: the singular locus of X has codimension at least 2.

Example A2.29. The ring $\mathbb{Z}$ is normal; so is any factorial domain (for example, any regular local ring). (Reason: Suppose $f = u/v$ is integral, satisfying an equation $f^n + a_1 f^{n-1} + \cdots + a_n = 0$, with $a_1 \in R$. If v is divisible by a prime p that does not divide u, then p divides all except the first term of the expression $u^n + a_1 v u^{n-1} + \cdots + a_n v^n = v^n(f^n + a_1 f^{n-1} + \cdots + a_n) = 0$, a contradiction.)

Example A2.30. Despite the simplicity of discrete valuation rings (see Example A2.22) there are a lot of nonisomorphic ones, even after avoiding the obvious differences of characteristic, residue class field $R/\mathfrak{m}$, and quotient field. For a concrete example, consider first the coordinate ring of a quartic affine plane curve, $R = \mathbb{K}[x,y]/(x^4 + y^4 - 1)$, where $\mathbb{K}$ is the field of complex numbers (or any algebraically closed field of characteristic not 2). The ring R has infinitely many maximal ideals, which have the form $(x-\alpha, y-\beta)$, where $\alpha \in \mathbb{K}$ is arbitrary and β is any fourth root of $1-\alpha^4$. But given one of these maximal ideals P, there are only finitely many maximal ideals Q such that $R_P \cong R_Q$. This follows at once from the theory of algebraic curves (see [Hartshorne 1977, Ch. I, §8], for example): any isomorphism $R_P \to R_Q$ induces an automorphism of the projective curve $x^4 + y^4 = z^4$ in $\mathbb{P}^2$ carrying the point corresponding to P to the point corresponding to Q. But there are only finitely many automorphisms of this curve (or, indeed, of any smooth curve of genus at least 2).

Example A2.31. The set of monomials in $x_1, \ldots, x_n$ corresponds to the set of lattice points $\mathbb{N}^n$ in the positive orthant (send each monomial to its vector of exponents). Let U be an subset of $\mathbb{N}^n$, and let $\mathbb{K}[U] \subset \mathbb{K}[x_1, \ldots, x_n]$ be the subring generated by the corresponding monomials. For simplicity we assume that the group generated by U is all of $\mathbb{Z}^n$, the group generated by $\mathbb{N}^n$. It is easy to see that any element of $\mathbb{N}^n$ that is in the convex hull of U, or even in the convex hull of the set generated by U under addition, is integral over $\mathbb{K}[U]$. In fact the integral closure of $\mathbb{K}[U]$ is $\mathbb{K}[\bar{U}]$, where $\bar{U}$ is the convex hull of the set generated by U using addition. For example take $U = \{x_1^4, x_1^3 x_2, x_1 x_2^3, x_2^4\}$ — all the monomials of degree 4 in two variables except the middle monomial $f := x_1^2 x_2^2$. The element f is in the quotient field of $\mathbb{K}[U]$ because $f = x_1^4 \cdot x_1 x_2^3 / x_1^3 x_2$. The equation $(2,2) = \frac{1}{2}((4,0) + (0,4))$, expressing the fact that f corresponds to a point in the convex hull of U, gives rise to the equation $f^2 - x_1^4 \cdot x_2^4 = 0$, so f is integral over $\mathbb{K}[U]$.

Example A2.32 (Resolution of singularities in codimension 1). Suppose that X is an affine variety over an algebraically closed field $\mathbb{K}$, with affine coordinate ring R. By Theorem A2.26 the normalization $\bar{R}$ corresponds to an affine variety Y, and the inclusion $R \subset \bar{R}$ corresponds to a map $g : Y \to X$. By Theorem A2.27 the map g is an isomorphism over the part of X that is smooth, or even normal. The map g is a finite morphism in the sense that the coordinate ring of $\bar{X}$ is a finitely generated *as a module* over the coordinate ring of X; this is a strong form of the condition that each fiber $g^{-1}(x)$ is a finite set.

Serre's Criterion in Theorem A2.28 implies that the coordinate ring of Y is smooth in codimension 1, and this means the singular locus of Y is of codimension at least 2.

Desingularization in codimension 1 is the most that can be hoped, in general, from a finite morphism. For example, the quadric cone $X \subset \mathbb{K}^3$ defined by the equation $x^2 + y^2 + z^2 = 0$ is normal, and it follows that any finite map $Y \to X$ that is isomorphic outside the singular point must be an isomorphism.

However, for any affine or projective variety X over a field it is conjectured that there is actually a *resolution of singularities*: that is, a *projective* map $\pi : Y \to X$ (this means that Y can be represented as a closed subset of $X \times \mathbb{P}^n$ for some projective space $\mathbb{P}^n$) where Y is a smooth variety, and the map π is an isomorphism over the part of X that is already smooth. In the example above, there is a desingularization (the *blowup* of the origin in X) that may be described as the subset of $X \times \mathbb{P}^2$, with coordinates x, y, z for X and u, v, w for $\mathbb{P}^2$, defined by the vanishing of the 2×2 minors of the matrix

$$\begin{pmatrix} x & y & z \\ u & v & w \end{pmatrix}$$

together with the equations $xu + yv + zw = 0$ and $u^2 + v^2 + w^2 = 0$. It is described algebraically by the *Rees algebra*

$$R \oplus I \oplus I^2 \oplus \cdots$$

where $R = \mathbb{K}[x, y, z]/(x^2 + y^2 + z^2)$ is the coordinate ring of X and $I = (x, y, z) \subset R$.

The existence of resolutions of singularities was proved in characteristic 0 by Hironaka. In positive characteristic it remains an active area of research.

A2E   The Cohen–Macaulay Property

Which curves in the projective plane pass through the common intersections of two given curves? The answer was given by the great geometer Max Noether (father of Emmy) [Noether 1873] in the course of his work algebraizing Riemann's amazing ideas about analytic functions, under the name of the "Fundamental Theorem of Algebraic Functions". However, it was gradually realized that Noether's proof was incomplete, and it was not in fact completed until work of Lasker in 1905. By the 1920s (see [Macaulay 1916] and [Macaulay 1934]), Macaulay had come to a much more general understanding of the situation for polynomial rings, and his ideas were studied and extended to arbitrary local rings by Cohen in the 1940s [Cohen 1946]. In modern language, the fundamental idea is that of a *Cohen–Macaulay ring*.

A curve in the projective plane is defined by the vanishing of a (square-free) homogeneous polynomial in three variables. Suppose that curves F, G and H are defined by the vanishing of f, g and h. For simplicity assume that F and G have no common component, so the intersection of F and G is finite. If h can be written as $h = af + bg$ for some a and b, then h vanishes wherever f and g vanish, so H passed through the intersection points of F and G. Noether's Fundamental Theorem is the converse: if H "passes through" the intersection of F and G, then h can be written as $h = af + bg$.

To understand Noether's Theorem we must know what it means for H to pass through the intersection of F and G. To make the theorem correct, the intersection, which may involve high degrees of tangency and singularity, must be interpreted subtly. We will give a modern explanation in a moment, but it is interesting first to phrase the condition in Noether's terms.

For Noether's applications it was necessary to define the intersection in a way that would only depend on data available locally around a point of intersection. Suppose, after a change of coordinates, that F and G both contain the point $p = (1, 0, 0)$. Noether's idea was to expand the functions $f(1, x, y)$, $g(1, x, y)$ and $h(1, x, y)$ as power series in x, y, and to say that H passes through the intersection of F and G locally at p if there are convergent power series $\alpha(x, y)$ and $\beta(x, y)$ such that

$$h(1, x, y) = \alpha(x, y) f(1, x, y) + \beta(x, y) g(1, x, y).$$

This condition was to hold (with different α and β!) at each point of intersection.

Noether's passage to convergent power series ensured that the condition "H passes through the intersection of F and G" depended only on data available locally near the points of intersection. Following [Lasker 1905] and using primary decomposition, we can reformulate the condition without leaving the context of homogeneous polynomials. We first choose a primary decomposition $(f, g) = \bigcap Q_i$. If p is a point of the intersection $F \cap G$, then the prime ideal P of forms vanishing at p is minimal over the ideal (f, g). By Theorem A2.1, P is an associated prime of (f, g). Thus one of the Q_i, say Q_1, is P-primary. We say that H passes through the intersection of F and G locally near p if $h \in Q_1$.

In this language, Noether's Fundamental Theorem becomes the statement that the only associated primes of (f, g) are the primes associated to the points of $F \cap G$. Since f and g have no common component, they generate an ideal of codimension at least 2, and by the Principal Ideal Theorem A2.11 the codimension of all the minimal primes of (f, g) is exactly 2. Thus the minimal primes of (f, g) correspond to the points of intersection, and Noether's Theorem means that there are no nonminimal, that is, embedded, associated primes of (f, g). This result was proved by Lasker in a more general form, *Lasker's Unmixedness Theorem*: if a sequence of c homogeneous elements in a polynomial ring generates an ideal I of codimension c, then every associated prime of I has codimension c. The modern version simply says that a polynomial ring over a field is Cohen–Macaulay. By Theorem A2.36, this is the same result.

Now for the definitions: a local ring R is *Cohen–Macaulay* if depth $R = \dim R$; it follows that the same is true for every localization of R (Theorem A2.33). More generally, an R-module M is *Cohen–Macaulay* if depth $M = \dim M$. For example, if S is a local ring and $R = S/I$ is a factor ring, then R is Cohen–Macaulay as a ring if and only if R is Cohen–Macaulay an S-module.

If R is not local, we say that R is Cohen–Macaulay if the localization R_P is Cohen–Macaulay for every maximal ideal P. If R is a homogeneous algebra with homogeneous maximal ideal $\mathfrak{m}$, then R is Cohen–Macaulay if and only if grade$(\mathfrak{m}) = \dim R$ (as can be proved from Theorem A2.15 and the existence of minimal graded free resolutions).

Globalizing, we say that a variety (or scheme) X is Cohen–Macaulay if each of its local rings $\mathcal{O}_{X,x}$ is a Cohen–Macaulay ring. More generally, a coherent sheaf F on X is Cohen–Macaulay if for each point $x \in X$ the stalk F_x is a Cohen–Macaulay module over the local ring $\mathcal{O}_{X,x}$.

If $X \subset \mathbb{P}^r$ is a projective variety (or scheme), we say that X is *arithmetically Cohen–Macaulay* if the homogeneous coordinate ring

$$S_X = \mathbb{K}[x_0, \ldots, x_r]/I(X)$$

is Cohen–Macaulay. The local rings of X are, up to adding a variable and its inverse, obtained from the homogeneous coordinate ring by localizing at certain primes. With Theorem A2.33 this implies that if X is arithmetically Cohen–Macaulay then X is Cohen–Macaulay. The "arithmetic" property is much stronger, as we shall see in the examples.

The Cohen–Macaulay property behaves well under localization and forming polynomial rings.

Theorem A2.33. *The localization of any Cohen–Macaulay ring at any prime ideal is again Cohen–Macaulay. A ring R is Cohen–Macaulay if and only if $R[x]$ is Cohen–Macaulay if and only if $R[\![x]\!]$ is Cohen–Macaulay if and only if $R[x, x^{-1}]$ is Cohen–Macaulay.*

The following result is an easy consequence of Theorems A2.19 and A2.15, and should be compared with Example A2.42 above.

Theorem A2.34. *Suppose that a local ring R is a finitely generated module over a regular local subring T. The ring R is Cohen–Macaulay as an R-module if and only if it is free as a T-module. A similar result holds in the homogeneous case.*

Sequences of c elements $f_1, \ldots, f_c$ in a ring R that generate ideals of codimension c have particularly nice properties. In the case when R is a local Cohen–Macaulay ring the situation is particularly simple.

Theorem A2.35. *If R is a local Cohen–Macaulay ring and $f_1, \ldots, f_c$ generate an ideal of codimension c then $f_1, \ldots, f_c$ is a regular sequence.*

Here is the property that started it all. We say that an ideal I of codimension c is *unmixed* if every associated prime of I has codimension exactly c.

Theorem A2.36. *A local ring is Cohen–Macaulay if and only if every ideal of codimension c that can be generated by c elements is unmixed, and similarly for a homogeneous algebra.*

Theorem A2.18 shows that a local ring is regular if its maximal ideal is generated by a regular sequence; here is the corresponding result for the Cohen–Macaulay property.

Theorem A2.37. *Let R be a local ring with maximal ideal $\mathfrak{m}$. The following conditions are equivalent:*

(a) R is Cohen–Macaulay.

(b) There is an ideal I of R that is generated by a regular sequence and contains a power of $\mathfrak{m}$.

The next consequence of the Cohen–Macaulay property is often taken as the definition. It is pleasingly simple, but as a definition it is not so easy to check.

Theorem A2.38. *A ring R is Cohen–Macaulay if and only if every ideal I of R has grade equal to its codimension.*

One way to prove that a ring is Cohen–Macaulay is to prove that it is a summand in a nice way. We will apply the easy first case of this result in Example A2.43.

Theorem A2.39. *Suppose that S is a Cohen–Macaulay ring and $R \subset S$ is a direct summand of S as R-modules. If either S is finitely generated as an R-module, or S is regular, then R is Cohen–Macaulay.*

The first statement follows from basic statements about depth and dimension [Eisenbud 1995, Proposition 9.1 and Corollary 17.8]. The second version, without finiteness, is far deeper. The version where S is regular was proved by Boutot [1987].

Example A2.40 (Complete intersections). Any regular local ring is Cohen–Macaulay (Theorem A2.18). If R is any Cohen–Macaulay ring, for example the power series ring $\mathbb{K}[x_1, \ldots, x_n]$, and $f_1, \ldots, f_c$ is a regular sequence in R, then $R/(f_1, \ldots, f_c)$ is Cohen–Macaulay; this follows from Theorem A2.13(a). For example, $\mathbb{K}[x_1, \ldots, x_n]/(x_1^{a_1}, \ldots, x_k^{a_k})$ is Cohen–Macaulay for any positive integers $k \leq n$ and $a_1, \ldots, a_k$.

Example A2.41. Any Artinian local ring is Cohen–Macaulay. So is any one-dimensional local domain. More generally, a one-dimensional local ring is Cohen–Macaulay if and only if the maximal ideal is not an associated prime of 0 (Theorem A2.1.2). For example, $\mathbb{K}[x, y]/(xy)$ is Cohen–Macaulay.

Example A2.42. The simplest examples of Cohen–Macaulay rings not included in the preceding cases are the homogeneous coordinate rings of set of points, studied in Chapter 3, and the homogeneous coordinate rings of rational normal curves, studied in 6.

Example A2.43. Suppose a finite group G acts on a ring S, and the order n of G is invertible in S. Let R be the subring of invariant elements of S. The *Reynolds operator*

$$s \mapsto \frac{1}{n} \sum_{g \in G} gs$$

is an R linear splitting of the inclusion map. Thus if S is Cohen–Macaulay, so is R by Theorem A2.39. Theorem A2.39 further shows that the ring of invariants of any linearly reductive group, acting linearly on a polynomial ring is a Cohen–Macaulay ring, a result first proved by Hochster and Roberts [1974].

Example A2.44. Perhaps the most imporant example of a ring of invariants under a finite group action is that where $S = \mathbb{K}[x_0, \ldots, x_r]$ is the polynomial ring on $r+1$ indeterminates and $G = (\mathbb{Z}/d)^{r+1}$ is the product of $r+1$ copies of the cyclic group of order d, whose i-th factor acts by multiplying x_i by a d-th root of unity. The invariant ring R is the *d-th Veronese subring* of S, consisting of all forms whose degree is a multiple of d.

Example A2.45. Most Cohen–Macaulay varieties in $\mathbb{P}^n$ (even smooth varieties) are not arithmetically Cohen–Macaulay. A first example is the union of two skew lines in $\mathbb{P}^3$. In suitable coordinates this scheme is represented by the homogeneous ideal $I := (x_0, x_1) \cap (x_2, x_3)$; that is, it has homogeneous coordinate ring $R := \mathbb{K}[x_0, x_1, x_2, x_3]/(x_0, x_1) \cap (x_2, x_3)$. To see that R is not Cohen–Macaulay, note that

$$R \subset R/(x_0, x_1) \times R/(x_2, x_3) = \mathbb{K}[x_2, x_3] \times \mathbb{K}[x_0, x_1],$$

so that $f_0 := x_0 - x_2$ is a nonzerodivisor on R. By the graded version of Theorem A2.14.1, it suffices to show that every element of the maximal ideal is a zerodivisor in $R/(f_0)$. As the reader may easily check, $I = (x_0 x_2, x_0 x_3, x_1 x_2, x_1 x_3)$, so $\bar{R} := R/(f_0) = \mathbb{K}[x_1, x_2, x_3]/(x_2^2, x_2 x_3, x_1 x_2, x_1 x_3)$. In particular, the image of x_2 is not zero in $\bar{R}$, but the maximal ideal annihilates x_2.

Example A2.46. Another geometric example easy to work out by hand is that of a smooth rational quartic curve in $\mathbb{P}^3$. We can define such a curve by giving its homogeneous coordinate ring. Let R be the subring of $\mathbb{K}[s, t]$ generated by the elements $f_0 = s^4, f_1 = s^3 t, f_2 = st^3, f_3 = t^4$. Since R is a domain, the element f_0 is certainly a nonzerodivisor, and as before it suffices to see that modulo the ideal $(f_0) = Rs^4$ the whole maximal ideal consists of zerodivisors. One checks at once that $s^6 t^2 \in R \setminus Rs^4$, but that $f_i s^6 t^2 \in Rs^4$ for every i, as required.

Many of the most interesting smooth projective varieties cannot be embedded in a projective space in any way as arithmetically Cohen–Macaulay varieties. Such is the case for all abelian varieties of dimension greater than 1 (and in general for any variety whose structure sheaf has nonvanishing intermediate cohomology.)

A2F The Koszul Complex

One of the most significant homological constructions is the Koszul complex. It is fundamental in many senses, perhaps not least because its construction depends only on the commutative and associative laws in R. It makes one of the essential bridges between regular sequences and homological methods in commutative algebra, and has been at the center of the action since the work of Auslander, Buchsbaum, and Serre in the 1950s. The construction itself was already exploited (implicitly) by Cayley; see [Gel'fand et al. 1994] for an exegesis. It enjoys the role of premier example in Hilbert's 1890 paper on syzygies. (The name Koszul seems to have been attached to the complex in the influential book [Cartan and Eilenberg 1956].) It is also the central construction in the Bernstein–Gelfand–Gelfand correspondence described briefly in Chapter 7. It appears in many other generalizations as well, for example in the Koszul duality associated with quantum groups (see [Manin 1988].)

I first learned about the Koszul complex from lectures of David Buchsbaum. He always began his explanation with the following special cases, and these still seem to me the best introduction.

Let R be a ring and let $x \in R$ be an element. The *Koszul complex of x* is the complex

$$\mathbf{K}(x): \quad 0 \xrightarrow{} \overset{0}{R} \xrightarrow{x} \overset{1}{R} \xrightarrow{} 0.$$

We give the *cohomological degree* of each term of $\mathbf{K}(x)$ above that term so that we can unambiguously refer to $\mathrm{H}^i(\mathbf{K}(x))$, the homology of $\mathbf{K}(X)$ at the term of cohomological degree i. Even this rather trivial complex has interesting homology: the element x is a nonzerodivisor if and only if $\mathrm{H}^0(\mathbf{K}(x))$ is 0. The homology $\mathrm{H}^1(\mathbf{K}(x)$ is always $R/(x)$, so that when x is a nonzerodivisor, $\mathbf{K}(x)$ is a free resolution of $R/(x)$.

If $y \in R$ is a second element, we can form the complex

$$\mathbf{K}(x) = \mathbf{K}(x,y): 0 \xrightarrow{} \overset{0}{R} \xrightarrow{\begin{pmatrix} x \\ y \end{pmatrix}} \overset{1}{R^2} \xrightarrow{(-y\ \ x)} \overset{2}{R} \xrightarrow{} 0.$$

Again, the homology tells us interesting things. First, $\mathrm{H}^0(\mathbf{K}(x,y))$ is the set of elements annihilated by both x and y. By Corollary A2.2, $\mathrm{H}^0(\mathbf{K}(x,y)) = 0$ if and only if the ideal (x,y) contains a nonzerodivisor. Supposing that x is a nonzerodivisor, we claim that $\mathrm{H}^1(\mathbf{K}(x,y)) = 0$ if and only if x,y is a regular sequence. By definition,

$$\mathrm{H}^1(\mathbf{K}(x,y)) = \frac{\{(a,b) \mid ay - bx = 0\}}{\{rx, ry \mid r \in R\}}.$$

The element a in the numerator can be chosen to be any element in the quotient ideal $(x) : y = \{s \in R \mid sy \in (x)\}$. Because x is a nonzerodivisor, the element b in the numerator is then determined uniquely by a. Thus the numerator is isomorphic to $(x) : y$, and $\mathrm{H}^1(\mathbf{K}(x,y)) \cong ((x) : y)/(x)$. It follows that $\mathrm{H}^1(\mathbf{K}(x,y)) = 0$ if and only if y is a nonzerodivisor modulo x, proving the claim. The module $\mathrm{H}^2(\mathbf{K}(x,y))$ is, in any case, isomorphic to $R/(x,y)$, so when x,y is a regular sequence the complex $\mathbf{K}(x,y)$ is a free resolution of $R/(x,y)$. This situation generalizes, as we shall see.

In general, the Koszul complex $\mathbf{K}(x)$ of an element x in a free module F is the complex with terms $K^i := \bigwedge^i F$ whose differentials $d : K^i \longrightarrow K^{i+1}$ are given by exterior multiplication by x. The formula $d^2 = 0$ follows because elements of F square to 0 in the exterior algebra. (Warning: our indexing is nonstandard — usually what we have called K^i is called K_{n-i}, where n is the rank of F, and certain signs are changed as well. Note also that we could defined a Koszul complex in exactly the same way without assuming that F is free — this makes it easy, for example, to define the Koszul complex of a section of a vector bundle.) If we identify F with R^n for some n, we may write x as a vector $x = (x_1, \ldots, x_n)$, and we will sometimes write $\mathbf{K}(x_1, \ldots, x_n)$ instead of $\mathbf{K}(x)$.

Here is a weak sense in which the Koszul complex is always "close to" exact.

Theorem A2.47. *Let* $x_1, \ldots, x_n$ *be a sequence of elements in a ring R. For every i, the homology $H^i(\mathbf{K}(x_1, \ldots, x_n))$ is anhilated by $(x_1, \ldots, x_n)$.*

The next result says that the Koszul complex can detect regular sequences inside an ideal.

Theorem A2.48. *Let* $x_1, \ldots, x_n$ *be a sequence of elements in a ring R. The grade of the ideal $(x_1, \ldots, x_n)$ is the smallest integer i such that*

$$H^i(\mathbf{K}(x_1, \ldots, x_n)) \neq 0.$$

In the local case, the Koszul complex detects whether a given sequence is regular.

Theorem A2.49. *Let* $x_1, \ldots, x_n$ *be a sequence of elements in the maximal ideal of a local ring R. The elements $x_1, \ldots, x_n$ form a regular sequence if and only if*

$$H^{n-1}(\mathbf{K}(x_1, \ldots, x_n)) = 0.$$

In this case the Koszul complex is the minimal free resolution of the module $R/(x_1, \ldots, x_n)$.

The Koszul complex is self-dual, and this is the basis for much of duality theory in algebraic geometry and commutative algebra. Here is how the duality is defined. Let F be a free R-module of rank n, and let e be a generator of $\bigwedge^n F \cong R$. Contraction with e defines an isomorphism $\phi_k \bigwedge^k F^* \to \bigwedge^{n-k} F$ for every $k = 0, \ldots, n$. The map ϕ_k has a simple description in terms of bases: if $e_1 \ldots, e_n$ is a basis of F such that $e = e_1 \wedge \cdots \wedge e_n$, and if $f_1, \ldots, f_n$ is the dual basis to $e_1 \ldots, e_n$, then

$$\phi_k(f_{i_1} \wedge \cdots \wedge f_{i_k}) = \pm e_{j_1} \wedge \cdots \wedge e_{j_{n-k}}$$

where $\{j_1, \ldots, j_{n-k}\}$ is the complement of $\{i_1, \ldots, i_k\}$ in $\{1, \ldots, n\}$, and the sign is that of the permutation $(i_1 \ldots i_k j_1 \ldots j_{n-k})$.

Theorem A2.50. *The contraction maps define an isomorphism of the complex* $\mathbf{K}(x_1, \ldots, x_n)$ *with its dual.*

Example A2.51. The Koszul complex can be built up inductively as a mapping cone. For example, using an element x_2 we can form the commutative diagram with two Koszul complexes $\mathbf{K}(x_1)$:

$$
\begin{array}{ccccccccc}
\mathbf{K}(x_1): & 0 & \longrightarrow & R & \xrightarrow{\ x_1\ } & R & \longrightarrow & 0 \\
 & & & \downarrow{\scriptstyle x_2} & & \downarrow{\scriptstyle x_2} & & \\
\mathbf{K}(x_1): & 0 & \longrightarrow & R & \xrightarrow[\ x_1\ ]{} & R & \longrightarrow & 0
\end{array}
$$

We regard the vertical maps as forming a map of complexes. The Koszul complex $\mathbf{K}(x_1, x_2)$ may be described as the mapping cone.

More generally, the complex $\mathbf{K}(x_1, \ldots, x_n)$ is (up to signs) the mapping cone of the map of complexes

$$\mathbf{K}(x_1, \ldots, x_{n-1}) \longrightarrow \mathbf{K}(x_1, \ldots, x_{n-1})$$

given by multiplication by x_n. It follows by induction that, when $x_1, \ldots, x_n$ is a regular sequence, $\mathbf{K}(x_1, \ldots, x_n)$ is a free resolution of $R/(x_1, \ldots, x_n)$. This is a weak version of Theorem A2.49.

Example A2.52. The Koszul complex may also be built up as a tensor product of complexes. The reader may check from the definitions that

$$\mathbf{K}(x_1, \ldots, x_n) = \mathbf{K}(x_1) \otimes \mathbf{K}(x_2) \otimes \cdots \otimes \mathbf{K}(x_n).$$

The treatment in Serre's book [Serre 2000] is based on this description.

A2G   Fitting Ideals and Other Determinantal Ideals

Matrices and determinants appear everywhere in commutative algebra. A linear transformation of vector spaces over a field has a well defined rank (the size of a maximal submatrix with nonvanishing determinant in a matrix representing the linear transformation) but no other invariants. By contrast linear transformations between free modules over a ring have as invariants a whole sequence of ideals, the *determinantal ideals* generated by all the minors (determinants of submatrices) of a given size. Here are some of the basic tools for handling them.

Let R be a ring and let A be a matrix with entries in R. The *ideal of $n \times n$ minors* of A, written $I_n(A)$, is the ideal in R generated by the $n \times n$ minors ($=$ determinants of $n \times n$ submatrices) of A. By convention we set $I_0(A) = R$, and of course $I_n(A) = 0$ if A is a $q \times p$ matrix and $n > p$ or $n > q$. It is easy to see that $I_n(A)$ depends only on the map of free modules ϕ defined by A — not on the choice of bases. We may thus write $I_n(\phi)$ in place of $I_n(A)$.

Let M be a finitely generated R-module, with free presentation

$$R^p \xrightarrow{\ \phi\ } R^q \longrightarrow M \longrightarrow 0.$$

Set $\text{Fitt}_j(M) = I_{q-j}(\phi)$. The peculiar-looking numbering makes the definition of $\text{Fitt}_j(M)$ independent of the choice of the number of generators chosen for M.

There is a close relation between the annihilator and the zeroth Fitting ideal:

Theorem A2.53. *If M is a module generated by n elements, then*

$$(\text{ann}\, M)^n \subset \text{Fitt}_0\, M \subset \text{ann}\, M.$$

Krull's Principal Ideal Theorem (Theorem A2.11) says that an ideal generated by n elements in a Noetherian ring can have codimension at most n; and when such an ideal has codimension n it is unmixed. An ideal generated by n elements is the ideal of 1×1 minors of a $1 \times n$ matrix. Macaulay generalized these statements to determinantal ideals in polynomial rings. The extension to any Noetherian ring was made by Eagon and Northcott [1962].

Theorem A2.54 (Macaulay's Generalized Principal Ideal Theorem). *If A is a $p \times q$ matrix with elements in a Noetherian ring R, and $I_t(A) \neq R$, then*

$$\text{codim}(I_t(A)) \leq (p-t+1)(q-t+1)$$

Let R be a local Cohen–Macaulay ring. Theorem A2.35 together with Example A2.40 show that if $f_1, \ldots, f_c$ is a sequence of elements that generates an ideal of the maximum possible codimension, c, then $R/(f_1, \ldots, f_c)$ is a Cohen–Macaulay ring. The next result, proved by Hochster and Eagon [1971], is the analogue for determinantal ideals.

Theorem A2.55. *If A is a $p \times q$ matrix with elements in a local Cohen–Macaulay ring R and $\text{codim}(I_t(A)) = (p-t+1)(q-t+1)$, then $R/I_t(A)$ is Cohen–Macaulay.*

Note that the determinantal ideals defining the rational normal curves (Example A2.58) have this maximal codimension.

Example A2.56. Suppose that $R = \mathbb{Z}$, $\mathbb{K}[x]$, or any other principal ideal domain. Let M be a finitely generated R-module. The structure theorem for such modules tells us that $M \cong R^n \oplus R/(a_1) \oplus \ldots \oplus R/(a_s)$ for uniquely determined nonnegative n and positive integers a_i such that a_i divides a_{i+1} for each i. The a_i are called the *elementary divisors* of M. The module M has a free presentation of the form $R^s \xrightarrow{\phi} R^{s+n}$ where ϕ is represented by a diagonal matrix whose diagonal entries are the a_i followed by a block of zeros. From this presentation we can immediately compute the Fitting ideals, and we find:

- $\text{Fitt}_j\, M = 0$ for $0 \leq j < n$.
- For $n \leq j$, the ideal $\text{Fitt}_j\, M$ is generated by all products of $j - n + 1$ of the a_i; in view of the divisibility relations of the a_i this means $\text{Fitt}_j\, M = (a_1 \cdots a_{j-n+1})$.

In particular the Fitting ideals determine n by the first relation above and the elementary divisors by the formulas

$$(a_1) = \text{Fitt}_n, \quad (a_2) = (\text{Fitt}_{n+1} : \text{Fitt}_n), \quad \ldots, \quad (a_s) = (\text{Fitt}_{n+s} : \text{Fitt}_{n+s-1}).$$

Thus the Fitting ideals give a way of generalizing to the setting of arbitrary rings the invariants involved in the structure theorem for modules over a principal ideal domain; this seems to have been why Fitting introduced them.

Example A2.57. Over more complicated rings cyclic modules (that is, modules of the form R/I) are still determined by their Fitting ideals ($\text{Fitt}_0(R/I) = I$); but other modules are generally not. For example, over $\mathbb{K}[x, y]$, the modules with presentation matrices

$$\begin{pmatrix} x & y & 0 \\ 0 & x & y \end{pmatrix} \quad \text{and} \quad \begin{pmatrix} x & y & 0 & 0 \\ 0 & 0 & x & y \end{pmatrix}$$

are not isomorphic (the second is annihilated by (x, y), the first only by $(x, y)^2$) but they have the same Fitting ideals: $\text{Fitt}_0 = (x, y)^2$, $\text{Fitt}_1 = (x, y)$, $\text{Fitt}_j = (1)$ for $j \geq 2$.

Example A2.58. A determinantal prime ideal of the "wrong" codimension Consider the smooth rational quartic curve X in $\mathbb{P}^3$ with parametrization

$$\mathbb{P}^1 \ni (s, t) \mapsto (s^4, s^3 t, s t^3, t^4) \in \mathbb{P}^3.$$

Using the "normal form" idea used for the rational normal curve in Proposition 6.1, it is not hard to show that the ideal $I(X)$ is generated by the 2×2 minors of the matrix

$$\begin{pmatrix} x_0 & x_2 & x_1^2 & x_1 x_3 \\ x_1 & x_3 & x_0 x_2 & x_2^2 \end{pmatrix}$$

The homogeneous coordinate ring $S_X = S/I(X)$ is not Cohen–Macaulay (Example A2.46). The ideal $I(X)$ is already generated by just four of the six minors:

$$I(X) = (x_0 x_3 - x_1 x_2,\ x_1 x_3^2 - x_2^3,\ x_0 x_2^2 - x_1^2 x_3,\ x_1^3 - x_0^2 x_2).$$

Compare this with the situation of Corollary A2.61.

A2H   The Eagon–Northcott Complex and Scrolls

Let A be a $g \times f$ matrix with entries in a ring R, and suppose for definiteness that $g \leq f$. The Eagon–Northcott complex of A [Eagon and Northcott 1962] bears the same relation to the determinantal ideal $I_g(A)$ of maximal minors of A that the Koszul complex bears to sequences of q elements; in fact the Koszul complex is the special case of the Eagon–Northcott complex in which $g = 1$. (A theory including the lower-order minors also exists, but it is far more complicated; it depends on rather representation theory, and is better-understood in characteristic 0 than in finite characteristic. See for example [Akin et al. 1982].) Because the material of this section is less standard than that in the rest of this appendix, we give more details.

Sets of points in $\mathbb{P}^2$ (Chapter 3) and rational normal scrolls (Chapter 6) are some of the interesting algebraic sets whose ideals have free resolutions given by Eagon–Northcott complexes.

Let R be a ring, and write $F = R^f$, $G = R^g$. The *Eagon–Northcott complex* of a map $\alpha : F \longrightarrow G$ (or of a matrix A representing α) is a complex

$$\mathbf{EN}(\alpha) : \ 0 \to (\mathrm{Sym}_{f-g}\, G)^* \otimes \bigwedge^f F \xrightarrow{d_{f-g+1}} (\mathrm{Sym}_{f-g-1}\, G)^* \otimes \bigwedge^{f-1} F \xrightarrow{d_{f-g}}$$

$$\cdots \longrightarrow (\mathrm{Sym}_2\, G)^* \otimes \bigwedge^{g+2} F \xrightarrow{d_3} G^* \otimes \bigwedge^{g+1} F \xrightarrow{d_2} \bigwedge^g F \xrightarrow{\bigwedge^g \alpha} \bigwedge^g G.$$

Here $\mathrm{Sym}_k\, G$ is the k-th symmetric power of G and the notation M^* denotes $\mathrm{Hom}_R(M, R)$. The maps d_j are defined as follows. First we define a diagonal map

$$\Delta : \ (\mathrm{Sym}_k\, G)^* \longrightarrow G^* \otimes (\mathrm{Sym}_{k-1}\, G)^*$$

as the dual of the multiplication map $G \otimes \mathrm{Sym}_{k-1}\, G \longrightarrow \mathrm{Sym}_k\, G$ in the symmetric algebra of G. Next we define an analogous diagonal map

$$\Delta : \ \bigwedge^k F \longrightarrow F \otimes \bigwedge^{k-1} F$$

as the dual of the multiplication in the exterior algebra of F^*. These diagonal maps can be defined as components of the maps of algebras induced by the diagonal map of modules $F \to F \oplus F$ sending f to (f, f). For the exterior algebra, for example, this is the composite

$$\bigwedge^k F \hookrightarrow \bigwedge F \longrightarrow \bigwedge(F \oplus F) = \bigwedge F \otimes \bigwedge F \to F \otimes \bigwedge^{k-1} F.$$

On decomposable elements, this diagonal has the simple form

$$f_1 \wedge \ldots \wedge f_k \mapsto \sum_i (-1)^{i-1} f_i \otimes f_1 \wedge \ldots \wedge \hat{f}_i \wedge \ldots \wedge f_k.$$

For $u \in (\mathrm{Sym}_{j-1}\, G)^*)$ we write $\Delta(u) = \sum_i u_i' \otimes u_i'' \in G^* \otimes (\mathrm{Sym}_{j-2}\, G)^*$ and similarly for $v \in \bigwedge^{g+j-1} F$ we write $\Delta(v) = \sum v_t' \otimes v_t'' \in F \otimes \bigwedge^{g+j-2} F$. Note that $\alpha^*(u_i') \in F^*$, so $[\alpha^*(u_i')](v_t') \in R$. We set

$$d_j : \ (\mathrm{Sym}_{j-1}\, G)^* \otimes \bigwedge^{g+j-1} F \to (\mathrm{Sym}_{j-2}\, G)^* \otimes \bigwedge^{g+j-2} F$$

$$u \otimes v \mapsto \sum_{s,t} [\alpha^*(u_s')](v_t') \cdot u_s'' \otimes v_t''.$$

That the Eagon–Northcott complex is a complex follows by a direct computation, or by an inductive construction of the complex as a mapping cone, similar to the one indicated above in the case of the Koszul complex. The most interesting part — the fact that d_2 composes with $\bigwedge^g \alpha$ to 0 — is a restatement of Cramer's Rule for solving linear equations; see Examples A2.67 and A2.68 below.

Rational Normal Scrolls

We give three definitions of rational normal scrolls, in order of increasing abstraction. See [Eisenbud and Harris 1987] for a proof of their equivalence. Fix nonnegative integers $a_1, \ldots, a_d$. Set $D = \sum a_i$ and $N = D + d - 1$.

As homogeneous ideals. Take the homogeneous coordinates on $\mathbb{P}^N$ to be

$$x_{1,0}, \ldots, x_{1,a_1}, \ x_{2,0}, \ldots, x_{2,a_2}, \ \cdots \ , x_{d,0}, \ldots, x_{d,a_d}.$$

Define a $2 \times D$ matrix of linear forms on $\mathbb{P}^N$ by

$$A(a_1, \ldots, a_d) = \begin{pmatrix} x_{1,0} & \cdots & x_{1,a_1-1} & x_{2,0} & \cdots & x_{2,a_2-1} & \cdots \\ x_{1,1} & \cdots & x_{1,a_1} & x_{2,1} & \cdots & x_{2,a_2} & \cdots \end{pmatrix}.$$

The rational normal scroll $S(a_1, \ldots, a_d)$ is the variety defined by the ideal of 2×2 minors of $I_2(A(a_1, \ldots, a_d))$. Each of the blocks

$$\begin{pmatrix} x_{i,0} & x_{i,1} & \cdots & x_{i,a_1-1} \\ x_{i,1} & x_{i,2} & \cdots & x_{i,a_1} \end{pmatrix}$$

used to construct $A(a_1, \ldots, a_d)$ is 1-generic by Proposition 6.3, and since the blocks involve different variables the whole matrix $A(a_1, \ldots, a_d)$ is 1-generic. It follows from Theorem 6.4 that the ideal of 2×2 minors $I_2(A(a_1, \ldots, a_d))$ is prime. (This could also be proved directly by the method of Proposition 6.1.)

As a union of planes. Let V_i be a vector space of dimension a_i. Regard $\mathbb{P}(V_i)$ as a subspace of $\mathbb{P}^N = \mathbb{P}(\bigoplus_i V_i)$. Consider in $\mathbb{P}(V_i)$ the parametrized rational normal curve

$$\lambda_i : \mathbb{P}^1 \longrightarrow \mathbb{P}(V_i)$$

represented in coordinates by

$$(s,t) \mapsto (s^{a_i}, s^{a_i-1}t, \ldots, t^{a_i}).$$

For each point $p \in \mathbb{P}^1$, let $L(p) \subset \mathbb{P}^N$ be the $(d-1)$-plane spanned by $\lambda_1(p), \ldots, \lambda_d(p)$. The rational normal scroll $S(a_1, \ldots, a_d)$ is the union $\bigcup_{p \in \mathbb{P}^1} L(p)$.

Structural definition. Let $\mathscr{E}$ be the vector bundle on $\mathbb{P}^1$ that is the direct sum $\mathscr{E} = \bigoplus_{i=1}^d \mathcal{O}(a_i)$. Consider the projectivized vector bundle $X := \mathbb{P}(\mathscr{E})$, which is a smooth d-dimensional variety mapping to $\mathbb{P}^1$ with fibers $\mathbb{P}^{d-1}$. Because all the a_i are nonnegative, the tautological bundle $\mathcal{O}_{\mathbb{P}(\mathscr{E})}(1)$ is generated by its global sections, which may be naturally identified with the $N+1$-dimensional vector space $H^0(\mathscr{E}) = \bigoplus_i H^0(\mathcal{O}_{\mathbb{P}^1}(a_i))$. These sections thus define a morphism $X \longrightarrow \mathbb{P}^N$. The rational normal scroll $S(a_1, \ldots, a_d)$ is the image of this morphism.

Here are generalizations of Theorems A2.47, A2.49 and Example A2.40.

Theorem A2.59. *Let $\alpha : F \to G$ with $\operatorname{rank} F \geq \operatorname{rank} G = g$ be a map of free R-modules. The homology of the Eagon–Northcott complex $\mathbf{EN}(\alpha)$ is annihilated by the ideal of $g \times g$ minors of α.*

The following result gives another (easier) proof of Theorem A2.55 in the case of maximal order minors. It can be deduced from Theorem A2.59 together with Theorem 3.3.

Theorem A2.60. *Let* $\alpha : F \to G$ *with* $\operatorname{rank} F = f \geq \operatorname{rank} G = g$ *be a map of free R-modules. The Eagon–Northcott complex* $\mathbf{EN}(\alpha)$ *is exact (and thus furnishes a free resolution of* $R/I_g(\alpha)$*) if and only if* $\operatorname{grade}(I_g(\alpha)) = f - g + 1$*, the greatest possible value. In this case the dual complex* $\operatorname{Hom}(\mathbf{EN}(\alpha), R)$ *is also a resolution.*

The following important consequence seems to use only a tiny part of Theorem A2.60, but I know of no other approach.

Corollary A2.61. *If* $\alpha : R^f \to R^g$ *is a matrix of elements in the maximal ideal of a local ring S such that* $\operatorname{grade}(I_g(\alpha)) = f - g + 1$*, then the* $\binom{f}{g}$ *maximal minors of* α *are minimal generators of the ideal they generate.*

Proof. The matrix of relations on these minors given by the Eagon–Northcott complex is zero modulo the maximal ideal of S. $\qquad\square$

We can apply the preceding theorems to the rational normal scrolls.

Corollary A2.62. *The ideal of* 2×2 *minors of the matrix* $A(a_1, \ldots, a_d)$ *has grade and codimension equal to* $D - 1$*, and thus the Eagon–Northcott complex* $\mathbf{EN}(A(a_1, \ldots, a_d))$ *is a free resolution of the homogeneous coordinate ring of the rational normal scroll* $S(a_1, \ldots, a_d)$*. In particular the homogeneous coordinate ring of a rational normal scroll is Cohen–Macaulay.*

The next results give some perspective on scrolls. The first is part of the Kronecker–Weierstrass classification of matrix pencils.

Theorem A2.63. *Suppose A is a* $2 \times D$ *matrix of linear forms over a polynomial ring whose ideal I of* 2×2 *minors has codimension* $D - 1$*. If I is a prime ideal then A is equivalent by row operations, column operations, and linear change of variables to one of the matrices* $A(a_1, \ldots, a_d)$ *with* $D = \sum a_i$*.*

Theorem A2.64. *If X is an irreducible subvariety of codimension c in* $\mathbb{P}^N$*, not contained in a hyperplane, then the degree of X is at least* $c + 1$*. Equality is achieved if and only if X is (up to a linear transformation of projective space) either*

- *a quadric hypersurface,*
- *a cone over the Veronese surface in* $\mathbb{P}^5$ *(whose defining ideal is the ideal of* 2×2 *minors of a generic symmetric* 2×2 *matrix), or*
- *a rational normal scroll* $S(a_1, \ldots, a_d)$ *with* $\sum a_i = c + 1$*.*

Consider a map $\alpha : F \longrightarrow G$, where F and G are free R-modules of ranks f and g respectively. The definition of the Eagon–Northcott complex is easier to understand if $g = 1$ or if f is close to g:

Example A2.65 (The Koszul complex). If $g = 1$ and we choose a generator for G, identifying G with R, then the symmetric powers $\mathrm{Sym}_k(G)$ and their duals may all be identified with R. If we suppress them in the tensor products defining the Eagon–Northcott complex, we get a complex of the form

$$0 \longrightarrow \textstyle\bigwedge^f F \longrightarrow \cdots \longrightarrow \textstyle\bigwedge^1 F \longrightarrow \{\textstyle\bigwedge^1 G = R\}.$$

Choosing a basis for F and writing $x_1, \ldots, x_f$ for the images of the basis elements in $G = R$, this complex is isomorphic to the Koszul complex $\mathbf{K}(x_1, \ldots, x_f)$.

Example A2.66. If $f = g$ then the Eagon–Northcott complex is reduced to

$$0 \longrightarrow \{R \cong \textstyle\bigwedge^f F\} \xrightarrow{\ \det \alpha\ } \{R \cong \textstyle\bigwedge^g G\}.$$

Example A2.67 (The Hilbert–Burch complex). Supose $f = g + 1$. If we choose an identification of $\bigwedge^f F$ with R then we may suppress the tensor factor $\bigwedge^f F$ from the notation, and also identify $\bigwedge^g F = \bigwedge^{f-1} F$ with F^*. If we also choose an identification of $\bigwedge^g G$ with R, then the Eagon–Northcott complex of α takes the form

$$0 \longrightarrow G^* \xrightarrow{\ \alpha^*\ } \{F^* = \textstyle\bigwedge^g F\} \xrightarrow{\ \bigwedge^g \alpha\ } \{\textstyle\bigwedge^g G = R\}.$$

This is the complex used for the Hilbert–Burch in Chapter 3.

Example A2.68. If α is represented by a matrix A, then the map at the far right of the Eagon–Northcott complex, $\bigwedge^g \alpha$, may be represented by the $1 \times \binom{f}{g}$ matrix whose entries are the $g \times g$ minors of α. The map d_2 admits a similarly transparent description: for every submatrix A' of A consisting of $g+1$ columns, there are g relations among the minors involving these columns that are given by A'^*, exactly as in the Hilbert–Burch complex, Example A2.67. The map d_2 is made by simply concatenating these relations.

Example A2.69. Suppose that α is represented by the 2×4 matrix

$$\begin{pmatrix} a & b & c & d \\ e & f & g & h \end{pmatrix}$$

so that $g = 2$, $f = 4$. There are six 2×2 minors, and for each of the four 2×3 submatrices of A there are two relations among the six, a total of eight, given as in A2.68. Since $(\mathrm{Sym}_2 G)^* \cong (\mathrm{Sym}_2(R^2))^* \cong R^3$, the the Eagon–Northcott complex takes the form

$$0 \longrightarrow R^3 \longrightarrow R^8 \longrightarrow R^6 \longrightarrow R .$$

The entries of the right-hand map are the 2×2 minors of A, which are quadratic in the entries of A, whereas the rest of the matrices (as in all the Eagon–Northcott complexes) have entries that are linear in the entries of A.

References

[Akin et al. 1982] K. Akin, D. A. Buchsbaum, and J. Weyman, "Schur functors and Schur complexes", *Adv. in Math.* **44**:3 (1982), 207–278.

[Altman and Kleiman 1970] A. Altman and S. Kleiman, *Introduction to Grothendieck duality theory*, Lecture Notes in Math. **146**, Springer, Berlin, 1970.

[Aprodu 2004] M. Aprodu, "Green–Lazarsfeld gonality conjecture for a generic curve of odd genus", 2004. Available at arXiv:math.AG/0401394. To appear in *Int. Math. Res. Notices.*

[Arbarello et al. 1985] E. Arbarello, M. Cornalba, P. A. Griffiths, and J. Harris, *Geometry of algebraic curves, I*, Grundlehren der math. Wissenschaften **267**, Springer, New York, 1985.

[Aure et al. 1997] A. Aure, W. Decker, K. Hulek, S. Popescu, and K. Ranestad, "Syzygies of abelian and bielliptic surfaces in $\mathbf{P}^4$", *Internat. J. Math.* **8**:7 (1997), 849–919.

[Avramov 1998] L. L. Avramov, "Infinite free resolutions", pp. 1–118 in *Six lectures on commutative algebra* (Bellaterra, 1996), Progr. Math. **166**, Birkhäuser, Basel, 1998.

[Bayer and Mumford 1993] D. Bayer and D. Mumford, "What can be computed in algebraic geometry?", pp. 1–48 in *Computational algebraic geometry and commutative algebra* (Cortona, 1991), Sympos. Math. **34**, Cambridge Univ. Press, Cambridge, 1993.

[Bayer and Stillman 1988] D. Bayer and M. Stillman, "On the complexity of computing syzygies: Computational aspects of commutative algebra", *J. Symbolic Comput.* **6**:2-3 (1988), 135–147.

[Bayer and Sturmfels 1998] D. Bayer and B. Sturmfels, "Cellular resolutions of monomial modules", *J. Reine Angew. Math.* **502** (1998), 123–140.

[Bayer et al. 1998] D. Bayer, I. Peeva, and B. Sturmfels, "Monomial resolutions", *Math. Res. Lett.* **5**:1-2 (1998), 31–46.

[Bernstein et al. 1978] I. N. Bernšteĭn, I. M. Gel'fand, and S. I. Gel'fand, "Algebraic vector bundles on $\mathbf{P}^n$ and problems of linear algebra", *Funktsional. Anal. i Prilozhen.* **12**:3 (1978), 66–67.

[Beĭlinson 1978] A. A. Beĭlinson, "Coherent sheaves on $\mathbf{P}^n$ and problems in linear algebra", *Funktsional. Anal. i Prilozhen.* **12**:3 (1978), 68–69.

[Boutot 1987] J.-F. Boutot, "Singularités rationnelles et quotients par les groupes réductifs", *Invent. Math.* **88**:1 (1987), 65–68.

[Bridgeland 2002] T. Bridgeland, "Flops and derived categories", *Invent. Math.* **147**:3 (2002), 613–632.

[Brill and Noether 1873] A. Brill and M. Noether, "Über die algebraischen Functionen und ihre Anwendung in der Geometrie", *Math. Ann.* **7** (1873), 269–310.

[Brodmann and Sharp 1998] M. P. Brodmann and R. Y. Sharp, *Local cohomology: an algebraic introduction with geometric applications*, Cambridge Studies in Advanced Mathematics **60**, Cambridge University Press, Cambridge, 1998.

[Bruns 1976] W. Bruns, ""Jede" endliche freie Auflösung ist freie Auflösung eines von drei Elementen erzeugten Ideals", *J. Algebra* **39**:2 (1976), 429–439.

[Bruns and Herzog 1998] W. Bruns and J. Herzog, *Cohen–Macaulay rings*, revised ed., Cambridge Studies in Advanced Mathematics **39**, Cambridge University Press, Cambridge, 1998.

[Buchsbaum and Eisenbud 1974] D. A. Buchsbaum and D. Eisenbud, "Some structure theorems for finite free resolutions", *Advances in Math.* **12** (1974), 84–139.

[Buchsbaum and Eisenbud 1977] D. A. Buchsbaum and D. Eisenbud, "Algebra structures for finite free resolutions, and some structure theorems for ideals of codimension 3", *Amer. J. Math.* **99**:3 (1977), 447–485.

[Cartan and Eilenberg 1956] H. Cartan and S. Eilenberg, *Homological algebra*, Princeton University Press, Princeton, N. J., 1956.

[Castelnuovo 1893] G. Castelnuovo, "Sui multipli di una serie lineare di gruppi di punti appartenente ad una curva algebrica", *Red. Circ. Mat. Palermo* **7** (1893), 89–110.

[Caviglia 2004] G. Caviglia, Ph.D. thesis, University of Kansas, 2004.

[Cayley 1847] A. Cayley, "On the theory of involution in geometry", *Cambridge Math. J.* **11** (1847), 52–61. See also *Collected Papers*, Vol. 1 (1889), 80–94, Cambridge Univ. Press, Cambridge.

[Ciliberto et al. 1986] C. Ciliberto, A. V. Geramita, and F. Orecchia, "Remarks on a theorem of Hilbert–Burch", pp. Exp. No. E, 25 in *The curves seminar at Queen's, IV* (Kingston, Ont., 1985–1986), Queen's Papers in Pure and Appl. Math. **76**, Queen's Univ., Kingston, ON, 1986.

[Cohen 1946] I. S. Cohen, "On the structure and ideal theory of complete local rings", *Trans. Amer. Math. Soc.* **59** (1946), 54–106.

[Conca 1998] A. Conca, "Straightening law and powers of determinantal ideals of Hankel matrices", *Adv. Math.* **138**:2 (1998), 263–292.

[Cox et al. 1997] D. Cox, J. Little, and D. O'Shea, *Ideals, varieties, and algorithms: An introduction to computational algebraic geometry and commutative algebra*, 2nd ed., Undergraduate Texts in Mathematics, Springer, New York, 1997.

[Decker and Eisenbud 2002] W. Decker and D. Eisenbud, "Sheaf algorithms using the exterior algebra", pp. 215–249 in *Computations in algebraic geometry with Macaulay 2*, Algorithms Comput. Math. **8**, Springer, Berlin, 2002.

[Decker and Schreyer 2000] W. Decker and F.-O. Schreyer, "Non-general type surfaces in $\mathbf{P}^4$: some remarks on bounds and constructions", *J. Symbolic Comput.* **29**:4-5 (2000), 545–582.

[Decker and Schreyer 2001] W. Decker and F.-O. Schreyer, "Computational algebraic geometry today", pp. 65–119 in *Applications of algebraic geometry to coding theory, physics and computation* (Eilat, 2001), edited by C. Ciliberto et al., NATO Sci. Ser. II Math. Phys. Chem. **36**, Kluwer, Dordrecht, 2001.

[Decker and Schreyer ≥ 2004] W. Decker and F.-O. Schreyer, *Algebraic curves and Gröbner bases*, Book in Preparation.

[Derksen and Sidman 2002] H. Derksen and J. Sidman, "A sharp bound for the Castelnuovo-Mumford regularity of subspace arrangements", *Adv. Math.* **172**:2 (2002), 151–157.

[Eagon and Northcott 1962] J. A. Eagon and D. G. Northcott, "Ideals defined by matrices and a certain complex associated with them", *Proc. Roy. Soc. Ser. A* **269** (1962), 188–204.

[Edmonds 1965] J. Edmonds, "Minimum partition of a matroid into independent subsets", *J. Res. Nat. Bur. Standards Sect. B* **69B** (1965), 67–72.

[Eisenbud 1980] D. Eisenbud, "Homological algebra on a complete intersection, with an application to group representations", *Trans. Amer. Math. Soc.* **260**:1 (1980), 35–64.

[Eisenbud 1988] D. Eisenbud, "Linear sections of determinantal varieties", *Amer. J. Math.* **110**:3 (1988), 541–575.

[Eisenbud 1995] D. Eisenbud, *Commutative algebra*, Graduate Texts in Math. **150**, Springer, New York, 1995.

[Eisenbud and Goto 1984] D. Eisenbud and S. Goto, "Linear free resolutions and minimal multiplicity", *J. Algebra* **88**:1 (1984), 89–133.

[Eisenbud and Harris 1983] D. Eisenbud and J. Harris, "A simpler proof of the Gieseker-Petri theorem on special divisors", *Invent. Math.* **74**:2 (1983), 269–280.

[Eisenbud and Harris 1987] D. Eisenbud and J. Harris, "On varieties of minimal degree, a centennial account", pp. 3–13 in *Algebraic geometry* (Bowdoin, 1985), edited by S. J. Bloch, Proc. Sympos. Pure Math. **46**, Amer. Math. Soc., Providence, RI, 1987.

[Eisenbud and Harris 2000] D. Eisenbud and J. Harris, *The geometry of schemes*, Graduate Texts in Math. **197**, Springer, New York, 2000.

[Eisenbud and Hochster 1979] D. Eisenbud and M. Hochster, "A Nullstellensatz with nilpotents and Zariski's main lemma on holomorphic functions", *J. Algebra* **58**:1 (1979), 157–161.

[Eisenbud and Koh 1991] D. Eisenbud and J. Koh, "Some linear syzygy conjectures", *Adv. Math.* **90**:1 (1991), 47–76.

[Eisenbud and Popescu 1999] D. Eisenbud and S. Popescu, "Gale duality and free resolutions of ideals of points", *Invent. Math.* **136**:2 (1999), 419–449.

[Eisenbud and Popescu 2000] D. Eisenbud and S. Popescu, "The projective geometry of the Gale transform", *J. Algebra* **230**:1 (2000), 127–173.

[Eisenbud and Sidman 2004] D. Eisenbud and J. Sidman, "The geometry of syzygies", in *Trends in commutative algebra*, edited by L. Avramov et al., Math. Sci. Res. Inst. Publ. **51**, Cambridge University Press, New York, 2004.

[Eisenbud and Weyman 2003] D. Eisenbud and J. Weyman, "A Fitting Lemma for *Z*/2-graded modules", *Trans. Amer. Math. Soc.* **355** (2003), 4451–4473.

[Eisenbud et al. 1981] D. Eisenbud, O. Riemenschneider, and F.-O. Schreyer, "Projective resolutions of Cohen-Macaulay algebras", *Math. Ann.* **257**:1 (1981), 85–98.

[Eisenbud et al. 1988] D. Eisenbud, J. Koh, and M. Stillman, "Determinantal equations for curves of high degree", *Amer. J. Math.* **110**:3 (1988), 513–539.

[Eisenbud et al. 1989] D. Eisenbud, H. Lange, G. Martens, and F.-O. Schreyer, "The Clifford dimension of a projective curve", *Compositio Math.* **72**:2 (1989), 173–204.

[Eisenbud et al. 2002] D. Eisenbud, S. Popescu, F.-O. Schreyer, and C. Walter, "Exterior algebra methods for the minimal resolution conjecture", *Duke Math. J.* **112**:2 (2002), 379–395.

[Eisenbud et al. 2003a] D. Eisenbud, G. Fløystad, and F.-O. Schreyer, "Sheaf Cohomology and Free Resolutions over Exterior Algebras", 2003. Available at arXiv:math.AG/0104203.

[Eisenbud et al. 2003b] D. Eisenbud, F.-O. Schreyer, and J. Weyman, "Resultants and Chow forms via exterior syzygies", *J. Amer. Math. Soc.* **16**:3 (2003), 537–579.

[Eisenbud et al. 2004] D. Eisenbud, C. Huneke, and B. Ulrich, "The regularity of Tor and graded Betti numbers", 2004. Available at arXiv:math.AC/0405373.

[Evans and Griffith 1979] E. G. Evans, Jr. and P. A. Griffith, "Local cohomology modules for normal domains", *J. London Math. Soc.* (2) **19**:2 (1979), 277–284.

[Evans and Griffith 1985] E. G. Evans and P. Griffith, *Syzygies*, London Mathematical Society Lecture Note Series **106**, Cambridge University Press, Cambridge, 1985.

[Fröberg 1999] R. Fröberg, "Koszul algebras", pp. 337–350 in *Advances in commutative ring theory* (Fez, 1997), Lecture Notes in Pure and Appl. Math. **205**, Dekker, New York, 1999.

[Fulton and Lazarsfeld 1983] W. Fulton and R. Lazarsfeld, "Positive polynomials for ample vector bundles", *Ann. of Math.* (2) **118**:1 (1983), 35–60.

[Gaeta 1951] F. Gaeta, "Sur la distribution des degrés des formes appartenant à la matrice de l'idéal homogène attaché à un groupe de *N* points génériques du plan", *C. R. Acad. Sci. Paris* **233** (1951), 912–913.

[Gallego and Purnaprajna 1998] F. J. Gallego and B. P. Purnaprajna, "Very ampleness and higher syzygies for Calabi–Yau threefolds", *Math. Ann.* **312**:1 (1998), 133–149.

[Gallego and Purnaprajna 1999] F. J. Gallego and B. P. Purnaprajna, "Projective normality and syzygies of algebraic surfaces", *J. Reine Angew. Math.* **506** (1999), 145–180. Erratum in **506** (1999), 145–180.

[Gelfand and Manin 2003] S. I. Gelfand and Y. I. Manin, *Methods of homological algebra*, Second ed., Springer Monographs in Mathematics, Springer, Berlin, 2003.

[Gel'fand et al. 1994] I. M. Gel'fand, M. M. Kapranov, and A. V. Zelevinsky, *Discriminants, resultants, and multidimensional determinants*, Mathematics: Theory & Applications, Birkhäuser, Boston, 1994.

[Geramita 1996] A. V. Geramita, "Inverse systems of fat points: Waring's problem, secant varieties of Veronese varieties and parameter spaces for Gorenstein ideals", pp. 2–114 in *The Curves Seminar at Queen's, X* (Kingston, ON, 1995), edited by A. V. Geramita, Queen's Papers in Pure and Appl. Math. **102**, Queen's Univ., Kingston, ON, 1996.

[Geramita et al. 1986] A. V. Geramita, D. Gregory, and L. Roberts, "Monomial ideals and points in projective space", *J. Pure Appl. Algebra* **40**:1 (1986), 33–62.

[Giaimo 2003] D. M. Giaimo, "On the Castelnuovo–Mumford regularity of connected curves", 2003. Available at arXiv:math.AG/0300051.

[Giaimo 2004] D. M. Giaimo, *On the Castelnuovo–Mumford regularity of curves and reduced schemes*, Ph.d. thesis, University of California, Berkeley, CA, 2004.

[Gieseker 1982] D. Gieseker, "Stable curves and special divisors: Petri's conjecture", *Invent. Math.* **66**:2 (1982), 251–275.

[Goto and Watanabe 1978a] S. Goto and K. Watanabe, "On graded rings. I", *J. Math. Soc. Japan* **30**:2 (1978), 179–213.

[Goto and Watanabe 1978b] S. Goto and K. Watanabe, "On graded rings. II. (Z^n-graded rings)", *Tokyo J. Math.* **1**:2 (1978), 237–261.

[Graham et al. 1995] R. L. Graham, M. Grötschel, and L. e. Lovász, *Handbook of combinatorics* (2 vols.), Elsevier, Amsterdam, 1995.

[Grayson and Stillman 1993–] D. R. Grayson and M. E. Stillman, "Macaulay 2, a software system for research in algebraic geometry", Available at http://www.math.uiuc.edu/Macaulay2/, 1993–.

[Green 1984a] M. L. Green, "Koszul cohomology and the geometry of projective varieties", *J. Differential Geom.* **19**:1 (1984), 125–171.

[Green 1984b] M. L. Green, "Koszul cohomology and the geometry of projective varieties. II", *J. Differential Geom.* **20**:1 (1984), 279–289.

[Green 1989] M. L. Green, "Koszul cohomology and geometry", pp. 177–200 in *Lectures on Riemann surfaces* (Trieste, 1987), World Sci. Publishing, Teaneck, NJ, 1989.

[Green 1999] M. L. Green, "The Eisenbud-Koh-Stillman conjecture on linear syzygies", *Invent. Math.* **136**:2 (1999), 411–418.

[Green and Lazarsfeld 1985] M. Green and R. Lazarsfeld, "On the projective normality of complete linear series on an algebraic curve", *Invent. Math.* **83**:1 (1985), 73–90.

[Green and Lazarsfeld 1988] M. Green and R. Lazarsfeld, "Some results on the syzygies of finite sets and algebraic curves", *Compositio Math.* **67**:3 (1988), 301–314.

[Greuel and Pfister 2002] G.-M. Greuel and G. Pfister, *A* **Singular** *introduction to commutative algebra*, Springer, Berlin, 2002.

[Griffiths and Harris 1978] P. Griffiths and J. Harris, *Principles of algebraic geometry*, Pure and Applied Mathematics, Wiley, New York, 1978.

[Gross and Popescu 2001] M. Gross and S. Popescu, "Calabi–Yau threefolds and moduli of abelian surfaces, I", *Compositio Math.* **127**:2 (2001), 169–228.

[Gruson and Peskine 1982] L. Gruson and C. Peskine, "Courbes de l'espace projectif: variétés de sécantes", pp. 1–31 in *Enumerative geometry and classical algebraic geometry* (Nice, 1981), edited by P. L. Barz and Y. Hervier, Progr. Math. **24**, Birkhäuser, Boston, 1982.

[Gruson et al. 1983] L. Gruson, R. Lazarsfeld, and C. Peskine, "On a theorem of Castelnuovo, and the equations defining space curves", *Invent. Math.* **72**:3 (1983), 491–506.

[Harris 1979] J. Harris, "Galois groups of enumerative problems", *Duke Math. J.* **46**:4 (1979), 685–724.

[Harris 1980] J. Harris, "The genus of space curves", *Math. Ann.* **249**:3 (1980), 191–204.

[Harris 1995] J. Harris, *Algebraic geometry*, Graduate Texts in Math. **133**, Springer, New York, 1995.

[Hartshorne 1966] R. Hartshorne, "Connectedness of the Hilbert scheme", *Inst. Hautes Études Sci. Publ. Math.* **29** (1966), 5–48.

[Hartshorne 1970] R. Hartshorne, *Ample subvarieties of algebraic varieties*, Lecture Notes in Math. **156**, Springer, Berlin, 1970. Notes written in collaboration with C. Musili.

[Hartshorne 1977] R. Hartshorne, *Algebraic geometry*, Graduate Texts in Math. **52**, Springer, New York, 1977.

[Hartshorne 1979] R. Hartshorne, "Complete intersections in characteristic $p > 0$", *Amer. J. Math.* **101**:2 (1979), 380–383.

[Hartshorne/Grothendieck 1967] R. Hartshorne, *Local cohomology: A seminar given by A. Grothendieck* (Harvard, 1961), vol. 41, Lecture Notes in Math., Springer, Berlin, 1967.

[Herzog and Trung 1992] J. Herzog and N. V. Trung, "Gröbner bases and multiplicity of determinantal and Pfaffian ideals", *Adv. Math.* **96**:1 (1992), 1–37.

[Hilbert 1970] D. Hilbert, *Gesammelte Abhandlungen. Band II: Algebra, Invariantentheorie, Geometrie*, Zweite Auflage, Springer, Berlin, 1970.

[Hilbert 1978] D. Hilbert, *Hilbert's invariant theory papers*, Lie Groups: History, Frontiers and Applications, VIII, Math Sci Press, Brookline, Mass., 1978.

[Hilbert 1993] D. Hilbert, *Theory of algebraic invariants*, Cambridge University Press, Cambridge, 1993.

[Hirschowitz and Ramanan 1998a] A. Hirschowitz and S. Ramanan, "New evidence for Green's conjecture on syzygies of canonical curves", *Ann. Sci. École Norm. Sup.* (4) **31**:2 (1998), 145–152.

[Hirschowitz and Ramanan 1998b] A. Hirschowitz and S. Ramanan, "New evidence for Green's conjecture on syzygies of canonical curves", *Ann. Sci. École Norm. Sup.* *(4)* **31**:2 (1998), 145–152.

[Hochster and Eagon 1971] M. Hochster and J. A. Eagon, "Cohen-Macaulay rings, invariant theory, and the generic perfection of determinantal loci", *Amer. J. Math.* **93** (1971), 1020–1058.

[Hochster and Roberts 1974] M. Hochster and J. L. Roberts, "Rings of invariants of reductive groups acting on regular rings are Cohen-Macaulay", *Advances in Math.* **13** (1974), 115–175.

[Huneke 1999] C. Huneke, "Hyman Bass and ubiquity: Gorenstein rings", pp. 55–78 in *Algebra, K-theory, groups, and education* (New York, 1997), edited by T.-Y. Lam and A. R. Magid, Contemp. Math. **243**, Amer. Math. Soc., Providence, RI, 1999.

[Kempf 1972] G. R. Kempf, "Schubert methods with an application to algebraic curves", Technical report, Mathematisch Centrum, Amsterdam, 1972.

[Kempf 1989] G. R. Kempf, "Projective coordinate rings of abelian varieties", pp. 225–235 in *Algebraic analysis, geometry, and number theory* (Baltimore, 1988), Johns Hopkins Univ. Press, Baltimore, MD, 1989.

[Kleiman and Laksov 1972] S. L. Kleiman and D. Laksov, "On the existence of special divisors", *Amer. J. Math.* **94** (1972), 431–436.

[Kleiman and Laksov 1974] S. L. Kleiman and D. Laksov, "Another proof of the existence of special divisors", *Acta Math.* **132** (1974), 163–176.

[Koh 1998] J. Koh, "Ideals generated by quadrics exhibiting double exponential degrees", *J. Algebra* **200**:1 (1998), 225–245.

[Koh and Stillman 1989] J. Koh and M. Stillman, "Linear syzygies and line bundles on an algebraic curve", *J. Algebra* **125**:1 (1989), 120–132.

[Kreuzer and Robbiano 2000] M. Kreuzer and L. Robbiano, *Computational commutative algebra. 1*, Springer, Berlin, 2000.

[Krull 1937] W. Krull, "Beiträge zur Arithmetik kommutativer Integritätsbereiche III", *Math. Z.* **42** (1937), 745–766.

[Kwak 1998] S. Kwak, "Castelnuovo regularity for smooth subvarieties of dimensions 3 and 4", *J. Algebraic Geom.* **7**:1 (1998), 195–206.

[Kwak 2000] S. Kwak, "Generic projections, the equations defining projective varieties and Castelnuovo regularity", *Math. Z.* **234**:3 (2000), 413–434.

[Lasker 1905] E. Lasker, "Zur Theorie der Moduln und Ideale", *Math. Ann.* **60** (1905), 20–116.

[Lazarsfeld 1987] R. Lazarsfeld, "A sharp Castelnuovo bound for smooth surfaces", *Duke Math. J.* **55**:2 (1987), 423–429.

[Leibniz 1962] G. W. Leibniz, *Mathematische Schriften. Bd. I.: Briefwechsel zwischen Leibniz und Oldenburg, Collins, Newton, Galloys, Vitale Giordano. Bd. II: Briefwechsel zwischen Leibniz, Hugens van Zulichem und dem Marquis de l'Hospital*, Herausgegeben von C. I. Gerhardt. Two volumes bound as one, Georg Olms Verlagsbuchhandlung, Hildesheim, 1962.

[Loose 1989] F. Loose, "On the graded Betti numbers of plane algebraic curves", *Manuscripta Math.* **64**:4 (1989), 503–514.

[L'vovsky 1996] S. L'vovsky, "On inflection points, monomial curves, and hypersurfaces containing projective curves", *Math. Ann.* **306**:4 (1996), 719–735.

[Lyubeznik 1989] G. Lyubeznik, "A survey of problems and results on the number of defining equations", pp. 375–390 in *Commutative algebra* (Berkeley, CA, 1987), edited by M. Hochter et al., Math. Sci. Res. Inst. Publ. **15**, Springer, New York, 1989.

[Lyubeznik 2002] G. Lyubeznik, "A partial survey of local cohomology", pp. 121–154 in *Local cohomology and its applications* (Guanajuato, 1999), edited by G. Lyubeznik, Lecture Notes in Pure and Appl. Math. **226**, Dekker, New York, 2002.

[Macaulay 1916] F. S. Macaulay, *The algebraic theory of modular systems*, University Press, Cambridge, 1916.

[Macaulay 1934] F. S. Macaulay, "Modern algebra and polynomial ideals", *Proc. Cam. Phil.Soc.* **30** (1934), 27–46.

[MacRae 1965] R. E. MacRae, "On an application of the Fitting invariants", *J. Algebra* **2** (1965), 153–169.

[Manin 1988] Y. I. Manin, *Quantum groups and noncommutative geometry*, Université de Montréal Centre de Recherches Mathématiques, Montreal, QC, 1988.

[Mattuck 1961] A. Mattuck, "Symmetric products and Jacobians", *Amer. J. Math.* **83** (1961), 189–206.

[Miller and Sturmfels 2004] E. Miller and B. Sturmfels, *Combinatorial commutative algebra*, Graduate Texts in Mathematics, Springer, New York, 2004.

[Mukai 1995] S. Mukai, "Curves and symmetric spaces. I", *Amer. J. Math.* **117**:6 (1995), 1627–1644.

[Mumford 1966] D. Mumford, *Lectures on curves on an algebraic surface*, With a section by G. M. Bergman. Annals of Mathematics Studies, No. 59, Princeton University Press, Princeton, N.J., 1966.

[Mumford 1970] D. Mumford, "Varieties defined by quadratic equations", pp. 29–100 in *Questions on Algebraic Varieties* (Varenna, 1969), Edizioni Cremonese, Rome, 1970.

[Noether 1873] M. Noether, "Über einen Satz aus der Theorie der algebraischen Functionen", *Math. Ann.* **6**:6 (1873), 351–359.

[Northcott 1976] D. G. Northcott, *Finite free resolutions*, Cambridge Tracts in Mathematics **71**, Cambridge University Press, Cambridge, 1976.

[Orlik and Terao 1992] P. Orlik and H. Terao, *Arrangements of hyperplanes*, Grundlehren der Math. Wissenschaften **300**, Springer, Berlin, 1992.

[Peeva and Sturmfels 1998] I. Peeva and B. Sturmfels, "Syzygies of codimension 2 lattice ideals", *Math. Z.* **229**:1 (1998), 163–194.

[Petri 1922] K. Petri, "Über die invariante Darstellung algebraischer Funktionen einer Veränderlichen", *Math. Ann.* **88** (1922), 242–289.

[Polishchuk and Zaslow 1998] A. Polishchuk and E. Zaslow, "Categorical mirror symmetry: the elliptic curve", *Adv. Theor. Math. Phys.* **2**:2 (1998), 443–470.

[Ranestad and Schreyer 2000] K. Ranestad and F.-O. Schreyer, "Varieties of sums of powers", *J. Reine Angew. Math.* **525** (2000), 147–181.

[Rathmann 1987] J. Rathmann, "The uniform position principle for curves in characteristic p", *Math. Ann.* **276**:4 (1987), 565–579.

[Ringel 1984] C. M. Ringel, *Tame algebras and integral quadratic forms*, Lecture Notes in Math. **1099**, Springer, Berlin, 1984.

[Room 1938] T. G. Room, *The geometry of determinantal loci*, Cambridge Univ. Press., Cambridge, 1938.

[Rubei 2001] E. Rubei, "On Koszul rings, syzygies, and abelian varieties", *Comm. Algebra* **29**:12 (2001), 5631–5640.

[Schenck 2003] H. Schenck, *Computational algebraic geometry*, London Math. Soc. Student Texts **58**, Cambridge Univ. Press, Cambridge, 2003.

[Schmitt and Vogel 1979] T. Schmitt and W. Vogel, "Note on set-theoretic intersections of subvarieties of projective space", *Math. Ann.* **245**:3 (1979), 247–253.

[Schreyer 1983] F.-O. Schreyer, *Syzygies of curves with special pencils*, Ph.D. thesis, Brandeis University, Waltham, MA, USA, 1983.

[Schreyer 1986] F.-O. Schreyer, "Syzygies of canonical curves and special linear series", *Math. Ann.* **275**:1 (1986), 105–137.

[Schreyer 1989] F.-O. Schreyer, "Green's conjecture for general p-gonal curves of large genus", pp. 254–260 in *Algebraic curves and projective geometry* (Trento, 1988), edited by E. Ballico and C. Ciliberto, Lecture Notes in Math. **1389**, Springer, Berlin, 1989.

[Schreyer 1991] F.-O. Schreyer, "A standard basis approach to syzygies of canonical curves", *J. Reine Angew. Math.* **421** (1991), 83–123.

[Schreyer 2001] F.-O. Schreyer, "Geometry and algebra of prime Fano 3-folds of genus 12", *Compositio Math.* **127**:3 (2001), 297–319.

[Schreyer 2003] F.-O. Schreyer, "Some topics in computational algebraic geometry", pp. 263–278 in *Advances in algebra and geometry, University of Hyderabad conference 2001*, Hindustan book agency, India, 2003.

[Serre 1988] J.-P. Serre, *Algebraic groups and class fields*, Graduate Texts in Mathematics **117**, Springer, New York, 1988.

[Serre 2000] J.-P. Serre, *Local algebra*, Springer Monographs in Mathematics, Springer, Berlin, 2000.

[Stückrad and Vogel 1982] J. Stückrad and W. Vogel, "On the number of equations defining an algebraic set of zeros in n-space", pp. 88–107 in *Seminar D.*

236   References

Eisenbud/B. Singh/W. Vogel, vol. 2, Teubner-Texte zur Math. **48**, Teubner, Leipzig, 1982.

[Stückrad and Vogel 1987] J. Stückrad and W. Vogel, "Castelnuovo bounds for certain subvarieties in $\mathbf{P}^n$", *Math. Ann.* **276**:2 (1987), 341–352.

[Sylvester 1853] J. J. Sylvester, "On a theory of syzygetic relations of two rational integral functions, comprising an application of the theory of Sturm's functions, and that of the greatest algebraic common measure", *Philos. Trans. Roy. Soc. London* **143** (1853), 407–548.

[Teixidor I Bigas 2002] M. Teixidor I Bigas, "Green's conjecture for the generic r-gonal curve of genus $g \geq 3r - 7$", *Duke Math. J.* **111**:2 (2002), 195–222.

[Voisin 1988] C. Voisin, "Courbes tétragonales et cohomologie de Koszul", *J. Reine Angew. Math.* **387** (1988), 111–121.

[Voisin 2002] C. Voisin, "Green's generic syzygy conjecture for curves of even genus lying on a $K3$ surface", *J. Eur. Math. Soc.* **4**:4 (2002), 363–404.

[Voisin 2003] C. Voisin, "Green's canonical syzygy conjecture for generic curves of odd genus", 2003. Available at arXiv:math.AG/0301359. To appear in *Compositio Math.*

[Yau and Chen 1996] Q. Yau and Z. Chen, "On the projective normality of the special line bundles on the complex projective curves", *Chinese J. Contemp. Math.* **17**:1 (1996), 17–26.

[Zariski 1947] O. Zariski, "The concept of a simple point of an abstract algebraic variety", *Trans. Amer. Math. Soc.* **62** (1947), 1–52.

Index

(continued from page ii)